Advances in Comparative and Environmental Physiology 21

Volumes already published

Vol. 10: Comparative Aspects of Mechanoreceptor Systems
Edited by *F. Ito* (1992)

Vol. 11: Mechanics of Animal Locomotion
Edited by *R. McN. Alexander* (1992)

Vol. 12: Muscle Contraction and Cell Motility: Molecular and Cellular Aspects.
Edited by *H. Sugi* (1992)

Vol. 13: Blood and Tissue Oxygen Carriers
Edited by *Ch.P. Mangum* (1993)

Vol. 14: Interaction of Cell Volume and Cell Function
Edited by *F. Lang* and *D. Häussinger* (1993)

Vol. 15: From the Contents: Salivary Gland Secretion – Nematocyst Discharge – Adaptations in Decapodan Crustaceans – Steroid-Sensitive Areas Mediating Reproductive Behaviors (1993)

Vol. 16: Ion Transport in Vertebrate Colon
Edited by *W. Clauss* (1993)

Vol. 17: Effects of High Pressure on Biological Systems
Edited by *A.G. Macdonald* (1993)

Vol. 18: Biomechanics of Feeding in Vertebrates
Edited by *V.L. Bels, M. Chardon* and *P. Vandewalle* (1994)

Vol. 19: Electrogenic Cl^- Transporters in Biological Membranes
Edited by *G.A. Gerencser* (1994)

Vol. 20: From the Contents: Motile Activities of Fish Chromatophores – Epithelial Transport of Heavy Metals – Heavy Metal Cytotoxicity in Marine Organisms – Comparative Pulmonary Morphology and Morphometry – Molecular Adaptations in Resistance to Penicillins (1994)

Vol. 21: Mechanisms of Systemic Regulation: Respiration and Circulation
Edited by *N. Heisler* (1995)

Vol. 22: Mechanisms of Systemic Regulation: Acid-Base Regulation, Ion Transfer and Metabolism
Edited by *N. Heisler* (1995)

Volumes in preparation

Comparative Immunology: Invertebrate Immune Responses.
Vol. 23: Cells and Molecular Products.
Vol. 24: Cell Activities and the Environment.
Edited by *E.L. Cooper*

Advances in

Comparative and Environmental Physiology 21

Mechanisms of Systemic Regulation: Respiration and Circulation

Guest Editor: N. Heisler

With Contributions by

G. Boutilier · W.W. Burggren · M.E. Feder · N. Heisler
J.W. Hicks · A. Ishimatsu · G.K. Iwama · G.M. Malvin
W.K. Milsom · M. Nikinmaa · S. Nilsson · B. Pelster
D.J. Randall · P. Scheid · D.P. Toews · L.A. Wentzell
A.L. Val

With 75 Figures

Springer

Guest Editor:

Prof. Dr. Norbert Heisler

Lehrstuhl für Tierphysiologie
Math. Naturwiss. Fak. I
Humboldt Universität zu Berlin
Abderhaldenhaus
Philippstr. 13
10115 Berlin
Germany

ISSN 0938-2673
ISBN-13:978-3-642-79668-5

CIP data applied for

Die Deutsche Bibliothek – CIP-Einheitsaufnahme
Mechanisms of systemic regulation: respiration and circulation/guest ed.: N. Heisler. With contributions by G. Boutilier ... – Berlin; Heidelberg; New York: Springer, 1995
(Advances in comparative and environmental physiology; 21)
ISBN-13:978-3-642-79668-5 e-ISBN-13:978-3-642-79666-1
DOI: 10.1007/978-3-642-79666-1

NE: Heilser, Norbert [Hrsg.]; Boutilier, G.; GT

Softcover reprint of the hardcover 1st edition 1995

Cover design: Springer-Verlag, Design & Productions

Typesetting: Thomson Press (India) Ltd., New Delhi

SPIN: 10051794 31/3137/SPS – 5 4 3 2 1 0 – Printed on acid-free paper

Foreword to the Series

The aim of the series is to provide comprehensive, integrated reviews giving sound, critical and provocative summaries of our present knowledge in environmental and comparative physiology, from the molecular to the organismic level.

Living organisms have evolved a widespread range of basic solutions to cope with the different problems, both organismal and environmental, with which they are faced. A clear understanding of these solutions is of course of fundamental interest for all biologists, zoologically or medically oriented. They can be best comprehended in the framework of the environmental and/or comparative approaches. These approaches demand either wide surveys of animal forms or a knowledge of the specific adaptive features of the species considered. This diversity of requirements, both at the conceptual and technological level, together with the fact that physiology and biochemistry have long been mainly devoted to the service of medicine, can account for the fact that these approaches emerged only slow amongst the other new, more rapidly growing disciplines of the biological sciences.

The field has now gained the international status it deserves and the organization of a series devoted to it appeared timely to me in view of its actual rapid development and of the interest it arouses for a growing number of biologists, physiologists, and biochemists, independently of their basic, major orientation.

Raymond Gilles

List of Editors

Preface

Various endogenous and environmental challenges of homoiostasis have resulted in the evolution of apparently quite different mechanisms for the same or similar functions in individual representatives of the animal kingdom. One of the prominent achievements of comparative physiology over the last few decades has been the description of regulatory features common to many studied species beyond the extreme diversity of their morphological forms. Delineation of functional principles universally applicable to the physiology and biochemistry of living systems became often possible through technical advances in the development of numerous new techniques, in many cases modified and adopted from other fields of science, but also by approaching certain problems using multifactorial analysis. The advance in technology has facilitated studies of minute functional details of mechanisms, which finally lead to better understanding of generally similar functions, covered by the multiple developments of Nature as a response to an extreme variety of different conditions. Improved understanding of specific mechanisms, however, has presented new problems at the level of system integration. The importance of the integrative aspect became particularly apparent during an international symposium on 'Mechanisms of Systemic Regulation in Lower Vertebrates: Respiration, Circulation, Ion Transfer and Metabolism' (organized in 1990 by Norbert Heisler and Johannes Piiper at the Max-Planck-Institut für experimentelle Medizin at Göttingen/Germany). Interaction, discussion and scientific cooperation among participants of the symposium initiated further progress in the understanding of the role of specific mechanisms in the light of organismic function, but at the same time indicated still largely fragmentary knowledge and existing deficits. The resulting progress in understanding homoiostatic control prompted the idea of publishing a number of concise reviews on selected aspects of this scientific field, which are presented in volumes 21 and 22 of this book series.

These reviews do not attempt to present encyclopedic completeness, but are intended to selectively delineate recent progress in some fields of comparative animal physiology, qualified by new approaches, interesting new data, novel interpretation or particular importance in terms of integrative merits. The large weight of the last point is reflected in the title of the two volumes. The general aim is to provide a synthesis of established ideas, models and the state of knowledge with new (sometimes controversial) ideas, approaches, results and interpretations.

The various chapters present aspects of cellular mechanisms and generalized functions related to import and export of substance, intracorporeal transfer and maintenance of the overall status of animals. Gas exchange with the environment is an essential function in homoiostatic regulation. Gas exchange by cutaneous mechanisms is energetically the most efficient for animals in its replacement of internal convective transport by pulmonary or branchial ventilation with diffusive mechanisms or exploitation of foreign convection of the gas exchange medium. However, certain characteristics limit the range in which cutaneous respiration can be utilized (Chap. 1). Elimination of the respiratory gas CO_2 is complicated in aquatic species by a number of factors, such as by the role of carbonic anhydrase. A particularly important factor is the activity of carbonic anhydrase on the mucosal surface and in the unstirred mucus layer of the branchial epithelium (Chap. 2). Gas exchange between body fluids and an internal gaseous space, the swimbladder, with its accumulation of high gas pressures, is an intriguing field. The involved mechanisms have been completely reevaluated over the last few years, indicating that buoyancy control is a more complex interaction between blood gas transport and metabolic function than thought before (Chap. 3). More than possible with cutaneous gas exchange, ventilation of specialized gas exchange structures like gills and lungs has to be closely adjusted and adapted to the special requirements of lower vertebrates by complex regulatory circuits, involving CO_2/pH chemoreceptors at various locations as sensory input devices (Chap. 4). Ionic distribution and membrane transfer and, consequently, oxygen transport in the nucleated erythrocytes of some fish species are governed by factors other than passive equilibria, greatly affected by the activity of plasma catecholamines (Chap. 5). Convective blood transport of oxygen in noncrocodilian reptiles is complicated by the anatomical arrangement with functional cross-over of oxygenated and deoxygenated blood flow pathways. Blood mixing is still relatively small, due to several mechanisms governing the complex central vascular blood flow pattern and greatly improving the efficiency of the system (Chaps. 6 and 7). Crocodilia have solved the problem of intracardiac blood flow separation, but have maintained the option for shunting under special conditions (Chap. 7). Cardiovascular function in amphibia is extremely diverse compared to reptiles on the basis of phylogenetic differentiation. The pattern is even more complicated by the fact that amphibia experience extreme anatomical reorganization during ontogenic development. As for phylogenetic differentiation and ontogenetic development, the regulatory range of cardiovascular function for amphibia is probably by far the widest in the animal kingdom (Chap. 8). In addition to the primary circulatory system for the transport of respiratory gases, nutrients and metabolic end products, lower vertebrates possess a secondary circulatory system, which has been neglected for a long time in terms of physiological function. Recently, first detailed studies of amphibia and fishes have presented data suggesting great importance of this system for homoiostatic control of the organism (Chaps. 9 and 10).

All of the chapters have been prepared by authorities in the respective fields and as such represent the state of the art of our present knowledge, sometimes spiced with controversial aspects delineated from various sides. With this high standard of the individual contributions, it is still hoped that this collection of reviews represents more than the sum of the parts. The volume is designed to attract readers from a variety of disciplines in biology and medicine, in addition to specialized fields in physiology, biochemistry and biophysics, providing solid information for established scientists as well as for students and postgraduates.

The editor would like to express his gratitude to the authors, who have provided an excellent selection of chapters and have made a great effort to make this volume comprehensible and wide ranging in a limited amount of print space. Numerous other people involved in the preparation of the manuscript and in the printing process, in particular my assistant Dr. Sabine Heinz and the editorial staff of Springer, in particular Ursula Gramm and Theodora Krammer, facilitated the publication process.

April 7, 1995 *Norbert Heisler*

Contents

Gas Exchange and Regulation of Respiration

Central Vascular Circulation

Chapter 6

Mechanisms of Intracardiac Shunting in Reptiles:
Pressure vs Washout Shunting
J.W. Hicks and *G.M. Malvin*

Chapter 7

Central Cardiovascular Dynamics in Reptiles
S. Nilsson

Chapter 8

Central Cardiovascular Function in Amphibians:
Qualitative Influences of Phylogeny, Ontogeny, and Season
W.W. Burggren

Veno-Lymphatic System

Chapter 9

The Role of the Lymphatic System for Water Balance and Acid-Base Regulation in the Amphibia
D.P. Toews and *L.A. Wentzell*

Chapter 10

Physiological Roles of the Secondary Circulatory System in Fish
A. Ishimatsu, G.K. Iwama and *N. Heisler*

Contents for Volume 22

Chapter 7

Regulation of Ion and Water Transport
by Hydrogen Ions in High Resistance Epithelia
B.J. Harvey and *V. Urbach*

Chapter 8

Intracellular Signals Controlling Ionic
and Acid-base Regulation in Avian Nasal Gland Cells
T.J. Shuttleworth

Hypoxia and Metabolic Adaptation

Chapter 9

Interrelationships Between Hypoxia
and Thermoregulation in Vertebrates
S.C. Wood

Chapter 10

Sulfide Tolerance in Marine Invertebrates
S. Völkel and *M.K. Grieshaber*

Gas Exchange and Regulation of Respiration

Chapter 1

The Regulation of Cutaneous Gas Exchange[1]

M.E. Feder[2]

Contents

1 Why Study Cutaneous Gas Exchange?

Although physiological studies usually focus on the human or mammalian lung to understand respiratory gas exchange, the laws of physics suggest that any permeable membrane can serve as a gas exchange organ and the phenomenology of natural history suggests that practically every suitable structure has done so in the past. Thus, the integument, limbs, head, pharynx, tail, gut, and even the cloaca have been specialized as gas exchange organs; air lungs, water lungs, external and internal gills, tracheae, and book lungs have evolved as organs primarily dedicated to gas exchange; the jelly coat, chorioallantois, eggshell, and placenta function similarly in a developmental sense; and medical technology has

[1] Dedicated to Professor Dr. Johannes Piiper on the occasion of his retirement

[2] Department of Organismal Biology & Anatomy and The Committee on Evolutionary Biology, The University of Chicago, 1025 East 57th Street, Chicago, IL 60637, USA

Advances in Comparative and Environmental Physiology, Vol. 21

exploited the peritoneal membrane and developed the extracorporeal membrane oxygenator. The broad perspective occasioned by this diversity transforms the analysis of respiratory gas exchange from a case-by-case study of particular taxa into a general study in biological design, constraint and regulation: What are the general rules governing the function of gas exchange organs? Along what axes has the evolutionary diversification of gas exchange organs been permitted or constrained? (Indeed, the development of general models of gas exchange and its diversity has been a premier accomplishment of Johannes Piiper and his colleagues.) According to this perspective, each particular gas exchange organ, including the human lung and metazoan integument inter alia, constitutes a test of the robustness of hypothesized general patterns and predictions.

Study of integumentary gas exchange may yield special insights to the general analysis of gas exchange organs in several ways:

- Unlike the "dedicated" gas exchange organs listed above, the integument has numerous functions in addition to gas exchange. These other functions may constrain both acute regulation of gas exchange and evolution of gas exchange organs. Accordingly, the skin can serve as a standard for comparison to help identify adaptive modifications in other gas exchange organs. For example, because the integument forms the outside of an organism, it must resist mechanical stresses and, in some cases, unbridled fluxes of mass and energy between organism and environment. As a result, the integument may be thicker than other gas exchange membranes. In amphibian skin, for example, the tissue barrier between respiratory medium and blood is 10 μm to more than 100 μm thick (Czopek 1965), making the amphibian skin the best available natural model of a diffusion-limited gas exchanger.
- Because the integument is external, it is amenable to direct manipulation and observation in situ.
- The integument was the primitive organ of gas exchange, has a long evolutionary history, and undergoes fossilization. We can relate organisms' occupancy of diverse habitats to their reliance on integumentary gas exchange.
- The skin relatively closely approximates diffusive exchange through a flat plane, a component of many models of gas exchange.

Not surprisingly, then, cutaneous gas exchange has received considerable attention in its own right and as a physiological or evolutionary standard of reference in general studies of gas exchange (Feder and Burggren 1985a, b; Feder et al. 1988a; Graham 1988). Although cutaneous gas exchange is by no means restricted to the vertebrate class Amphibia (Feder and Burggren 1985a; Graham 1988), amphibians as a group may rely more heavily upon cutaneous gas exchange than other vertebrates (Foxon 1964; Feder and Burggren 1985a) and the utility of amphibians as experimental models has resulted in their widespread exploitation in this regard (Feder 1992). Accordingly, the bulk of research on cutaneous gas exchange has used amphibians as subjects, although it usually addresses broader issues. The present work will examine major themes and recent findings in the regulation of cutaneous gas exchange in amphibians; readers interested in a more comprehensive review of the older literature should consult the references cited above.

2 Major Themes

2.1 August Krogh and the Modern Era of Investigation

Although the physiological study of cutaneous gas exchange dates back to Spallanzani in 1797 (Krogh 1904), the modern age of investigation began (as did so much of comparative physiology) with August Krogh. Krogh's doctoral dissertation (1904) set forth a variety of issues, some of which still continue to attract attention. Krogh characterized the roles of the skin and lungs in O_2 and CO_2 exchange throughout the year and in different species, and related differences between skin and lungs to the surface area and gas partial pressures of the blood perfusing each. Krogh investigated the interaction of O_2 uptake and CO_2 excretion, and examined the regulatory role of the nervous system in detail. Finally, Krogh considered how cutaneous gas exchange might be regulated, and concluded (p. 415) that "the cutaneous absorption of oxygen cannot be regulated at all by the organism,... [and] is effected by plain physical powers–by diffusion."

2.2 Multiple Gas Exchangers and the Role of the Skin

Many vertebrates deploy multiple organs of gas exchange in various combinations (e.g. lungs and gills, lungs and skin, skin and gills, lungs and skin and gills), while others rely predominantly or exclusively on but a single gas exchanger. Much work has sought to determine how an organism might divide its gas exchange among its various gas exchange organs, under what circumstances particular organs are favored, and whether organisms can redistribute gas exchange among the various organs to mitigate unfavorable environments or major increases in demands for gas exchange. Works by Dolk and Postma (1927), Guimond and Hutchison (1976), Nonotte and Kirsch (1978), Burggren and West (1982), West and Burggren (1984), and Boutilier et al. (1986), among many others, exemplify this theme. Common (but by no means universal) patterns when multiple gas exchange organs are present are: The skin figures more prominently in CO_2 excretion than in O_2 uptake. The skin tends to be most important to gas exchange at the relatively low levels of gas exchange in resting animals; other gas exchange organs tend to support increased rates of gas exchange during or after activity. Similarly, the skin often accounts for the greatest proportion of total gas exchange at times of year when the gas exchange requirement is low. In the face of rising temperatures, however, the proportion of overall gas exchange attributable to the skin may decrease, remain constant, or even increase (Burggren et al. 1983). Aquatic air-breathing animals can partially reallocate gas exchange among skin and other gas exchange organs in relation to the suitability of the respiratory media for gas exchange; e.g. blood flow to the skin increases when the lungs are hypoxic and decreases when the skin is hypoxic (West and Burggren 1984; Boutilier et al. 1986).

Several genera of salamanders have attracted particular attention with respect to cutaneous gas exchange (*Rhyacotriton*: Dicamptodontidae; *Onychodactylus* and *Ranodon*: Hynobiidae; the entire family Plethodontidae; and *Chioglossa* and *Salamandrina*:Salamandridae). In each instance, ancestral forms with lungs independently underwent evolutionary loss of lungs, and their extant descendants rely exclusively on the skin and lining of the buccopharyngeal cavity for gas exchange as adults. Diving animals and animals overwintering underwater also may rely exclusively on the skin for gas exchange, albeit temporarily (Pinder 1987; Boutilier 1990; Pinder and Feder 1990).

2.3 The Evolution of Gas Exchange

Some issues of long-standing interest to biological scientists are how dedicated gas exchangers (e.g. lungs, gills) evolved (presumably from integumentary gas exchange), and how vertebrates accomplished the transition from water-breathing aquatic habits to air-breathing terrestrial habits. Integumentary gas exchange figures prominently in both issues (Graham 1988). A common theme is that the skin was a key gas exchange organ during the transition from water to land in vertebrates (Rahn and Howell 1976; Randall et al. 1981; see also Sect. 4). Living amphibians, however, which are the source of most data bearing on the evolution of cutaneous gas exchange, are highly derived forms that may be inappropriate as models for the primitive amphibians that first invaded the land (Gans 1970a,b).

2.4 Cutaneous Surface Area as a Limit to Energy Metabolism

Some seeming advantages to internal gas exchange organs are that the diffusion barrier between the blood and the respiratory medium can be very thin, and that their surface area can be greatly expanded within a given volume to accommodate evolutionary variation in gas exchange requirements. The outer surface of animals, when serving as a gas exchange organ, would seem to be much less malleable in evolutionary adjustment of its area. These expectations have led numerous investigators to examine the allometry of respiratory surface area in skin-breathing animals and its consequences. A common working hypothesis has been that the allometric coefficient for surface area (i.e. the slope of the line relating the logarithm of surface area to the logarithm of body mass) should be less for exclusively skin-breathing species than for species with multiple gas exchange organs (Czopek 1965; Whitford and Hutchison 1967), as should the allometric coefficient for organismal metabolism. The comparative allometry of surface area and metabolism also bears on a second theme, whether respiratory surface area or cellular oxidative capacity is "the limit" to metabolism (Ultsch

1973, 1976). In addition to the foregoing work, numerous authors (e.g. Beckenbach 1975; Feder 1976; Withers 1980; among many others) have examined allometry of gas exchange in various amphibians. These studies are consistent with the prediction that respiratory surface area places an upper limit on the magnitude of cutaneous gas exchange and the lowest Po_2 at which cutaneous gas exchange can be regulated (Feder and Burggren 1985a). Unequivocal attribution of these limits to surface area per se is somewhat clouded: Allometric variation in cutaneous conductance is not due solely to variation in surface area, but also to variation in the thickness and material composition of the skin overlying the cutaneous capillaries as well as the radii and spacing of these capillaries (Ultsch 1973; Feder 1988; Malvin 1988). Regardless of the source of allometric variation in cutaneous conductance, the physiological limits it imposes do not translate readily into ecological or evolutionary restrictions to small body sizes or cool temperatures in exclusively skin-breathing animals (Feder and Burggren 1985a). For example, lungless plethodontid salamanders reach their greatest abundance and diversity in the tropics, where some routinely encounter high temperatures and attain large body sizes (Feder 1976; Feder et al. 1982).

2.5 The Regulation of Gas Exchange

The final theme, which is a focus of ongoing research in several laboratories, concerns not long-term or evolutionary adjustments in the overall capacity for gas exchange, but short-term responses to acute variation in the gas exchange requirement or the respiratory environment (Feder and Burggren 1985b). This theme examines to what extent cutaneous gas exchange can be modulated in the short-term, and the means available for doing so. The remainder of this chapter is devoted to this issue. The chapter will examine first the possible importance of adjustments in perfusion and diffusing capacity within the skin itself in regulating gas exchange, and then the possible importance of skin ventilation in regulating cutaneous gas exchange by dissipating unstirred layers or boundary layers outside the skin itself. The former topic concerns skin in both air and water; the latter is primarily relevant to skin in water only.

3 Relative Importance of Diffusion and Perfusion in Regulating Cutaneous Gas Exchange

3.1 Models

As developed by Johannes Piiper and colleagues (Piiper and Scheid 1975, 1977, 1981; Piiper 1982, 1988), the context of these studies is typically to examine the portions of total respiratory conductance (or resistance) attributable to diffusion versus perfusion. As an element of nonarterialized blood with an initial Po_2 of $P_{b(in)}$ enters the skin and transits a cutaneous capillary, its Po_2 ($P_{b(x)}$) will increase

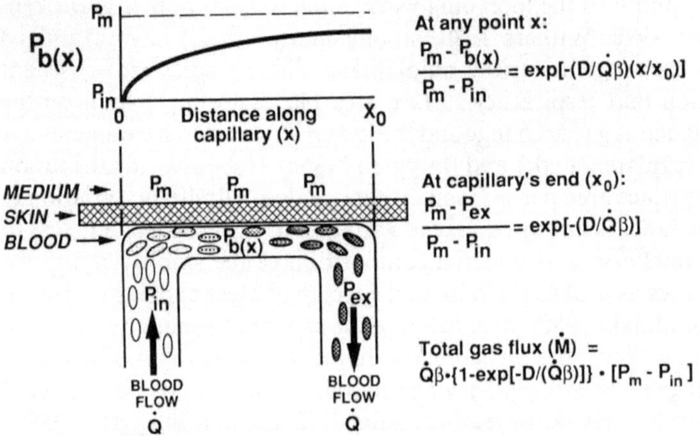

Fig. 1. Oxygen uptake along a cutaneous capillary. *Lower left.* Schematic of oxygen partial pressures (Po_2). As blood enters a cutaneous capillary at a Po_2 of $P_{b(in)}$, its Po_2 ($P_{b(x)}$) increases as a function of distance (x), approaching the partial pressure of the medium (P_m) as a limit. Typically, blood exits the cutaneous circulation at a Po_2 of $P_{ex,}$ without having fully equilibrated with the respiratory medium. *Upper left* Increase in blood Po_2 as a function of distance along a capillary, from its start ($x = 0$) to its end ($x = x_0$). *Right* Equations relating Po_2 and oxygen uptake ($\dot{M}o_2$) to the ratio of diffusive conductance: perfusive conductance. See text for explanation. (After Piiper and Scheid 1975, 1981; Piiper 1982, 1988)

with increasing distance (x) or residence time, approaching the Po_2 of the external medium (P_m) as a limit (Fig. 1). The efficacy of the skin as a gas exchange organ can be characterized by the equilibration deficit (Piiper 1982), the ratio of the amount by which the blood has failed to equilibrate with the respiratory medium at any distance (x) along the capillary (i.e., $P_m - P_{b(x)}$) to the total possible increase in the Po_2 of the blood (i.e. $P_m - P_{b(in)}$):

$$\text{Equilibration Deficit} = \frac{P_m - P_{b(x)}}{P_m - P_{b(in)}}.$$

A large equilibration deficit implies that the blood fails to equilibrate with the respiratory medium as it passes through the skin. As Piiper and Scheid (1975, 1977, 1981) have shown, the equilibration deficit is a joint function of the diffusive conductance (D, the diffusing capacity) and the perfusive conductance [$\dot{Q}\beta$, the product of blood flow ($\dot{Q}$) and O_2 capacitance (β)]:

$$\frac{P_m - P_{b(x)}}{P_m - P_{b(in)}} = \exp \frac{D}{\dot{Q}\beta}.$$

The total oxygen flux ($\dot{M}o_2$) will also be a function of these variables (Piiper 1988):

$$\dot{M}o_2 = \dot{Q}\beta \cdot \{1 - \exp[-D/(\dot{Q}\beta)]\} \cdot [P_m - P_{b(in)}].$$

Thus, the ratio $D/(\dot{Q}\beta)$ emerges as a critical parameter governing the extent of equilibration and oxygen uptake through the skin. As $D/(\dot{Q}\beta)$ decreases, pro-

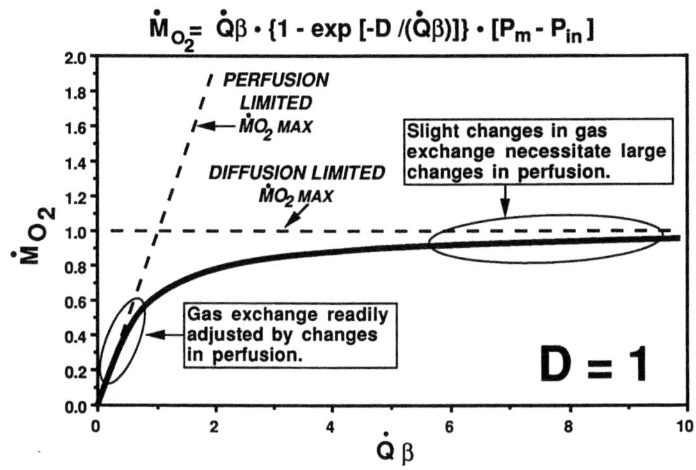

Fig. 2. Theoretical expectation of relationship of oxygen uptake ($\dot{M}o_2$) to perfusion under joint diffusion and perfusion limitation. Diffusive conductance or diffusing capacity is 1; a unit difference between the oxygen partial pressure of the medium and that of the blood ($P_m - P_{in}$) is expected to yield a unit $\dot{M}o_2$ in the absence of any perfusion limitation; this $\dot{M}o_2$ is labeled *diffusion limited $\dot{M}o_2$ max.* In the absence of any diffusion limitation, $\dot{M}o_2$ would be expected to increase according to the line labeled *perfusion limited $\dot{M}o_2$ max.* The expected relationship (*solid line*), whose equation is given at the top, approaches the perfusion limited $\dot{M}o_2$ max at low perfusive conductance ($\dot{Q}\beta$) and approaches the diffusion limited $\dot{M}o_2$ max at high perfusive conductance. (After Piiper and Scheid 1975, 1981; Piiper 1982, 1988)

gressively longer capillaries (or red cell residence times in the skin) are required to achieve full equilibration of the blood with the respiratory medium; low $D/(\dot{Q}\beta)$ signifies that equilibration is incomplete (Piiper 1982, 1988).

The ratio $D/(\dot{Q}\beta)$ also delimits the regulatory options available to adjust rates of gas exchange (Fig. 2). When D is large relative to $(\dot{Q}\beta)$, changes in $\dot{Q}$ will markedly affect gas flux [the gas exchange organ is said to be perfusion limited; L_{perf} (Piiper 1988) is large]. Because $\dot{Q}$ is easily regulated, gas flux is readily regulated as well. When D is small relative to $(\dot{Q}\beta)$, changes in D will have much greater effect on gas flux than changes in $\dot{Q}$, and large changes in $\dot{Q}$ will be required to achieve even small changes in gas flux [the gas exchange organ is said to be diffusion limited; L_{diff} (Piiper 1988) is large]. D is a function of several physical and morphological variables:

$$D = d\alpha A/x,$$

where A and x are the surface area and thickness of the diffusion barrier, respectively, and *d* and α are the diffusivity and solubility of the diffusing species in the diffusion barrier. Unlike $\dot{Q}$, which is easily regulated (e.g. by changing heart rate), the variables that D comprises are seemingly fixed in the short term and incapable of being regulated acutely to modulate gas exchange. Thus, a low $D/(\dot{Q}\beta)$ ratio may represent both a substantial equilibration deficit and problematic regulation of gas exchange.

The foregoing theoretical discussion is relevant to cutaneous gas exchange because the $D/(\dot{Q}\beta)$ ratio of skin is lower than for other gas exchange organs, and low enough to be problematic for the regulation of gas exchange (Piiper 1982). Several complementary studies of an exclusively skin-breathing plethodontid salamander, *Desmognathus fuscus*, yield $D/(\dot{Q}\beta)$ ratios of about 0.2 (Gatz et al. 1975; Piiper et al. 1976). Several other studies yield descriptive or experimental data that are consistent with this conclusion (Gottlieb and Jackson 1976; Mackenzie and Jackson 1978; Jackson and Braun 1979; Moalli et al. 1980; Feder et al. 1988b; Malvin and Hlastala 1989). Krogh, in fact, reached the same conclusion in 1904! The skin, therefore, has been considered "passive" and incapable of being regulated in its role as a gas exchanger. The massive diffusion limitation is evident in that "arterialized" blood exits the skin of amphibians with a Po_2 of ca. 13.3 kPa ($\sim$ 100 mmHg) less than that of the respiratory medium (Piiper 1982).

3.2 Evidence for Regulation and Possible Mechanisms

Despite this large diffusion limitation, at least some amphibians clearly demonstrate a capacity to increase or decrease cutaneous gas exchange in some circumstances. For example, larvae of ranid frogs can undergo three fold variations in cutaneous oxygen uptake with changes in temperature (Burggren et al. 1983). In air, adult frogs can adjust excretion of indicator gases (and hence presumably exchange of oxygen) in inverse relationship to the Po_2 surrounding the skin and/or the Po_2 of lung gas (Malvin and Hlastala 1986a, b, 1989). In water, the cutaneous oxygen uptake of frogs is to some degree independent of ambient Po_2 (Pinder 1987). Up-modulation of cutaneous gas exchange is clearly evident in plethodontid salamanders, which lack lungs and gills as adults and respire via the skin and buccopharyngeal lining. During exercise, the rate of oxygen consumption in these salamanders can increase by more than an order of magnitude within minutes (Withers 1980; Feder 1986, 1987; Full 1986; Full et al. 1988; Gatten et al. 1992). Much recent work has attempted to reconcile these phenomena with the conclusion that cutaneous gas exchange is passive and incapable of being regulated. Possible explanations include (Fig. 3):

1. Diffusive conductance or diffusing capacity (D) can be increased acutely such that the $D/(\dot{Q}/\beta)$ ratio increases, lessening diffusion limitation and enhancing the efficacy of perfusion modulation as a regulatory mechanism (Feder and Burggren 1985b). In support of this hypothesis, Pinder (1987) reported that diffusing capacity may double when submerged bullfrogs (in simulated overwintering conditions) are exposed to hypoxia, and Feder et al. (1988b) reported patterns of inert gas equilibration consistent with a lessening of diffusion limitation during exercise in a large plethodontid salamander. By contrast, Malvin and Hlastala (1986a, b, 1989) could find no evidence for changes in diffusing capacity, and attributed regulation to other mechanisms (see below). If D is regulated, how might such changes be effected? Burggren and Moalli (1984) proposed that capillary recruitment, which acutely increases the surface area

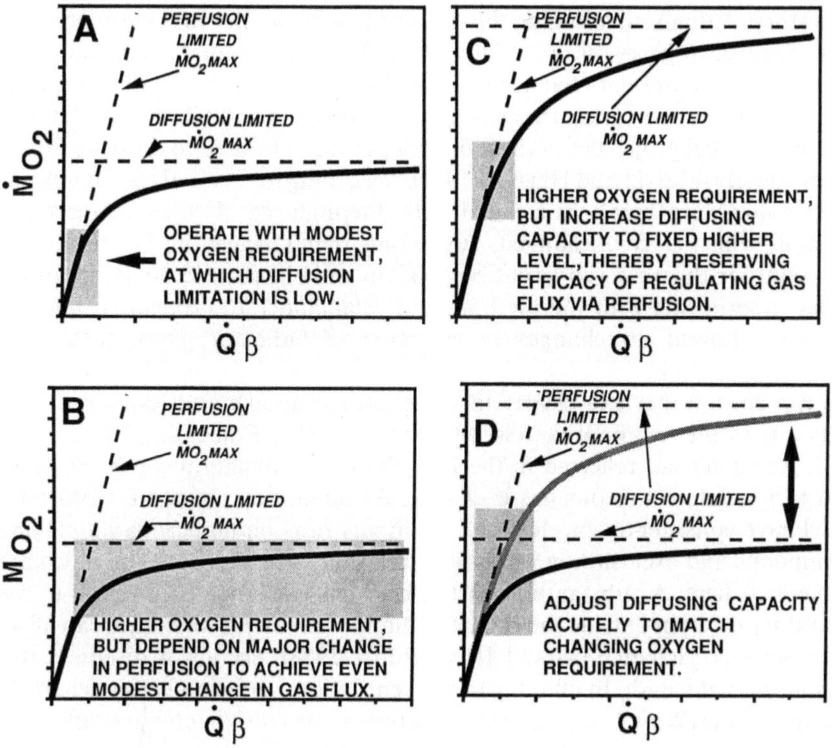

Fig. 3. Options for regulating cutaneous uptake of oxygen under mixed diffusion and perfusion limitation. $\dot{M}o_2 - \dot{Q}\beta$ relationships, diffusion limited $\dot{M}o_2$ max, and perfusion limited $\dot{M}o_2$ max are plotted as in Fig. 2. In each case, the *height* of the *shaded rectangle* signifies the relative and absolute magnitude of oxygen flux, and the *width* of the *shaded rectangle* signifies the corresponding change in perfusion that is necessary to achieve such modulation of oxygen flux

through which oxygen may enter the cutaneous circulation, could affect D. Moreover, they reported correspondence between bulk blood flow to the skin, the number of recruited capillaries per unit area, and gross rates of gas exchange. These variables also undergo correlated change in response to skin ventilation (Burggren and Feder 1986; Feder and Pinder 1988). Malvin and Hlastala (1987), however, reported regulated changes in gas flux in the absence of capillary recruitment. Another possible mechanism is a redistribution of blood flow between deep and superficial levels of the skin (Feder et al. 1988b; Piiper 1988) or between regions of an animal in which the skin is thick or thin (Talbot 1991). This possible mechanism has not been examined experimentally.

2. Because diffusion limitation, while predominant, is not exclusive, changes in bulk blood flow to the skin are sufficient to account for variation in cutaneous gas exchange (Malvin 1988). In work with anesthetized ranid frogs, Malvin and Hlastala (1986a, b, 1989) demonstrated $\pm$ 30% changes in excretion of indicator gases (halothane, freon, acetylene, CO_2) in response to changes in the Po_2 of the

respiratory medium surrounding the skin or in the lungs. Because the concentrations of indicator gases on both sides of the skin are presumably constant, this variation must represent changes in total skin conductance. Capillary recruitment did not change during these experiments (Malvin and Hlastala 1987). The differing physical properties of the indicator gases can be used to estimate D and $\dot{Q}$ individually (Malvin and Hlastala 1989). According to this analysis, significant changes in $\dot{Q}$ are highly correlated with Po_2-induced changes in inert gas excretion while D is unchanged, suggesting that changes in $\dot{Q}$ but not in D account for the observed variation (Malvin and Hlastala 1989). By contrast, similar experiments with the plethodontid salamander *Desmognathus quadramaculatus* showed no changes in excretion of indicator gases, total conductance, $\dot{Q}$ or D in response to variation in Po_2 (Malvin and Hlastala 1989). This is a species that can clearly increase oxygen uptake and lessen diffusion limitation during treadmill exercise (Feder et al. 1988a; Full et al. 1988).

3. Variables not reflected in the $D/(\dot{Q}\beta)$ ratio formulation can be manipulated to regulate cutaneous gas exchange. As the authors of the $D/(\dot{Q}\beta)$ ratio model have consistently emphasized, the model rests on numerous simplifying assumptions and exclusion of potentially relevant variables, some of which are enumerated here. A substantial resistance to gas exchange may reside in the respiratory medium outside the skin, and diminution of this resistance can affect cutaneous gas exchange (Model III of Piiper 1988). Section 4 discusses this phenomenon at length. In analogy to the ventilation: perfusion heterogeneity in the human lung (West 1977), regional variation in the $D/(\dot{Q}\beta)$ ratio may affect the efficiency of gas exchange (Model IV of Piiper 1988). For example, the dorsal and ventral skin of the bullfrog are sufficiently different in D, $\dot{Q}$, and partial pressures of entering blood to result in ca. 20% differences in O_2 and CO_2 exchange within an individual frog (Talbot 1991; see also Model VI of Piiper 1988). Back-diffusion of gases via countercurrent exchange between ascending and descending portions of cutaneous capillaries may also reduce the efficiency of cutaneous gas exchange (Model V of Piiper 1988). Several microcirculatory variables, including the radius of capillaries, pulsatility of blood flow, distance between red cells within a capillary, and resistance of the plasma to diffusion, should affect rates of gas flux (Federspiel and Sarelius 1984; Malvin 1988). For a constant number of recruited cutaneous capillaries, cutaneous gas exchange will be most efficient if their spacing remains below a "critical density", which can be predicted from first principles (Malvin 1988); thus, the spacing of perfused capillaries can affect cutaneous gas exchange. Most theoretical and experimental analyses of $D/(\dot{Q}\beta)$ ratios have assumed a constant β; i.e. a constant and linear Co_2-Po_2 relationship in the blood. Kobayashi et al. (1991) have recently reconsidered the $D/(\dot{Q}\beta)$ ratio in light of the more typical sigmoidal (i.e. nonlinear) shape of hemoglobin oxygen equilibrium curves, and concluded that the $D/(\dot{Q}\beta)$ ratio for amphibian skin had previously been underestimated. A second source of variation in β, acute change in hematocrit, has been demonstrated to occur in frogs (Pinder and Smits 1987) and can substantially affect cutaneous gas exchange (Malvin and Wood 1991). Finally, the red cell residence time in the capillary,

which is critical to the magnitude of the equilibration deficit, will be a joint function of the bulk flow of red cells to the skin and the total volume of red cells in the cutaneous capillaries, the latter a function of capillary recruitment and/or vasoconstriction. Red cell residence times are yet to be characterized.

4. The apparent discrepancy between diffusion limitation and capacity for regulating cutaneous gas exchange reflects the prevailing rate of gas exchange. For a gas exchange organ with a given $D/(\dot{Q}\beta)$ ratio and partial pressure gradient between the blood and respiratory medium, diffusion limitation will be less (and perfusion limitation greater) if rates of gas exchange are low. All else equal, skin-breathing species with modest gas exchange requirements or circumstances in which gas exchange requirements are reduced (e.g. cool temperatures, inactivity) might be expected to exhibit greater capacities for the regulation of cutaneous gas exchange than if the gas exchange requirement is high or elevated. A case in point concerns exclusively skin-breathing salamanders, whose ability to increase oxygen consumption during activity is apparently inconsistent with a large diffusion limitation (see above). The resting rates of oxygen consumption in these and other salamanders are exceptionally low for vertebrates (Feder 1976). When these salamanders exercise, a considerable factorial scope (i.e. maximal O_2 consumption: standard O_2 consumption) is possible, because the standard O_2 consumption is so low (Feder 1988). The maximum rates of O_2 consumption are no greater than expected from the diffusing capacity and medium-blood partial pressure gradient (Feder 1988), and are not particularly large by comparison with those of other vertebrates (Gatten et al. 1992). Similarly, instances in which the capacity for regulation of cutaneous gas exchange is large concern hypothermic frogs with low rates of oxygen consumption (Pinder 1987) or normoxic frogs also relying upon lungs (e.g. Burggren and Moalli 1984). When the skin is pressed to its limit (e.g. in apneic normothermic frogs, in which the skin must bear a large gas exchange requirement normally assigned to the lungs), the capacity for regulation is modest and D has little scope for adjustment (Malvin and Hlastala 1986a, b, 1989; Talbot 1991).

3.3 An Isolated Perfused Skin Preparation

Most of the foregoing modeling and experimental work relies on the measurement of cutaneous gas fluxes in intact animals. However, as Section 3.2 discusses, numerous interacting variables can affect cutaneous gas exchange, and can putatively confound analyses. Moreover, a complex set of regulatory controls, whose function is only just beginning to be understood, is at work (Malvin and Riedel 1990; West and Van Vliet 1992). These issues have led my colleagues and myself to develop an isolated perfused frog skin preparation, in which relevant variables are entirely under experimental control and can be manipulated individually and jointly. The preparation has been described in detail elsewhere; Pinder et al. (1990, 1991) and Clemens and Feder (1992) provide a detailed account of experimental methods and results.

As discussed in Section 3.1, a parameter of major interest is the $D/(\dot{Q}\beta)$ ratio. This ratio can be equated with L_{diff}, a second index of diffusion limitation that represents the difference between actual gas exchange and gas exchange in the absence of any perfusion limitation (i.e. $\dot{Q}\beta \rightarrow \infty$), as follows:

$$L_{diff} = \exp[-D/(\dot{Q}\beta].$$

We experimentally manipulated the $D/(\dot{Q}\beta)$ ratio for oxygen uptake of the preparation by changing the perfusion rate (i.e. $\dot{Q}$) of an autologous red cell suspension at fixed $P_{b(in)}$, with the partial pressure of the medium (P_m) held constant. We chose $P_{b(in)}$ so as to maintain the perfusate on the most linear portion of the hemoglobin oxygen equilibrium curve and to minimize the errors discussed by Kobayashi et al. (1991). The rate at which O_2 is taken up by the perfusate ($\dot{M}o_2$) can be estimated as follows:

$$\dot{M}o_2 = \dot{Q} \cdot C_{Hb} (So_{2\,outflow} - So_{2\,inflow}),$$

where C_{Hb} is the molar hemoglobin concentration of the perfusate (assuming a molecular weight equal to 16 700) and $So_{2\,inflow}$ and $So_{2\,outflow}$ are the percentage O_2 saturations of the perfusate as it enters and exits the preparation, respectively. D can then be calculated as:

$$D = \dot{M}o_2/(P_m - P_{b(in)}).$$

Throughout the experimental range of $\dot{Q}$, $\dot{M}o_2$ increases approximately fivefold. The summary equation relating $\dot{M}o_2$ to D and $\dot{Q}\beta$ is:

$$\dot{M}o_2 = 0.500\dot{Q} \cdot C_{Hb}\{1 - \exp[-4.73/(0.500\dot{Q} \cdot C_{Hb})]\},$$

where $\dot{Q}$ is bulk perfusate flow rate in $\mu l\,cm^{-2}\,min^{-1}$, 0.500 is the average $So_{2\,inflow}$, and 4.73 is the estimated $\dot{M}o_2$ at maximum diffusing capacity.

As $\dot{Q}$ was increased experimentally, D increased as well (see Fig. 2 in Pinder et al. 1991). This relationship is consistent with increasing capillary recruitment (Burggren and Moalli 1984), which in turn is consistent with decreases in the vascular resistance of the preparation (Pinder et al. 1991). In Fig. 4, the summary equation relating $\dot{M}o_2$ to D and $\dot{Q}\beta$ has been decomposed into individual curves for each of four bulk perfusion flow rates and their corresponding D. Although the change in D is convincing evidence that diffusional aspects of cutaneous gas exchange can be regulated, the increase in D was at best modest, averaging 21% as $\dot{Q}$ was increased from 4 to 16 μl perfusate $cm^{-2}\,min^{-1}$. Nonetheless, the increase in D was responsible for a significant fraction of the increase in $\dot{M}o_2$ as perfusion rates were increased from moderate to high levels. $\dot{M}o_2$ increased from 2.7 to 4.3 nmol $cm^{-2}\,min^{-1}$ as perfusion rate was increased from ca. 4 to 16 μl perfusate $cm^{-2}\,min^{-1}$, an increase of ca. 60% (Fig. 4). Approximately 60% of this increase would have resulted from increasing perfusive conductance even if D did not change as perfusion was increased (arrow B in Fig. 4); 40% of this increase was thus due to increasing D (arrow A in Fig. 4).

At the lowest $\dot{Q}$ for which D was computed (4 μl perfusate $cm^{-2}\,min^{-1}$), L_{diff} was 38%, $D/(\dot{Q}\beta)$ was 0.96, and changes in $\dot{Q}$ can somewhat regulate $\dot{M}o_2$. As

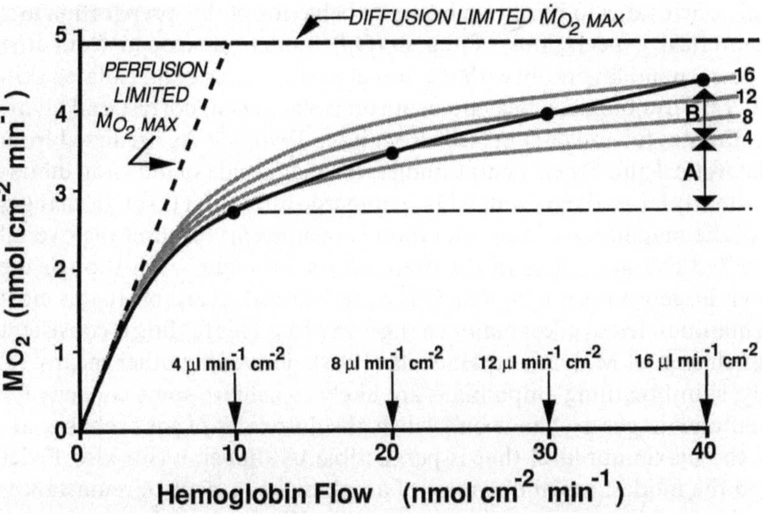

Fig. 4. Actual effect of perfusion on oxygen uptake ($\dot{M}o_2$) of an isolated perfused frog skin preparation (cf. Figs. 2 and 3, which represent theoretical predictions). The Po_2 of the gaseous respiratory medium was about 20 kPa; the Po_2 of the perfusate entering the preparation was about 4 kPa. The preparation was perfused at 4 flow rates, 4–16 µl min^{-1} cm^{-2}, with the corresponding hemoglobin flows shown on the abscissa. In the absence of any diffusion limitation, $\dot{M}o_2$ would be expected to increase according to the line labeled *perfusion limited* $\dot{M}o_2$ *max*. In the absence of any perfusion limitation and with maximum diffusing capacity, the $\dot{M}o_2$ would be expected to be fixed at the level labeled *diffusion limited* $\dot{M}o_2$ *max*. As perfusion rate was increased, diffusing capacity increased: the *four shaded curves* depict the $\dot{M}o_2$-$\dot{Q}\beta$ relationship for each of the four corresponding perfusion rates. $\dot{M}o_2$ also increased with increasing perfusion (*solid circles*; points connected with solid line fitted by eye). Thus, *arrow A* represents the increase in $\dot{M}o_2$ expected with increasing perfusion if D were fixed at the lowest level, and *arrow B* represents the additional increase in $\dot{M}o_2$ due to increases in D. Derived from the data of Figs. 2 and 3 in Pinder et al. (1991)

$\dot{Q}$ was increased, D changed as well, which progressively altered the $\dot{M}o_2$–$\dot{Q}\beta$–D relationship from the lowest curve to the highest curve shown in Fig. 4. Despite this change from one curve to the next, however, $\dot{M}o_2$ continued to occupy progressively higher positions along any given curve as $\dot{Q}\beta$ was increased. The higher position, which approached the maximum $\dot{M}o_2$ that is permitted by diffusion, represents an increasing degree of diffusion limitation. At the highest experimental $\dot{Q}$, 16 µl perfusate cm^{-2} min^{-1}, L_{diff} was 79%, $D/(\dot{Q}\beta)$ was 0.24, and fairly large changes in $\dot{Q}$ would be required to effect even modest changes in $\dot{M}o_2$.

Some additional studies (Clemens and Feder 1992) corroborate these findings. If P_m is manipulated while $\dot{Q}$ is either jointly manipulated or held constant, the $\dot{M}o_2$ of the preparation is strongly dependent upon P_m. Exposing the isolated skin to hypoxic Po_2, for example, reduces the $\dot{M}o_2$ in direct proportion to the change in $(P_m - P_{b(in)})$. Experimental increases in $\dot{Q}$ cannot compensate for or offset this reduction. If P_m is made hyperoxic, by contrast, $\dot{M}o_2$ does not increase in proportion to the change in $(P_m - P_{b(in)})$, because of the reduction in β at high Po_2 (see also Kobayashi et al. 1991).

General conclusions to be drawn from the behavior of this preparation are: (1) The theoretical expectations of the $D/(\dot{Q}\beta)$ model, developed from first principles, correspond favorably with the actual performance of an isolated skin preparation. (2) Physiological measurements on isolated skin correspond favorably with estimates for intact skin (see also Vitalis 1990). (3) As predicted from previous theoretical and experimental studies, the magnitude of diffusion limitation is consistently large. Even though D can undergo alteration through changes in perfusion, the magnitude of this alteration is insufficient to offset the overall diffusion-limited characteristics of the preparation. (4) Thus, even though the preparation is largely diffusion limited, $\dot{Q}$ is nonetheless the variable that is most likely to be manipulated to effect major changes in $\dot{M}o_2$. This finding is consistent with the conclusion of Malvin and Hlastala (1989), derived by other means. (5) Accordingly, skin-breathing amphibians are likely to achieve some capacity for regulating cutaneous gas exchange only when absolute rates of gas exchange are well below the maximum level that is permissible by diffusion (see also Feder 1988). Given the modest metabolic rates of amphibians in many circumstances (e.g. Pinder 1987), this tactic is sometimes feasible. When rates of gas exchange approach the diffusion-limited maximum, however, regulation of cutaneous gas exchange becomes extremely problematic for amphibians.

4 Importance of Ventilating the Skin in Regulating Cutaneous Gas Exchange in Water

For internal gas exchange organs, ventilation can be an important mechanism of regulation because ventilation is critical in transporting the respiratory medium to and from the gas exchange surface. The integument, by contrast, usually contacts an "infinite pool" (Piiper and Scheid 1975, 1977) of respiratory medium, seemingly suggesting that ventilation may be less significant to cutaneous gas exchange than to other forms of gas exchange. In reality, as Feder and Burggren (1985a, b) proposed and work during the past two decades has shown, skin ventilation can markedly affect gas exchange in aquatic amphibians by dissipating an unstirred layer or boundary layer surrounding the skin.

The boundary layer opposes transmembrane exchange, and its resistance must be overcome as respiratory gases exchange between an organism and the bulk medium around it (Vogel 1983; Barry and Diamond 1984; Feder and Pinder 1988). The boundary layer surrounding the skin of amphibians immersed in water is 30–4000 μm thick (Dainty and House 1966a, b; Pinder and Feder 1990; Booth and Feder 1991; Feder and Booth 1992). Flow is zero immediately adjacent to the skin, and increases gradually with distance from the skin until it approaches the "free stream" velocity (Vogel 1983). If cutaneous uptake of dissolved O_2 exceeds replenishment of O_2 via bulk flow or diffusion from the bulk medium, then the Po_2 of the boundary layer will decline and, in turn, depress cutaneous oxygen uptake (Fig. 5; Vitalis 1990; Booth and Feder 1991). The

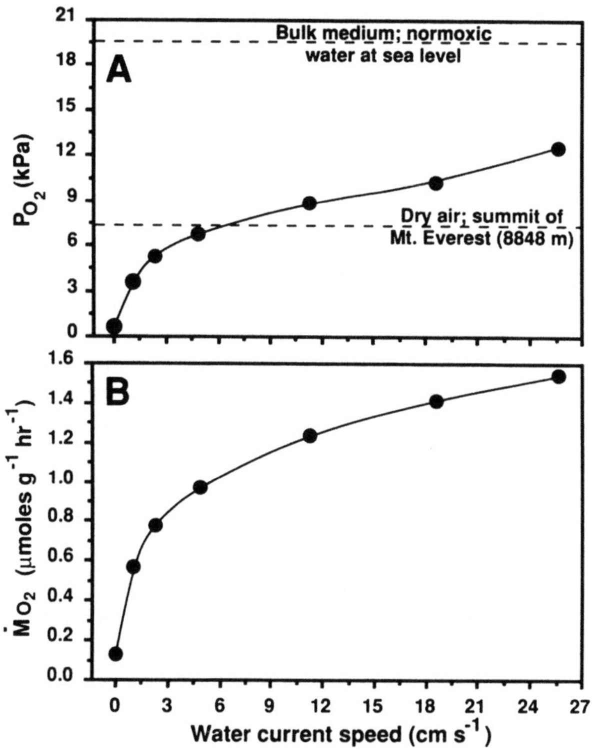

Fig. 5. Effect of current speed of normoxic water on **A** the Po_2 at the skin-medium interface and **B** oxygen consumption ($\dot{M}o_2$) of an exclusively skin-breathing aquatic salamander, *Desmognathus quadramaculatus* at 15 °C. Data and Po_2-$\dot{M}o_2$ relationship are from Booth and Feder (1991). In nature, this species is frequently at current speeds of 5 cm s^{-1} or less, at which oxygen consumption ought to be depressed. In **A**, the *broken lines* represent the Po_2 of the bulk medium (normoxic water at sea level) and of dry air at the summit of Mt. Everest (8848 m elevation) (Dejours 1982) as a frame of reference

severity of this depression ought to depend upon the flow velocity of the water surrounding a skin-breathing animal as well as on its size and the physical characteristics of its skin (Feder and Pinder 1988). The thickness of the fluid dynamic boundary layer scales as flow velocity$^{-0.5}$ (Vogel 1983); thus, the boundary layer will be especially thick in still or slowly moving water. Several studies with oxygen microelectrodes (Pinder and Feder 1990; Booth and Feder 1991; Feder and Booth 1992) have now revealed the actual Po_2 at the skin-medium interface (Po_2s-m) (Fig. 5). In still normoxic water, Po_2s-m ranges between 2–4 kPa (15–30 mmHg) for various regions of representative animals, and for various species of different body sizes at temperatures of 5–20 °C (Feder and Booth 1991). Sometimes the Po_2s-m is even lower (Fig. 5). To place these values in some perspective, they are less than the Po_2 at the summit of Mt. Everest (8848 m altitude) and are less than experimental Po_2s used in many experimental studies of hypoxia. Thus, exposure to still normoxic water is evidently tantamount to a profound hypoxic stress.

The consequences of large boundary layers are severalfold. The physical resistance of the boundary layer to gas flux can be nearly an order of magnitude greater than that of the skin itself (Pinder and Feder 1990). As stated above, the hypoxia in the boundary layer can directly reduce oxygen uptake by the skin (Fig. 5; Booth and Feder 1991; see also Beckenbach 1975; Withers 1980; Vitalis 1990) and blood oxygen carriage to the rest of the body (Burggren and Feder 1986; Feder and Pinder 1988; Pinder and Feder 1990). In many diving amphibians, dive length is a joint function of depletion of oxygen stored at the start of the dive and cutaneous oxygen uptake while immersed (Boutilier 1990). Hypoxic boundary layers, which reduce the Po_2s-m and hence putatively cutaneous oxygen uptake, ought thus to shorten dives. Feder and Booth (1992) tested this hypothesis by measuring the decline in pulmonary Po_2 in anesthetized *Xenopus* undergoing forced immersion in moving and in unstirred water. Voluntarily diving *Xenopus* typically begin dives at 16 kPa pulmonary Po_2 and end them at Po_2's averaging 4 kPa (Boutilier and Shelton 1986). In the experiment, *Xenopus* required 18.7 min to reduce pulmonary Po_2 from 16 to 4 kPa in flowing water but only 13.8 min to achieve a comparable reduction in unstirred water, i.e. a 26% reduction in dive time. In experiments with awake, unrestrained amphibians, cessation of stirring reduced the length of voluntary dives by 59% in *Xenopus* and 74% in the aquatic salamander *Siren* (Feder and Booth 1992). Such reductions in dive time, in turn, may have major effects on evolutionary fitness by compromising courtship of mates or bottom-feeding (Halliday and Sweatman 1976; Feder and Moran 1985), or by increasing vulnerability to aquatic and aerial predators (Feder 1983; Kramer 1988).

Because boundary layer thickness scales as flow velocity$^{-0.5}$, even slight ventilation of the skin ought to have a disproportionate effect on boundary layer thickness. Accordingly, potential solutions to the boundary layer problem are for an animal to inhabit water in which the flow rate is high, or to entrain water flow past the skin through ventilatory movements or routine locomotion. Indeed, experimental exposure to moving water increases cutaneous oxygen uptake (Burggren and Feder 1986; Pinder and Burggren 1986; Pinder and Feder 1990; Booth and Feder 1991) and blood oxygen transport (Burggren and Feder 1986; Feder and Pinder 1988), retards depletion of lung oxygen stores (Feder and Booth 1992), and prolongs submergence in diving amphibians (Feder and Booth 1992). Moreover, blood flow to the skin is rapidly adjusted when the skin is ventilated (Burggren and Feder 1986; Feder and Pinder 1988). Evidence is presently lacking for a behavioral preference for aquatic microhabitats with rapid flow velocities, and measurements of actual flow velocities in such microhabitats are few. In fact, *Desmognathus quadramaculatus*, a large plethodontid (lungless) salamander found under rocks in montane streams, seemingly avoids high flows. These salamanders tend to occupy rocks underneath which flow is ≤ 5 cm s^{-1}, a speed slow enough to depress the resting metabolic rate by $\geq 40\%$ (Fig. 5), even though higher flow velocities are available nearby (Booth and Feder 1991). By contrast, evidence is growing that amphibians exhibit specific ventilatory behaviors (Guimond and Hutchison 1973; Hutchison et al. 1976; Boutilier et al. 1980; Boutilier and Toews 1981; Pinder and Feder 1990) or simply move more frequently (Feder and Booth 1992) in response to unstirred water, aquatic hypoxia, or both. For example, exclusively skin-breathing frogs submerged in cold water in simulated overwintering conditions move their limbs when current flow is reduced; this limb movement is sufficient to dissipate the boundary layer and maintain the rate of O_2 uptake by the skin (Pinder and Feder 1990).

Hypoxic boundary layers bear on a still unresolved controversy concerning the evolution of lunglessness. The repeated evolution of lunglessness in salamanders (see Sect. 2.2), particularly in the Plethodontidae, has long been thought to have been associated with life in or near torrential streams (Wilder and Dunn 1920). In such habitats, lungs would be a liability, because of their buoyancy and might be unnecessary, because of the cool, well-oxygenated conditions of mountain streams. Ruben and Boucot (1989) challenged this scenario, asserting that the ancestral habitat of plethodontids was neither montane nor cool at the time lunglessness evolved. The foregoing work on hypoxic boundary layers, however, seemingly suggests that lunglessness is unlikely to have evolved in water at all. In currents rapid enough to dissipate boundary layers, lungless salamanders are readily dislodged and swept away. In slower currents, routine oxygen consumption is depressed (Fig. 5). Indeed, only a small minority of lungless species (e.g. *Desmognathus quadramaculatus*, *Leurognathus marmoratus*) remain in water for long periods. These few exceptional species, however, are evidently able to tolerate the local hypoxia brought on by boundary layer formation, and can survive for months in still normoxic water (unpubl. data). Whether this tolerance is due to metabolic depression or anaerobic metabolism is presently under investigation. Nonetheless, occupancy of terrestrial habitats facilitates cutaneous gas exchange by entirely eliminating hypoxic boundary layers as a problem. The physical properties of air make it far more conducive to gas exchange than is water, and oxygen boundary layers in air are likely to be negligible (Feder and Burggren 1985a). Even so, terrestrial habitats pose a serious problem to cutaneous exchange in terms of water loss.

5 Conclusions and Future Directions

A conclusion of the 1970s (and of Krogh in 1904) was that cutaneous gas exchange is primarily governed by diffusion and hence is passive and poorly regulated. During the 1980s, several mechanisms were hypothesized to allow the regulation of cutaneous gas exchange despite a large diffusion limitation. One class of mechanisms, alterations in diffusive and perfusive conductance of the skin, has now been demonstrated in several experimental systems. These mechanisms, however, appear to allow only modest increases in cutaneous gas flux except when gas exchange rates are initially low. A second kind of mechanism, skin ventilation in water, can markedly affect cutaneous gas fluxes, but assessment of its biological relevance must await measurements of actual current flow rates in nature (Feder and Booth 1992).

Many aspects of cutaneous gas exchange could justifiably receive attention in the future. One such aspect is microcirculatory variation in the skin. Because the skin is external, its microcirculation is accessible in situ to sophisticated analytical instrumentation (e.g. Malvin and Wood 1991). For example, real-time analysis of red cell position and velocity, perhaps performed with confocal fluorescence videomicroscopy, could resolve many outstanding questions concerning the regulation of cutaneous gas exchange (see Sect. 3.2). Accordingly, skin offers particular promise as a model system for many aspects of microcirculatory study. Another issue deserving future attention is the interaction of cutaneous gas exchange, cutaneous exchange of ions and water, and acid-base balance (Feder et al. 1991; Boutilier et al. 1992). Each of these processes must be regulated without compromising the others, and all occur in the same organ. Undoubtedly, the regulation of cutaneous gas exchange will continue to attract as much attention in the future as it has in the past.

Acknowledgments. Research was supported by NSF grant DCB-8718264. Juan Markin assisted with the preparation of the manuscript.

References

Barry PH, Diamond JM (1984) Effects of unstirred layers on membrane phenomenon. Physiol Rev 64: 763–872

Beckenbach AT (1975) Influence of body size and temperature on the critical oxygen tension of some plethodontid salamanders. Physiol Zool 48: 338–347

Booth DT, Feder ME (1991) Formation of hypoxic boundary layers and their biological implications in a skin-breathing aquatic salamander, *Desmognathus quadramaculatus*. Physiol Zool 64: 1307–1321

Boutilier RG (1990) Control and co-ordination of gas exchange in bimodal breathers. In: Boutilier RG (ed) Advances in comparative and environmental physiology 6. Springer, Berlin Heidelberg New York, pp 279–345

Boutilier RG, Shelton G (1986) Gas exchange, storage and transport in voluntarily diving *Xenopus laevis*. J Exp Biol 126: 133–155

Boutilier RG, Toews DP (1981) Respiratory, circulatory and acid-base adjustments to hypercapnia in a strictly aquatic and predominantly skin-breathing urodele, *Cryptobranchus alleganiensis*. Respir Physiol 46: 177–192

Boutilier RG, McDonald DG, Toews DP (1980) The effect of enforced activity on ventilation, circulation, and blood acid-base balance in the aquatic gill-less urodele, *Cryptobranchus alleganiensis*: a comparison with the semiterrestrial anuran, *Bufo marinus*. J Exp Biol 84: 289–302

Boutilier RG, Glass ML, Heisler N (1986) The relative distribution of pulmocutaneous blood flow in *Rana catesbeiana*: effects of pulmonary or cutaneous hypoxia. J Exp Biol 126: 33–40

Boutilier RG, Stiffler DF, Toews DP (1992) Exchange of respiratory gases, ions, and water in amphibious and aquatic amphibians. In: Feder ME, Burggren WW (eds) Environmental physiology of the amphibians. Univ Chicago Press, Chicago, pp 81–124

Burggren WW, Feder ME (1986) Effect of experimental ventilation of the skin on cutaneous gas exchange in amphibians. J Exp Biol 121: 445–449
Burggren WW, Moalli R (1984) 'Active' regulation of cutaneous gas exchange by capillary recruitment in amphibians: experimental evidence and a revised model for skin respiration. Respir Physiol 55: 379–392
Burggren WW, West NH (1982) Changing respiratory importance of gills, lungs and skin during metamorphosis in the bullfrog, *Rana catesbeiana*. Respir Physiol 47: 151–164
Burggren WW, Feder ME, Pinder A (1983) Temperature and the balance between aerial and aquatic respiration in larvae of *Rana berlandieri* and *Rana catesbeiana*. Physiol Zool 56: 263–273
Clemens DC, Feder ME (1992) Dependence of oxygen uptake on ambient Po_2 in isolated perfused frog skin. J Comp Physiol 171: 646–650
Czopek J (1965) Quantitative studies on the morphology of respiratory surfaces in amphibians. Acta Anat 62: 296–323
Dainty J, House CR (1966a) 'Unstirred layers' in frog skin. J Physiol 182: 66–78
Dainty J, House CR (1966b) An examination of the evidence for membrane pores in frog skin. J Physiol 185: 172–184
Dejours P (1982) Mount Everest and beyond; breathing air. In: Taylor CR, Johansen K, Bolis L (eds) A companion to animal physiology. Cambridge Univ Press, Cambridge, pp 17–30
Dolk HE, Postma N (1927) Über die Haut-und die Lungenatmung von *Rana temporaria*. Z Vergl Physiol 5: 417–444
Feder ME (1976) Lunglessness, body size, and metabolic rate in salamanders. Physiol Zool 49: 398–406
Feder ME (1983) The relation of air-breathing and locomotion to predation on tadpoles (*Rana berlandieri*) by turtles. Physiol Zool 56: 522–531
Feder ME (1986) Effect of thermal acclimation on locomotor energetics and locomotor performance in a lungless salamander, *Desmognathus ochrophaeus*. J Exp Biol 121: 271–283
Feder ME (1987) Effect of thermal acclimation on locomotor energetics and locomotor performance in a tropical salamander, *Bolitoglossa subpalmata*. Physiol Zool 60: 18–26
Feder ME (1988) Exercising with and without lungs. II. Experimental elimination of pulmonary and buccopharyngeal gas exchange in individual salamanders (*Ambystoma tigrinum*). J Exp Biol 138: 487–497
Feder ME (1992) A perspective on environmental physiology of the amphibians. In: Feder ME, Burggren WW (eds) Environmental physiology of the amphibians. Univ Chicago Press, Chicago, pp 1–6
Feder ME, Booth DT (1992) Hypoxic boundary layers surrounding skin-breathing aquatic amphibians: occurrence, consequences, and organismal responses. J Exp Biol 166: 237–261
Feder ME, Burggren WW (1985a) Cutaneous gas exchange in vertebrates: design, patterns, control, and implications. Biol Rev 60: 1–45
Feder ME, Burggren WW (1985b) The regulation of cutaneous gas exchange in vertebrates. In: Gilles R (ed) Circulation, respiration, and metabolism. Springer, Berlin Heidelberg New York, pp 101–113
Feder ME, Moran CM (1985) Effect of water depth on costs of aerial respiration and its alternatives in tadpoles of *Rana pipiens*. Can J Zool 63: 643–648
Feder ME, Pinder AW (1988) Ventilation and its effect on "infinite pool" exchangers. Am Zool 28: 973–983
Feder ME, Papenfuss TJ, Wake DB (1982) Body size and elevation in neotropical salamanders. Copeia 1982: 186–188
Feder ME, Burggerg, WW, Graham JB (eds) (1988a). Cutaneous exchange of gases and ions. Am Zool 28: 937–1045
Feder ME, Full RJ, Piiper J (1988b) Elimination kinetics of acetylene and Freon 22 in resting and active lungless salamanders. Respir Physiol 72: 229–240
Feder ME, Robbins T, Talbot CR (1991) Effect of bulk flow on cutaneous ion uptake in the bullfrog, *Rana catesbeiana*. Am Zool 31: 121A
Federspiel WJ, Sarelius IH (1984) An examination of the contribution of red cell spacing to the uniformity of oxygen flux at the capillary wall. Microvasc Res 27: 273–285

Foxon GEH (1964) Blood and respiration. In: Moore JA (ed) Physiology of the Amphibia. Academic Press, New York, pp 151–209
Full RJ (1986) Locomotion without lungs: energetics and performance of a lungless salamander. Am J Physiol 251: R775–R780
Full RJ, Anderson BD, Finnerty CJ, Feder ME (1988) Exercising with and without lungs. I. The effects of metabolic cost, maximal oxygen transport, and body size on terrestrial locomotion in salamander species. J Exp Biol 138: 471–485
Gans C (1970a) Respiration in early tetrapods—the frog is a red herring. Evolution 24: 740–751
Gans C (1970b) Strategy and sequence in the evolution of the external gas exchangers of ectothermal vertebrates. Forma Functio 3: 61–104
Gatten RE, Miller K, Full RJ (1992) Energetics of amphibians at rest and during locomotion. In: Feder ME, Burggren WW (eds) Environmental physiology of the amphibians. Univ Chicago Press, Chicago, pp 314–377
Gatz RN, Crawford EC, Piiper J (1975) Kinetics of inert gas equilibration in an exclusively skin-breathing salamander, *Desmognathus fuscus*. Respir Physiol 24: 15–29
Gottlieb G, Jackson DC (1976) Importance of pulmonary ventilation in respiratory control in the bullfrog. Am J Physiol 230: 608–613
Graham JB (1988) Ecological and evolutionary aspects of integumentary respiration: body size, diffusion, and the Invertebrata. Am Zool 28: 1031–1045
Guimond RW, Hutchison VH (1973) Aquatic respiration: an unusual strategy in the Hellbender *Cryptobranchus alleganiensis alleganiensis* (Daudin). Science 182: 1263–1265
Guimond RW, Hutchison VH (1976) Gas exchange of the giant salamanders of North America. In: Hughes GM (ed) Respiration of amphibious vertebrates. Academic Press, New York, pp 313–338
Halliday TR, Sweatman HPA (1976) To breathe or not to breathe: the newt's problem. Anim Behav 24: 551–561
Hutchison VH, Haines HB, Engbretson G (1976) Aquatic life at high altitude: respiratory adaptation in the Lake Titicaca frog, *Telmatobius culeus*. Respir Physiol 27: 115–129
Jackson DC, Braun BA (1979) Respiratory control in bullfrogs: cutaneous versus pulmonary response to selective CO_2 exposure. J. Comp Physiol 129: 339–342
Kobayashi H, Pelster B, Piiper J, Scheid P (1991) Diffusion and perfusion limitation in alveolar O_2 exchange: shape of the blood O_2 equilibrium curve. Respir Physiol 83: 23–34
Kramer DL (1988) The behavioral ecology of air breathing by aquatic animals. Can J Zool 66: 89–94
Krogh A (1904) On the cutaneous and pulmonary respiration of the frog. Skand Archiv Physiol 15: 328–419
Mackenzie J, Jackson DC (1978) The effect of temperature on cutaneous CO_2 loss and conductance in the bullfrog. Respir Physiol 32: 313–323
Malvin GM (1988) Microvascular regulation of cutaneous gas exchange in amphibians. Am Zool 28: 999–1007
Malvin GM, Hlastala MP (1986a) Regulation of cutaneous gas exchange by environmental O_2 and CO_2 in the frog. Respir Physiol 65: 99–111
Malvin GM, Hlastala MP (1986b) Effect of lung volume and O_2 and CO_2 content on cutaneous gas exchange in frogs. Am J Physiol 251: R941–R946
Malvin GM, Hlastala MP (1987) Capillary recruitment is not the primary mechanism regulating cutaneous gas exchange in the frog. Fed Proc 46: 792
Malvin GM, Hlastala MP (1989) Effects of environmental O_2 on blood flow and diffusing capacity in amphibian skin. Respir Physiol 76: 229–242
Malvin GM, Riedel C (1990) Autonomic regulation of cutaneous vascular resistance in the bullfrog *Rana catesbeiana*. J Exp Biol 152: 425–439
Malvin GM, Wood SC (1991) Tissue diffusing capacity in frog skin: effect of capillary red cell density. FASEB J 5: A684
Moalli R, Meyers RS, Ultsch GR, Jackson DC (1980) Acid-base control in a predominantly skin-breathing salamander, *Cryptobranchus alleganiensis*. Respir Physiol 43: 1–11
Nonotte G, Kirsch R (1978) Cutaneous respiration in seven sea-water teleosts. Respir Physiol 35: 111–118

Piiper J (1982) A model for evaluating diffusion limitation in gas-exchange organs of vertebrates. In: Taylor CR, Johansen K, Bolis L (eds) A companion to animal physiology. Cambridge Univ Press, Cambridge, pp 49–64

Piiper J (1988) Models for cutaneous gas exchange and transport. Am Zool 28: 963–972

Piiper J, Scheid P (1975) Gas transport efficacy of gills, lungs, and skin: theory and experimental data. Respir Physiol 23: 209–221

Piiper J, Scheid P (1977) Comparative physiology of respiration: functional analysis of gas exchange organs in vertebrates. In: Widdicombe JG (ed) International review of physiology, vol 14. Respiratory physiology II. Univ Park Press, Baltimore, pp 219–253

Piiper J, Scheid P (1981) Model for capillary-alveolar equilibration with special reference to O_2 uptake in hypoxia. Respir Physiol 51: 287–290

Piiper J, Gatz R, Crawford EC Jr (1976) Gas transport characteristics in an exclusively skin-breathing salamander, *Desmognathus fuscus* (Plethodontidae). In: Hughes GM (ed) Respiration of amphibious vertebrates. Academic Press, New York, pp 339–356

Pinder AW (1987) Cutaneous diffusing capacity increases during hypoxia in cold submerged bullfrogs (*Rana catesbeiana*). Respir Physiol 70: 85–95

Pinder AW, Burggren WW (1986) Ventilation and partitioning of oxygen uptake in the frog *Rana pipiens:* effects of hypoxia and activity. J Exp Biol 126: 453–468

Pinder AW, Feder ME (1990) Effect of boundary layers on cutaneous gas exchange. J Exp Biol 143: 67–80

Pinder AW, Smits AW (1987) Oxygen carrying capacity varies inversely with oxygen saturation of arterial blood in bullfrogs submerged at 10 °C. Physiologist 30: 189

Pinder AW, Clemens D, Feder ME (1990) An isolated perfused frog skin preparation for the study of gas exchange. In: Piiper J, Meyer M (eds) Oxygen transport to tissue, vol 12. Plenum Press, New York, pp 719–724

Pinder AW, Clemens DC, Feder ME (1991) An isolated perfused frog skin preparation for the study of gas exchange. I. Effect of bulk flow on oxygen uptake and diffusing capacity. Respir Physiol 85: 1–14

Rahn H, Howell BJ (1976) Bimodal gas exchange. In: Hughes GM (ed) Respiration of amphibious vertebrates. Academic Press, New York, pp 271–286

Randall DJ, Burggren WW, Farrell AP, Haswell MS (1981). The evolution of air-breathing in vertebrates. Cambridge Univ Press, Cambridge

Ruben JA, Boucot AJ (1989) The origin of lungless salamanders (Amphibia: Plethodontidae). Am Nat 134: 161–169

Talbot CR (1991) Regional variation in cutaneous gas exchange and its regulation in the bullfrog. Respir Physiol 89: 261–272

Ultsch GR (1973) A theoretical and experimental investigation of the relationships between metabolic rate, body size, and oxygen exchange capacity. Respir Physiol 18: 143–160

Ultsch GR (1976) Respiratory surface area as a factor controlling the standard rate of O_2 consumption of aquatic salamanders. Respir Physiol 26: 357–369

Vitalis TZ (1990) Pulmonary and cutaneous oxygen uptake and oxygen consumption of isolated skin in the frog *Rana pipiens*. Respir Physiol 81: 391–400

Vogel S (1983) Life in moving fluids. Princetion Univ Press, Princeton

West JB (1977) Ventilation/blood flow and gas exchange, 3rd ed. Blackwell, Oxford

West NH, Burggren WW (1984) Control of pulmonary and cutaneous blood flow in the toad, *Bufo marinus*. Am J Physiol 247: R884–R994

West NH, Van Vliet BN (1992) Sensory mechanisms regulating the cardiovascular and respiratory systems. In: Feder ME, Burggren WW (eds) Environmental physiology of the amphibians. Univ Chicago Press, Chicago, pp 151–182

Whitford WG, Hutchison VH (1967) Body size and metabolic rate in salamanders. Physiol Zool 40: 127–133

Wilder IW, Dunn ER (1920) The correlation of lunglessness in salamanders with a mountain brook habitat. Copeia 1920: 63–68

Withers PC (1980) Oxygen consumption of plethodontid salamanders during rest, activity, and recovery. Copeia 1980: 781–786

Chapter 2

The Role of Carbonic Anhydrase in Aquatic Gas Exchange

D.J. Randall[1] and *A.L. Val*[2]

Contents

1 Introduction

Carbonic anhydrase (carbonate hydrolyase, E.C. 4.2.1.1) is a multigene family zinc metalloenzyme (Bergenhem and Carlsson 1990) catalyzing the interconversion between CO_2 and HCO_3^- in almost all groups of organisms. Since the realization 60 years ago that the velocity at which HCO_3^- dehydrates to CO_2 was not great enough to allow the liberation of CO_2 into the lung, this enzyme has received much attention. Many physiological and biochemical processes have been associated with carbonic anhydrase activity including ion, gas and fluid transport, calcification, photosynthesis, formation of gastric acid, pancreatic

[1] Department of Zoology, University of British Columbia, 6270 University Boulevard, Vancouver, B.C., Canada V6T 2A9
[2] Instituto Nacional de Pesquisas da Amazônia (INPA), Alameda Cosme Ferreira 1756, 69083 Manaus, Amazonas, Brasil

Advances in Comparative and Environmental Physiology, Vol. 21

juice, saliva, cerebral-spinal fluid, aqueous humor, and synthesis of glucose, urea, and lipids (Tashian et al. 1990).

Thus, carbonic anhydrase is a multigene family and a multifunctional enzyme. The aim of this chapter is to review the role of carbonic anhydrase in aquatic gas exchange.

1.1 Isozymes

Seven genes have been described in vertebrates encoding the carbonic anhydrase isozymes which are termed as CAI, CAII, CAIII, CAIV, CAV, CAVI, and CAVII. In addition, CA "Y" from mouse liver cDNA might represent another (eighth) gene (Hewett-Emmett and Tashian 1991) and also a transmembrane protein (D8) has been described in *Vaccinia virus* (Tashian 1989). Although these isozymes have exhibited activity as carboxylesterases, aldeyde hydratases, and phosphatases in vitro, their physiological role seems to be restricted to the interconversion of CO_2 and HCO_3^-.

All these isozymes have practically the same structure as human CAI and CAII which are single chain peptides containing 259 or 260 amino acid residues and having a molecular weight of 29 000 (Deutsch 1984). The great homology detected for almost all isozymes indicates that they are the evolutionary products of a single primordial gene. Although they present great structural similarities, including the residues comprising the active site, they have exhibited strong differences in their catalytic activity and binding properties to inhibitors. As far as we know, carbonic anhydrase has one of the highest turnover numbers compared with all other studied enzymes. The isozyme CAII has a CO_2 hydration turnover number of $1 \times 10^{-6}\ s^{-1}$ at pH 9 and 25 °C. Human isozymes I and III are less active, exhibiting turnover numbers of $2 \times 10^{-5}\ s^{-1}$ and $4 \times 10^{-3}\ s^{-1}$ respectively, under similar conditions (Lindskog et al. 1984).

1.2 Carbonic Anhydrase in Fish

Hagfish have a single form of carbonic anhydrase in their red blood cell. Carlsson et al. (1980) showed that it has a similar inhibitor susceptibility to type CAII present in mammals. Maren et al. (1980), however, showed that the turnover number is comparable with isozyme CAI. Sanyal (1984) assumed that hagfish have CAI type in its red blood cells.

The tiger shark (*Galeocerdo cuvieri*) has a higher molecular weight CA than normal (38 000–41 000, Maynard and Coleman 1971) which is due to intermolecular linkage between glutathione and cysteine rather than a longer polypeptide chain (Bergenhem et al. 1986). According to Bergenhem and Carlsson (1990), both red cell and muscle types of CA of tiger shark have the same pI and similar sensitivities to sulfonamide inhibition, indicating that they are encoded by only one gene. In addition, these authors point out, based on the inhibition characteristics of the CA present in shark, that it is not the CAIII isozyme. Analysis of the amino acid sequence has indicated that tiger shark CA resembles the CAII type present in the human red blood cell (Bergenhem and Carlsson 1990). Based on their comparisons of substituted residues, Tashian et al. (1990) concluded that tiger shark CA is similar to isozyme CAI.

Only one type of carbonic anhydrase has been detected in teleost fish. The available data on turnover number and catalytic properties indicate that it resembles the isozyme CAII present in mammals (Sanyal 1984). Although some differences exist in amino acid composition (Deutsch 1984) and in the presence of only one sulfhydryl group in the teleost CA compared with four in hagfish and many in sharks (Sanyal 1984), the active site seems to be essentially similar in all these animals.

Gills of decapod crustaceans contain carbonic anhydrase, but no activity has been detected in the hemolymph (Randall and Wood 1981; Burnett and McMahon 1985). The external membrane-bound pool constitutes 70% of the total activity detected and consists of an isozyme sensitive to membrane impermeable dextran-bound inhibitors injected into the hemolymph (Burnett and McMahon 1985). The intracellular pool is acetazolamide-sensitive (Henry and Cameron 1983). Thus, there are two separate pools of carbonic anhydrase in crab gills, as pointed out by Bottcher et al. (1990).

1.3 Tissue Distribution

Carbonic anhydrase activity has been detected in almost all mammalian tissues. This wide distribution seems to be primarily due to the presence of the high-activity isozyme CAII and to some extent, the isozyme CAI (Tashian 1989). All other isozymes are expressed in specific tissues of mammals, either singly or in groups.

The cytoplasmic isozyme CAII seems to be widely distributed among fish tissues. Carbonic anhydrase activity has been detected, for example, in red blood cell, gill, skeletal muscle, rectal gland, and eyes of shark (Swenson et al. 1984; Bergenhem and Carlsson 1990). In addition, recently, Flugel and Lutjen-Drecoll (1991) have reported the presence of CA activity in the endometrial tissue from a late-term pregnant spiny dogfish uterus. There is some variation in carbonic anhydrase activity in diffferent tissues. Maren and Friedland (1978) reported a higher K_1 for anionic inhibitors for the rectal gland compared to the enzyme from eye and blood. Bergenhem and Carlsson (1990), however, suggested that the differences detected could be the result of different reduction states of the enzyme.

Henry et al. (1988) assayed carbonic anhydrase in lysed red blood cells and gills of *Ictalurus punctatus*. They observed that the gill CA activity is mainly due to the cytoplasmic fraction (56.3% of the CO_2 generated per minute per gram of tissue) and that the red cell CA activity was more than fourfold greater than that in the gills. Although only low activity was detected, CA is present in gill subcellular fractions (microsomes and mitochondria) of this catfish species.

Rahim et al. (1988) showed that trout gills contain carbonic anhydrase activity, located both within the cell and on the apical surface exposed to the water flow. Wright et al. (1986) reported high levels of carbonic anhydrase activity in the body mucus, and concluded from circumstantial evidence that the gill mucus also contained carbonic anhydrase activity.

Kidney and liver of teleost fish also exhibit CA activity. Toadfish livers (*Opsanus beta*) exhibit significant CA activity for both soluble and mitochondrial fractions which are involved with bicarbonate supply for urea synthesis (Walsh et al. 1989). Kidney CA in rainbow trout, through their effects on HCO_3^-/Cl^- and H^+/Na^+ exchanges, may maintain the rates of sodium and chloride uptake (Houston and McCarty 1978).

Carbonic anhydrase has been localized in gills, epithelial lining of the branchial chamber, muscle, heart, lungs, and antennal gland of crustaceans (Randall and Wood 1981; Burnett et al. 1985; Wheatly and Henry 1987; Morris and Greenaway 1990). As mentioned above, there are two pools of carbonic anhydrase in crabs. One of them is located on the outer surface of the basolateral membrane where it is involved with CO_2 excretion (Henry 1988). Otherwise, the intracellular pool of CA in the gill of land crabs provides HCO_3^- for exchange with Cl^- having a ionoregulatory role (Morris and Greenaway 1990). In fact, CA activity in stenohaline crab species is one-third to one-fifth the levels found in euryhaline species acclimated to the same salinity (Henry et al. 1990).

2 Biochemical Properties

The enzyme carbonic anhydrase, particularly the isozyme CAII, is an extremely efficient catalyst. A k_{cat} of $10^6\ s^{-1}$ and a K_m for CO_2 of 8.3 mM have been detected in the hydration reaction, whereas in the dehydration reaction these values are k_{cat} of $6 \times 10^5\ s^{-1}$ and K_m for HCO_3^- of 32 mM (Fersht 1985). Many studies have been carried out in order to solve the catalytic mechanism of carbonic anhydrase.

2.1 Kinetic Properties

Carbonic anhydrase is a metalloenzyme and in all cases this metal ion is zinc (Zn^{2+}). Zinc seems to play an important role in the catalytic mechanism of the enzyme in both directions. As pointed out by Fersht (1985), the catalytic mechanism of this enzyme "invokes a zinc-bound water molecule that ionizes with pK_a of 7 to give the nucleophilic zinc-bound hydroxide ion at high pH". Merz (1990) showed that there is easy access of CO_2 to the deep water pocket in the enzyme and a maximal exposure of a hydroxide oxygen lone pair to the CO_2 carbon atom. The zinc hydrogen bonding network, however, is still the subject of much speculation (Eriksson and Liljas 1991).

The kinetic mechanism as shown by Fersht (1985) has been used to explain many of the characteristics of carbonic anhydrase. The mechanism involves four steps, described as follows for the hydration direction: the negatively charged water oxygen binds to Zn^{2+} (step 1) with the subsequent release of H^+ and bound CO_2 (step 2). This provides the condition for one oxygen atom of carbon dioxide to bind loosely to zinc with a simultaneous hydroxide ion transfer to CO_2 (step 3); finally, there is the release of HCO_3^- and enzyme recovery (step 4). This mechanism has been considered insufficient to explain some characteristics when inhibitors are present. Lindskog et al. (1984) proposed a kinetic mechanism which they called "alkaline ping-pong". It includes the assumption that protons can be transferred from the inhibitor-enzyme complex to His-64. In this scheme, since there is no proton release from the enzyme to the medium, it is possible to explain why the dehydration reaction cannot proceed in the presence of an inhibitor.

The hydration of CO_2 and the dehydration of HCO_3^- exhibit opposite pH profiles when catalyzed by carbonic anhydrase. At pH 5, for example, dehydration occurs at a maximum rate while hydration occurs at a minimum rate; the opposite occurs at pH 9. Near pH 7 the enzyme activity in both directions is quite similar (Coleman 1984). This aspect is probably important for CO_2 transfer in animals since changes in cell pH and buffer concentrations will influence carbonic anhydrase activity.

Using a stopped-flow method, Sanyal et al. (1982) reported on the red blood cell carbonic anhydrase activity of the teleost *Archosargus probatocephalus*. In this case, the turnover number for CO_2 hydration was 77 000 s^{-1} at pH 7.5 and 1 °C, increasing up to 149 000 s^{-1} at 25 °C, i.e. three and five times smaller than that for human CAII under the same conditions, respectively. The reported K_m ($\approx$25 mM) for the above-mentioned teleost carbonic anhydrase is significantly higher than human CAII K_m (8.3 mM).

2.2 Inhibitors

Heterocyclic sulfonamides containing the thiadiazole ring are, in general, strong inhibitors of carbonic anhydrase, particularly the isozyme CAII. The human isozymes CAI and CAIII exhibit different sensitivities to this class of compounds (Sanyal 1984). Thus, many studies have been conducted using acetazolamide, also known as diamox (2-acetylamino-1,3,4-thiadiazole-5-sulfonamide) as a carbonic anhydrase inhibitor in fish (Dimberg 1988; Henry et al. 1988), frogs (Scheid and Siffert 1985), and crustaceans (Burnett et al. 1985). Sulfonamide compounds also include methazolamide, chlorzolamide, benzolamide, sulfanilamide, and ethoxzolamide which have different properties in inhibiting carbonic anhydrase. Acetazolamide and methazolamide, for example, are less potent than ethoxzolamide in inhibiting skeletal muscle CA in frogs (Scheid and Siffert 1985). Dimberg (1988) showed that acetazolamide and methazolamide have different effects on the blood CO_2 balance and that the former is more diffusible than the latter. In fact, we observed that the carbonic anhydrase in intact red blood cells of rainbow trout is less inhibited by acetazolamide than carbonic anhydrase from lysed cells under identical conditions.

Cyanate is also a potent inhibitor of CA exhibiting a K_i of 5×10^{-7} M for cat isozyme CAIII (Sanyal 1984; Geers and Gros 1988). Sanyal (1984) reported the cyanate inhibition constants for CA in hagfish (3×10^{-6} M), dogfish (3×10^{-5} M), and frog (6×10^{-6} M). Sanyal (1984) also showed that the constants for chloride inhibition for this group of animals are very high (7–30×10^{-2} M). Although cyanate is a potent inhibitor, it has not been used in experiments of CO_2 transfer, because of its parallel effects on respiratory metabolism and hemoglobin. Watson et al. (1982) reported that acephate, an insecticide that may easily enter aquatic ecosystems, inhibits both gill and red blood cell CA of rainbow trout, K_i values, however, are very high (207 and 48 mM for gill and rbc CA, respectively).

Inhibition of CA by a factor present in plasma has been demonstrated for dog and rabbit (Hill 1986) and pig and sheep (Leiner et al. 1962). In these animals, a high inhibition level has been reported. C_{50} values (% plasma that inhibits 50% of CA activity) were reported for red blood cell CA of dog (0.44%) and rabbit (1.02%) by Hill (1986). The physiological role and/or significance of plasma inhibitor(s) of carbonic anhydrase are unclear. Recently, Galar and Marroquin (1990) have shown that cAMP inhibits CA of frog skin. It is possible that cAMP could be involved with the inhibition of red blood cell CA by plasma in the above-mentioned animals since cAMP seems to be actively extruded from erythrocytes, at least of birds (Heasley and Brunton 1985). Haswell and Randall (1976) concluded that there were high levels of carbonic anhydrase inhibitor in trout plasma. It was shown subsequently, however, that this conclusion was invalid (Heming and Randall 1982). There may be inhibitors of CA in fish plasma, but there is no information available.

2.3 Role in Intermediary Metabolism

Bicarbonate ion and protons are necessary for a variety of metabolic pathways. Thus, carbonic anhydrase affects many different metabolic processes which are bicarbonate ion or proton dependent. Gluconeogenesis is affected by carbonic anhydrase inhibition since bicarbonate availability determines the rate of pyruvate carboxylation in the generation of glucose from lactate. For example, exercised alligators that received a CA inhibitor were hypoglycemic relative to the control group that did not receive the CA inhibitor (Herbert et al. 1983). Thus, in this case, CA inhibition not only stops bicarbonate ion production but also prevents glycogen synthesis, because exhaustion of bicarbonate makes pyruvate carboxylation the rate-limiting step.

Excess acetate is converted to fatty acids by incorporation of one carbon from bicarbonate to form malonate and then two of the three carbons of malonyl are added to the growing fatty acid chain. Herbert and Coulson (1984) showed that the inhibition of liver carbonic anhydrase blocks fatty acid synthesis, malonate formation being the rate-limiting step.

Coulson and Herbert (1984), who studied the effects of carbonic anhydrase inhibition in alligator, showed that ammonia excretion was unimpaired but that bicarbonate concentration decreased and chloride increased in the blood. Instead of ammonium bicarbonate excretion, the alligator excreted ammonium chloride and as a result the urine became acidic and the plasma alkaline. Interactions between ammonia and carbon dixide excretion were recently considered (Randall 1990; Heisler 1990) and will be discussed later in this chapter.

3 Gas Transfer

Carbonic anhydrase has been shown to affect many physiological functions in vertebrates, most studies being carried out on mammals. Relatively few experiments have been carried out on fish and most of these have been concerned with gas transfer, the subject of this review.

Fish transfer oxygen from water to blood across the gills, transport it to the tissues where it is utilized and is associated with the production of ammonia and carbon dioxide. Carbon dioxide is produced in the tissues, diffuses into the blood, and is then transported to the respiratory surface and excreted. Membranes are generally permeable to CO_2 but are not very permeable to bicarbonate. Carbonic anhydrase catalyzes the interconversion of carbon dioxide and bicarbonate, augmenting CO_2 transfer across membranes and facilitating CO_2 transfer through the body (Geers and Gros 1991). Carbonic anhydrase has a direct effect on carbon dioxide transfer in fish but only an indirect effect on ammonia and oxygen transfer via changes in pH. The ammonia:ammonium ion ratio is dependent on pH, as is oxygen binding to hemoglobin. Carbonic anhydrase through catalysis of the CO_2/bicarbonate hydration/dehydration reactions effects pH and, therefore, ammonia:ammonium ion ratios and oxygen binding to hemoglobin.

3.1 Muscle Carbonic Anhydrase

It would appear that fish muscle membranes are not very permeable to bicarbonate (Tang et al. 1992). As a result, molecular carbon dioxide, rather than bicarbonate, diffuses from muscle into the blood. Tang et al. (1992) came to this conclusion, because trout muscle P_{CO_2} was similar to that of venous blood but bicarbonate concentrations were far from having electrochemical equilibrium between muscle and plasma. Carbonic anhydrase activity has been demonstrated in fish muscle (Sanyal et al. 1982) and may play a role in carbon dioxide-facilitated diffusion and muscle homeostasis and contraction (Geers and Gros 1991), as in other vertebrates, but little is known of the functions of carbonic anhydrase in fish muscle. It is possibly involved in enhanced ammonia excretion from muscle

(Tang et al. 1992). If carbon dioxide is hydrated after leaving the muscle, the extracellular fluid surrounding the muscle will be acidified. This acidified layer around the muscle could, in turn, trap ammonia as ammonium ions and enhance ammonia transfer from the muscle. Bicarbonate infusion into the blood resulted in alkaline conditions in the plasma and ammonia retention in the muscle (Tang et al. 1992). These observations are consistent with an interaction between carbon dioxide and ammonia transfer from the muscle, possibly based on the development of an acid boundary layer resulting from carbon dioxide hydration close to the outside of the muscle. It is not clear, however, whether there is carbonic anhydrase activity available to plasma flowing through the muscle capillaries of fish. The presence or absence of such activity could have consequences for the role of catecholamines in regulating erythrocytic pH and the unloading of oxygen to the muscle (see below).

3.2 Gill Epithelium Carbonic Anhydrase

Carbon dioxide is transported in the blood mainly as bicarbonate, as in other vertebrates. At the gills, bicarbonate is dehydrated to carbon dioxide which then diffuses into the gill epithelium. Perfused gills, in the absence of any carbonic anhydrase in the perfusate, excrete carbon dioxide in proportion to the carbon dioxide, but not the bicarbonate, gradient across the gill epithelium. Under these conditions, if there is no carbon dioxide gradient, there will be no carbon dioxide excretion, even in the face of enormous bicarbonate gradients, indicating that the epithelium is not permeable to bicarbonate, at least on the basal surface and that, unlike the situation in terrestrial vertebrates, there is no carbonic anhydrase activity available to plasma as it flows through the gas exchange organ (Perry et al. 1982).

Carbonic anhydrase in the gill epithelium is found within the cell but is concentrated on the apical membrane (Rahim et al. 1988) and basal regions are devoid of the enzyme, consistent with the observation that CA activity is not available to the plasma during passage through the gills. Body mucus contains carbonic anhydrase activity (Wright et al. 1986) and it was presumed, based on other circumstantial evidence, that mucus on the gill surface also contains CA activity. It was concluded (Wright et al. 1986) that molecular CO_2 excreted across the gills was rapidly hydrated, acidifying the water in the gill lamellar channel (Wright et al. 1986; Randall et al. 1991).

Carbon dioxide hydration at the gill surface will enhance CO_2 transfer across the gills. Certainly, the chemical composition of the water, including pH, effects $PaCO_2$ in trout (Lin and Randall 1990). Proton excretion occurs across fish gills and is pH dependent, as is CO_2 hydration. In some instances, because of the magnitude of proton excretion, bicarbonate dehydration occurs in the water in the face of CO_2 excretion across the gills.

Water passing over the gills of fish is acidified if the pH is above about 5.3 but is alkalinized at lower pH values. The alkalinization at low pH can be accounted for by assuming all ammonia is excreted as NH_3 and then binds a proton raising water pH (Lin and Randall 1990). The degree of acidification at high pH, however, is too large to be accounted for by CO_2 hydration, even if it is assumed that all carbon dioxide excretion occurs as molecular CO_2. Thus, at high pH, there must be acid excretion in excess of that due to carbon dioxide hydration. In trout, acidification is not due to carbon dioxide hydration to any degree because bicarbonate levels decrease as water passes over the gills (Lin and Randall 1991), indicating that acid excretion causes bicarbonate dehydration even though there is simultaneous carbon dioxide excretion. That is, carbon dioxide levels in water passing over the gills increase due to both carbon dioxide excretion and bicarbonate dehydration as a result of acid excretion by the fish.

Lin and Randall (1991) concluded that trout gills may possess a proton ATPase on the apical membrane similar to that found in frog skin (Ehrenfeld et al. 1985). Acetazolamide inhibits acid excretion across trout gills, indicating that gill epithelial carbonic anhydrase may play some role in proton excretion across the apical membrane (Lin and Randall 1991), presumably by increasing the rate of proton supply through CO_2 hydration. Dimberg and Hogland (1987) reported that long-term acclimation of trout to hypercapnia resulted in an increase in branchial CA. In addition, Lin (unpubl. observ.) showed that hypercapnia increases proton ATPase activity within the gills, but this activity is not directly affected by acetazolamide. Thus, trout exposed to hypercapnia have increased carbonic anhydrase and proton ATPase activity. CA inhibitors reduce acid excretion, not by affecting proton ATPase directly, but possibly by reducing the supply of protons by inhibiting CO_2 hydration. Swenson et al. (1984) observed a differential effect of CA inhibition by methazolamide (inhibition of both gill and RBC CA) and benzolamide (inhibition of RBC CA only) in dogfish on excretion of a bicarbonate load. They concluded that gill epithelial CA played a role in CO_2 excretion. Their results, however, can also be explained by an effect on acid excretion across the gills.

The development of an acid boundary layer, whether by CO_2 hydration or acid excretion, has been shown to affect ammonia. Removal of the acid boundary layer by reducing CO_2 excretion or by adding buffer to the water reduced ammonia excretion (Wright et al. 1989). The increased ammonia excretion could be due to increased ammonia transfer from the gill surface, resulting from the combined movement of both ammonia and ammonium ions or due to changes in the NH_3 gradient across the gill epithelium. Randall et al. (1991) concluded that changes in the NH_3 gradient were more important in augmenting ammonia excretion under the conditions of the experiments of Wright et al. (1989).

3.3 Erythrocytic Carbonic Anhydrase

Erythrocytes are required for bicarbonate dehydration flowing through the gills, with the subsequent diffusion of molecular CO_2 through the gill epithelium. Thus, as blood flows through the gills, there will be some diffusion of molecular carbon dioxide from the plasma but no appreciable bicarbonate dehydration in the plasma as the transit time for blood flow through the gills is less than a second (Randall 1982). Oxygenation of the blood will result in the release of protons from hemoglobin (Albers et al. 1983) which will drive catalyzed bicarbonate dehydration and carbon dioxide production in the fish red blood cell. Red cell bicarbonate levels will fall and bicarbonate will enter the erythrocyte from the plasma via the band III anion exchanger. Hence, carbon dioxide excretion will be closely coupled to oxygen uptake across the fish gill. Red blood cell carbonic anhydrase activity is high, perhaps to ensure that the protons released during hemoglobin oxygenation are utilized in bicarbonate dehydration rather than being bound by nonbicarbonate buffers in the cell.

Some teleost fish have very low hematocrits or even no erythrocytes, e.g. ice fish. It is not clear how these animals excrete CO_2 at the low temperatures at which they exist and at which the uncatalyzed dehydration reaction is very slow. Possibly, the dehydration reaction is catalyzed by branchial carbonic anhydrase (Swenson 1990), similar to the situation in crabs (Burnett and McMahon 1985).

Many teleosts respond to hypoxemia by increasing the levels of circulating catecholamines and subsequently RBC intracellular pH (Nikinmaa 1990). This is of importance in maintaining blood oxygen content during acidosis. The blood of many teleost fish has a Root shift, whereby acid conditions result in a reduction in hemoglobin oxygen saturation even at high PaO_2. The Root shift is important in gas secretion into the swimbladder in response to local acidosis created by the gas gland, but if the acidosis is general, then oxygen delivery could be compromised. During conditions of extreme hypoxia or acidosis, catecholamine levels in the blood increase and sodium/proton exchange across the red blood cell membrane is stimulated by catecholamines acting via a beta adrenoreceptor mechanism. This increase in red blood cell pH occurs, because the rate of proton extrusion exceeds the rate of entry of acid via the Jacob-Stewart cycle. In this cycle, acid is transferred back into the erythrocyte, because the extruded proton causes bicarbonate dehydration and a rise in plasma carbon dioxide levels, carbon dioxide then diffuses back into the red blood cell, the subsequent hydration acidifying the cell. Plasma bicarbonate dehydration is the rate-limiting step in this process (Forster and Steen 1969). Thus, the pH of the red blood cell will only rise if the rate of proton extrusion via sodium/proton exchange exceeds the rate of plasma bicarbonate dehydration, which is uncatalyzed. The addition of carbonic anhydrase to the plasma will short-circuit the catecholamine-induced elevation in red blood cell pH (Nikinmaa 1990). Thus, plasma bicarbonate dehydration reactions are uncatalyzed, at least during blood flow through the gills, due to the absence of any carbonic anhydrase activity. This, in turn, permits catecholamines to exert an effect on RBC pH. The situation in fish muscle is unknown. If carbonic anhydrase is present in the muscle capillary endothelium, then the catecholamine-induced elevation in RBC pH would be short-circuited and this would drive oxygen from the red blood cell as a result of the decrease in intracellular pH. Carbon dioxide flux through the red blood cell, however, could also be affected and it is not clear what the overall effect would be on gas transport.

As blood flows through the gills, molecular carbon dioxide will be lost from he plasma by diffusion and bicarbonate will be lost due to transfer into the red lood cell. There will be essentially no interconversion of CO_2 and bicarbonate in lasma as blood flows quickly through the gills. A disequilibrium between carbon lioxide and bicarbonate will exist depending on the relative amounts of these two ompounds lost from the plasma. For example, if there is little oxygenation of the lood, the bicarbonate flux into the red blood cell will be reduced and bicarbonte levels in the plasma will remain high compared with plasma carbon dioxide evels. As blood flows away from the gills, bicarbonate will be dehydrated to arbon dioxide and $PaCO_2$ will rise (Gilmour et al. 1994). Infusion of carbonic nhydrase into the plasma will equilibrate plasma CO_2 and bicarbonate as blood lows through the gills. During hypoxia when the degree of hemoglobin oxygenaion, and hence bicarbonate flux through the red cell, is reduced then carbonic nhydrase infusion causes a drop in $PaCO_2$ (Lessard et al. 1995). Catecholamine elease results in the efflux of protons from the red blood cell and this has the otential to increase the disequilibrium between plasma bicarbonate and CO_2 by educing bicarbonate flux through the erythrocyte and/or to decrease plasma pH. arbonic anhydrase infusion following exhaustive exercise results in a reduction n $PaCO_2$ (Perry and Wood 1989), probably in part, because circulating catcholamines are elevated (Primmett et al. 1986).

.4 Crustacean Carbonic Anhydrase

rustacean hemolymph is devoid of carbonic anhydrase activity but the basal nembrane of the gills has high levels of activity that catalyzes bicarbonate lehydration (Randall and Wood 1981; Henry 1988). CO_2 diffuses across the gill pithelium of crabs and cytoplasmic CA is involved in CO_2 hydration to supply icarbonate and protons as counterions for ion exchange across the apical nembrane (Perry and Laurent 1990).

Concluding Remarks

n conclusion, the low-activity isozyme CAI predominates in hagfish and sharks, vhile the high-activity CAII type seems to be present in teleosts, as well as in mphibians and reptiles (Bundy and Cheng 1976; Hall and Shraer 1979; Scheid nd Siffert 1985). Red blood cells of turtles, however, also contain the low-activity ype CAI isozyme (Tashian et al. 1981). From an evolutionary point of view, it ppears that gene duplication and the mutational events encoding all other CA sozymes detected in mammals occurred after the appearance of terrestrial ertebrates.

There are high levels of carbonic anhydrase activity in fish muscle, erythocytes, and the gill epithelium. The plasma is devoid of carbonic anhydrase activity and this is of significance, because catecholamine-induced elevations of red blood cell pH depend on the sodium/proton exchange working against the uncatalyzed bicarbonate dehydration reaction. The absence of carbonic anhydrase activity in the plasma, however, results in CO_2/bicarbonate disequilibria in plasma flowing away from the gills and perhaps in other regions of the circulation.

The gill epithelial cytosolic carbonic anhydrase activity is involved in CO_2 hydration reactions, increasing the availability of protons and bicarbonate for ion-exchange processes. The apical regions of the gill epithelium and the mucus contain carbonic anhydrase activity and this catalyzes reactions in the water close to the gill surface. Carbon dioxide excretion and its subsequent catalyzed hydration due to the presence of carbonic anhydrase may result in an acid boundary layer around muscle and at the gill surface. In both instances, the formation of this acid boundary layer may enhance ammonia by a process of ammonium ion trapping.

Crustacean hemolymph is devoid of carbonic anhydrase activity. Catalyzed hemolymph bicarbonate dehydration, however, occurs at the surface of the basal membrane of the gills. As in fish, crustacean gill cytosolic carbonic anhydrase activity is involved in supplying protons and bicarbonate for ion-exchange processes in acid-base regulation.

Acknowledgments. This review is based upon work supported by the NSERC Canada. ALV has been supported by CNPq/Brasil.

References

Albers C, Goetz K-H, Hughes GM (1983) Effect of acclimation temperature on intraerythrocytic acid-base balance and nucleoside triphosphates in the carp, *Cyprinus carpio*. Respir Physiol 54: 145–159

Bergenhem N, Carlsson U (1990) Partial amino acid sequence of erythrocyte carbonic anhydrase from tiger shark. Comp Biochem Physiol 95B: 205–213

Bergenhem N, Carlsson U, Strid L (1986) The existence of glutathione and cysteine disulfide-linked to erythrocyte carbonic anhydrase from tiger shark. Biochim Biophys Acta 871: 55–60

Bottcher K, Siebers D, Becker W (1990) Localization of carbonic anhydrase in the gills of *Carcinus maenas*. Comp Biochem Physiol 96B: 243–246

Bundy HF, Cheng B (1976) Amphibian carbonic anhydrase: purification and partial characterization of the enzyme from erythrocytes of *Rana catesbeiana*. Comp Biochem Physiol 55B: 265–271

Burnett LE, McMahon BR (1985) Facilitation of CO_2 excretion by carbonic anhydrase located on the surface of the basal membrane of crab gill epithelium. Resp Physiol 62: 341–348

Burnett LE, Dunn TN, Infantino RL (1985) The function of carbonic anhydrase in crustacean gills. In: Gilles R, Gilles-Baillien M (eds) Transport processes, iono- and osmoregulation. Springer, Berlin Heidelberg New York, pp 159–168

Carlsson U, Kjellstrom B, Antonosson B (1980) Purification and properties of cyclostome carbonic anhydrase from erythrocytes of hagfish. Biochim Biophys Acta 612: 160–170

Coleman JE (1994) Carbonic anhydrase: zinc and the mechanism of catalysis. Ann N Y Acad Sci 429: 26–48

Coulson RA, Herbert JD (1984) A role for carbonic anhydrase in intermediary metabolism. Ann N Y Acad Sci 429: 505–515
Deutsch HF (1984) Primary structures and genetic changes in mammalian carbonic anhydrase isozymes. Ann N Y Acad Sci 429: 183–194
Dimberg K (1988) Inhibition of carbonic anhydrase in vivo in freshwater-adapted rainbow trout: differential effects of acetazolamide and metazolamide on blood CO_2 levels. Comp Biochem Physiol 91A(2): 253–258
Dimberg K, Hogland LB (1987) Carbonic anhydrase activity in the blood and gills of rainbow trout during long term hypercapnia in hard, bicarbonate rich freshwater. J Comp Physiol 157: 405–412
Ehrenfeld J, Garcia-Romeu F, Harvey BJ (1985) Electrogenic active proton pump in *Rana esculenta* skin and its role in sodium ion transport. J Physiol Lond 359: 331–355
Eriksson AE, Liljas A (1991) X-Ray Crystallographic studies of carbonic anhydrase isozymes I, II, and III. In: Dodgson SJ, Tashian RE, Gros G, Carter ND (eds) The carbonic anhydrases. Plenum Press, New York, pp 33–48
Fersht A (1985) Enzyme structure and mechanism. Freeman, New York
Forster RE, Steen JB (1969) The rate of the "Root shift" in eel red cells and eel hemoglobin solutions. J Physiol 204: 259–282
Flugel C, Lutjen-Drecoll E (1991) Distribution of carbonic anhydrase in the uterus of late-term pregnant spiny dogfish (*Squalus acanthias*). J Exp Biol 158: 531–537
Galar I, Marroquin MC (1990) Cyclic AMP effects on chloride transport and carbonic anhydrase activity in frog skin. Can J Physiol Pharmacol 68: 1269–1274
Geers C, Gros G (1988) Carbonic anhydrase inhibition affects contraction of directly stimulated rat soleus. Life Sci 42: 37–45
Geers C, Gros G (1991) Muscle carbonic anhydrases: function in muscle contraction and in the homeostasis of muscle pH and pCO_2. In: Dodgson SJ, Tashian RE, Gros G, Carter ND (eds) The carbonic anhydrases. Plenum Press, New York, pp 227–240
Gilmour KM, Randall DJ, Perry SF (1994) Acid-base disequilibrium in the arterial blood of rainbow trout. RESP Physiol 96: 259–272
Hall GE, Shraer R (1979) Purification and partial characterization of high- and low-activity carbonic anhydrase isoenzymes from *Malaclemys terrapin centrata*. Comp Biochem Physiol 63B: 561–567
Haswell MS, Randall DJ (1976) Carbonic anhydrase inhibitor in trout plasma. Respir Physiol 28: 17–27
Heasley LE, Brunton LL (1985) Prostaglandin A_1 metabolism and inhibition of cyclic AMP extrusion by avian erythrocytes. J Biol Chem 260: 11514–11519
Heisler N (1990) Acid-base regulation: interrelationships between gaseous and ionic exchange. In: Boutilier RG (ed) Advances in comparative and environmental physiology, vol 6. Springer, Berlin Heidelberg New York, pp 211–252
Heming TA, Randall DJ (1982) Fish erythrocytes are bicarbonate permeable: problems with determining carbonic anhydrase activity using the modified boat technique. J Exp Zool 219: 125–128
Henry RP (1988) Subcellular distribution of carbonic anhydrase activity in the gills of blue crab, *Callinectes sapidus*. J Exp Zool 245: 1–8
Henry RP, Cameron JN (1983) The role of carbonic anhydrase in respiration, ion regulation and acid-base balance in the aquatic crab *Callinectes sapidus* and the terrestrial crab *Gecarcirnus lateralis*. J Exp Biol 103: 205–223
Henry RP, Smatresk NJ, Cameron JN (1988) The distribution of branchial carbonic anhydrase and the effects of gill and erythrocyte carbonic anhydrase inhibition in the channel catfish *Ictalurus punctatus*. J Exp Biol 134: 201–218
Henry RP, Perry HM, Trigg CB, Handley HL, Krarup A (1990) Physiology of two species of deep-water crabs, *Chaceon fenneri* and *C. quinquedens*: gill morphology, and hemolymph ionic and nitrogen concentrations. J. Crust Biol 103(3): 375–381
Herbert JD, Coulson RA (1984) A role for carbonic anhydrase in de novo fatty acid synthesis in liver. Ann N Y Acad Sci 429: 525–527

Herbert JD, Coulson RA, Hernandez T (1983) Inhibition of pyruvate carboxylation in alligators and chameleons by carbonic anhydrase inhibitors. Comp Biochem Physiol 75A: 185–192

Hewett-Emmett D, Tashian RE (1991) Structure and evolutionary origins of the carbonic anhydrase multigene family. In: Dodgson SJ, Tashian RE, Gros G, Carter ND (eds) The carbonic anhydrases. Plenum Press, New York, pp 15–32

Hill EP (1986) Inhibition of carbonic anhydrase by plasma of dogs and rabbits. J Appl Physiol 60(1): 191–197

Houston AH, McCarty LS (1978) Carbonic anhydrase (acetazolamide-sensitive esterase) activity in the blood, gill and kidney of the thermally acclimated rainbow trout, *Salmo gairdneri*. J Exp Biol 73: 15–27

Leiner M, Beckand H, Eckert H (1962) Über die Kohlensäure-Dehydratase in den einzelnen Wirbeltierklassen. I. Der Zinkgehalt in den einzelnen Fermenten und die Wirkung des Inhibitors aus dem Schafblut auf die einzelnen Enzyme. Hoppe-Seyler's Physiol Chemie 327: 144–165

Lessard J, Val AL, Aota S, Randall DJ (1995) Why is there no carbonic anhydrase activity available to fish plasma? J Exp Biol 198: 31–38

Lin H, Randall DJ (1990) The effect of varying water pH on the acidification of expired water in rainbow trout. J Exp Biol 149: 149–160

Lin H, Randall DJ (1991) Evidence for the presence of an electrogenic proton pump on the trout gill epithelium. J Exp Biol 161: 119–134

Lindskog S, Engberg P, Forsman C, Ibrahim SA, Jonsson BH, Simonsson I, Tibell L (1984) Kinetics and mechanism of carbonic anhydrase isoenzymes. Ann N Y Acad Sci 429: 61–75

Merz Jr KM (1990) Insights into the function of the zinc hydroxide-Thr199-Glu 106 hydrogen bonding network in carbonic anhydrase. J Mol Biol 214: 799–802

Maren TH, Friedland BR (1978) Further studies on the phylogeny of vertebrate carbonic anhydrase in red cells and secretory organs. Bull Mount Desert Island Biol Lab 18: 79–82

Maren TH, Friedland BR, Rittmaster RS (1980) Kinetic properties of primitive vertebrate carbonic anhydrases. Comp Biochem Physiol 67B: 69–74

Maynard JR, Coleman JE (1971) Elasmobranch carbonic anhydrase purification and properties of the enzyme from two species of shark. J Biol Chem 246: 4455–4464

Morris S, Greenaway P (1990) Adaptations to a terrestrial existence by the robber crab, *Birgus latro* L. V. The activity of carbonic anhydrase in gills and lungs. J. Comp Physiol B 160: 217–221

Nikinmaa N (1990) Vertebrate red blood cells. Zoophysiology 28. Springer, Berlin Heidelberg New York

Perry SF, Laurent P (1990) The role of carbonic anhydrase in carbon dioxide excretion, acid-base balance and ionic regulation in aquatic gill breathers. Comp Physiol 6. Karger, Basel, pp 39–57

Perry SF, Wood CM (1989) Control and co-ordination of gas transfer in fishes. Can J Zool 67: 2961–2970

Perry SF, Davie PS, Daxboeck C, Randall DJ (1982) A comparison of CO_2 excretion in a spontaneously ventilating blood-prefused trout preparation and saline-perfused gill preparation: contribution of the branchial epithelium and the red blood cell. J Exp Biol 101: 47–60

Primmett DRN, Randall DJ, Mazeaud M, Boutilier RG (1986) The role of catecholamines in erythrocyte pH regulation and oxygen transport in rainbow trout *Salmo gairdneri* during exercise. J Exp Biol 122: 139–148

Rahim SM, Delaunoy JP, Laurent P (1988) Identification and immunocytochemical localization of two different carbonic anhydrase isoenzymes in teleostean fish erythrocytes and gill epithelia. Histochemistry 89: 451–459

Randall DJ (1982) The control of respiration and circulation in fish during exercise and hypoxia. J Exp Biol 100: 275–288

Randall DJ (1990) Control and co-ordination of gas exchange in water breathers. In: Boutilier RG (ed) Advances in comparative and environmental physiology, vol 6. Springer, Berlin Heidelberg New York, pp 253–278

Randall DJ, Wood CM (1981) Carbon dioxide excretion in the land crab (*Cardisoma carniflex*). J Exp Zool 218: 37–44
Randall DJ, Lin H, Wright PA (1991) Gill water flow and the chemistry of the boundary layer. Physiol Zool 64: 26–38
Sanyal G (1984) Comparative carbon dioxide hydration kinetics and inhibition of carbonic anhydrase isozymes in vertebrates. Ann N Y Acad Sci 429: 165–178
Sanyal G, Pessah NI, Swenson ER, Maren TH (1982) The isolation of carbonic anhydrase from the muscle of *Squalus acanthias* and *Scomber scombrus*: inhibition studies. Bull Mount Desert Island Biol Lab 24: 66–68
Sanyal G, Swenson ER, Maren TH (1982) The carbon dioxide hydration activity of purified teleost red cell carbonic anhydrase: inhibition by sulfonalamides and anions. Comp Biochem Physiol 73B: 937–944
Scheid P, Siffert W (1985) Effects of inhibiting carbonic anhydrase on isometric contraction of frog skeletal muscle. J Physiol 361: 91–101
Swenson ER (1990) Kinetics of oxygen and carbon dioxide exchange. In: Boutilier RG (ed) Advances in comparative and environmental physiology, vol 6. Springer, Berlin Heidelberg New York, pp 162–210
Swenson ER, Maren TH, Hildesley MA (1984) Further studies on elasmobranch CO_2 and HCO_3^- excretion: roles of branchial erythrocyte carbonic anhydrase in *Squalus acanthias*. Bull Mount Desert Island Biol Lab 24: 72–75
Tang Y, Lin H, Randall DJ (1992) Compartmental distribution of ammonia and carbon dioxide in rainbow trout at rest and following exercise, and the effect of bicarbonate infusion. J Exp Biol 169: 235–249
Tashian R (1989) The carbonic anhydrases: widening perspectives on their evolution, expression and function. Bioessays 10: 186–192
Tashian RE, Stroup SK, Hall GE (1981) Primary sequence of turtle low-activity red cell carbonic anhydrase: homology with mammalian low-activity CAI isozymes. Fed Proc 40: 1677
Tashian RE, Venta PJ, Nicewander PH, Hewett-Emmett D (1990) Evolution, structure, and expression of the carbonic anhydrase multigene family. In: Ratazzi GC, Scandaliso JG, Whitt GS (eds) Isozymes: structure, function and use in biology and medicine. Wiley-Liss, New York, pp 159–175
Walsh PJ, Parent JJ, Henry RP (1989) Carbonic anhydrase supplies bicarbonate for urea synthesis in toadfish (*Opsanus beta*) hepatocytes. Physiol Zool 62(6): 1257–1272
Watson TA, Tilley PAG, McKeown BA, Geen GH (1982) In vitro effects of acephate on carbonic anhydrase activity in the blood and gills of rainbow trout, *Salmo gairdneri*. J Environ Sci Health B17(1): 63–75
Wheatly MG, Henry RP (1987) Branchial and antennal gland Na^+/K^+-dependent ATPase and carbonic anhydrase activity during salinity acclimation of the eryhaline crayfish *Pacifastus leniusculus*. J Exp Biol 133: 73–86
Wright PA, Heming TA, Randall DJ (1986) Downstream pH changes in water flowing over the gills of rainbow trout. J Exp Biol 126: 499–512
Wright PA, Randall DJ, Perry SF (1989) Fish gill water boundary layer: a site of linkage between carbon dioxide and ammonia excretion. J Comp Physiol 158B: 627–635

Chapter 3

Gas Exchange in the Fish Swimbladder

P. Scheid and *B. Pelster*

Contents

1 Introduction

Neutral buoyancy is of great advantage to any object in water – a submarine as well as an animal – since it will help effectively in saving energy to avoid sinking or rising. Depending on their composition, fish are usually somewhat denser than the surrounding water, largely because of their skeleton and protein masses (densities around 2–3 and 1.3 g·cm^{-3}, respectively), which overrides the "float-

Institut für Physiologie, Ruhr-Universität Bochum, D-44780 Bochum, Germany

Advances in Comparative and Environmental Physiology, Vol. 21

ing" mass of fat (density 0.9 $g \cdot cm^{-3}$). Thus, for marine fish, overall tissue density is typically some 5% greater than that of seawater (8% for fresh water fish), which would result in a sinking force of some 5% of their weight. To overcome this force in water a formidable energy expenditure is neccessary. For an animal on land, this "sinking" force, 100% of the weight, is totally counteracted by its resting on solid ground. There are several ways in which animals try to achieve neutral buoyancy (Denton 1961). Fish have adopted the option of incorporating a gas-filled chamber, the swimbladder, but due to its compressible walls the gases must be released or deposited when fish migrate between different depths.

An air-filled chamber is an excellent device to reduce the density of fish without adding much weight. In fact, the density of air is only 0.00125 $g \cdot cm^{-3}$ at 1 atm, and still only 0.7 $g \cdot cm^{-3}$ at 700 atm, a pressure experienced in the sea at a depth of 7000 m below the surface. These values are not basically different for gases other than air. There are, nonetheless, particular problems with the swimbladder as a floating device.

A given swimbladder volume will grant the fish neutral buoyancy only at one particular depth, provided the gas-filled swimbladder obeys Boyle's law, which it apparently does (Denton 1961). Any elevation would result in the expansion of the swimbladder gas, and the increased swimbladder volume would add a lifting force, which would increase with elevation. By the same token, any descent from the level of neutral buoyancy would reduce swimbladder volume, creating a sinking force which would increase with depth. Thus, like a Cartesian diver, the level of buoyancy has a labile equilibrium; any deviation upward will lift the fish with increasing speed, and vice versa.

Since fish are found at various depths, partly following a diurnal rhythm, the possibility for rapid adjustment of swimbladder volume must exist, i.e. deposition and resorption of gas from the swimbladder. Moreover, it must be possible to prevent the loss of gases from inside the swimbladder in the face of high partial pressure head.

We will briefly discuss the "typical" anatomy of the swimbladder, and will then review the mechanisms by which gases are deposited, reabsorbed and trapped in the swimbladder. Particular emphasis has recently been placed on the mechanisms of gas deposition. We shall describe the classical model for gas deposition, some recent additions to it and some still open questions.

Comprehensive reviews on various aspects of the fish swimbladder can be found elsewhere (Denton 1961; Alexander 1966; Fänge 1966, 1983; Scheid et al. 1990; Pelster and Scheid 1992a).

2 Swimbladder Architecture

Many marine fish have a swimbladder, i.e. mostly those inhabiting the upper 1000 m in free ocean water and species found below about 2000 m; the depth record being 7000 m.

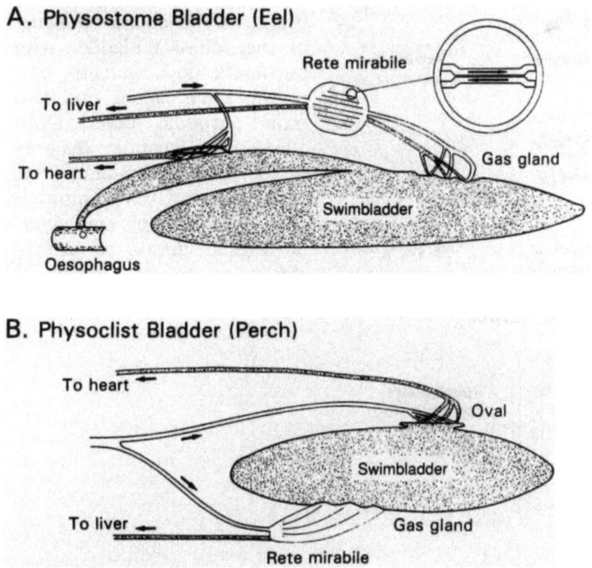

Fig. 1. General anatomy of the swimbladder of physostome fishes (**A**) and physoclist fishes (**B**). (After Denton 1961)

The swimbladder may be viewed as a gas-containing sac. Despite differences in details, it is functionally very similar in most fish. (For a comprehensive discussion of the various swimbladder types, see Fänge 1953; Steen 1970). Embryonically, the bladder originates as an outgrowth of the foregut. In physostome fishes the connection to the digestive tract persists, in physoclist fishes it is lost (Fig. 1). Here, the swimbladder of the eel will be considered, since it has served as an experimental animal in many studies on swimbladder function.

The swimbladder of the eel (Fig. 2A) consists of an unpaired, elongated sac which is connected to the esophagus through the pneumatic duct. The opening to the esophagus is, however, functionally closed and, therefore, not of significance for gas filling or emptying of the bladder by this route. Swimbladder and pneumatic duct serve quite different functions; the former as the site for gas deposition, the latter for gas resorption. The opening between both can be occluded.

The swimbladder is lined by a cuboidal epithelium, whereas the epithelium of the pneumatic duct is flat. Both structures are supplied with blood through a common artery. However, whereas the capillaries of the pneumatic duct are drained by veins that enter directly into the large veins, there is a countercurrent capillary arrangement in the red bodies of the swimbladder, the retia mirabilia. Thus, arterial blood enters the arterial capillary supply of the rete, an arrangement of 30000–50000 arterial capillaries, before it reassembles into (post-rete) arterial vessels that feed the capillaries of the swimbladder wall. Blood draining

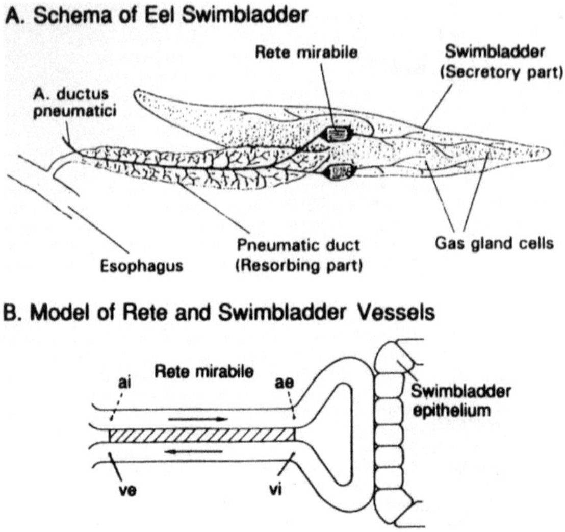

Fig. 2. **A** Schematic drawing of the eel swimbladder with pneumatic duct, secretory part of the swimbladder and two retia mirabilia (after Dorn 1961). **B** Schematic diagram of the vascular system of the swimbladder with rete mirabile and swim-bladder epithelium. *ai* arterial influx; *ae* arterial efflux; *vi* venous influx; *ve* venous efflux of the rete

from them into veins in the swimbladder wall feed, in turn, the venous capillaries in the rete, again about 20000–40000 capillaries that run strictly parallel, intermingled in a checkerboard arrangement with the arterial capillaries. The diffusion distance between these capillaries ranges between 1–2 µm (Stray-Pedersen and Nicolaysen 1975). Venous blood continues on to the liver (again a capillary network) before it returns to the main circulation. The rete thus presents the basis for a hairpin countercurrent system, similar to that of the kidney, and this is of functional significance for the mechanism of gas deposition. A schematic diagram of the vascular system of the swimbladder is shown in Fig. 2B.

3 Swimbladder Gas

3.1 Composition of Swimbladder Gas

The total pressure of the swimbladder gas at any depth is in equilibrium with the hydrostatic pressure. Thus, at 2000 m this pressure would be about 200 atm. Analysis of swimbladder gas from fish at various depths has shown that CO_2, O_2, N_2 and other inert gases occur in varying fractions. In shallow water, the fractional composition resembles that of air, whereas in deeper regions, O_2 becomes the major component (Fig. 3). There is, furthermore, a time dependence of this fraction: freshly deposited gas contains up to 25% CO_2, but CO_2 is later replaced by O_2 and inert gases (Wittenberg et al. 1964; Steen 1970).

What is important, though, is that at any depth the partial pressure of any gas component exceeds that in the surrounding water. Indeed, the partial pressures of

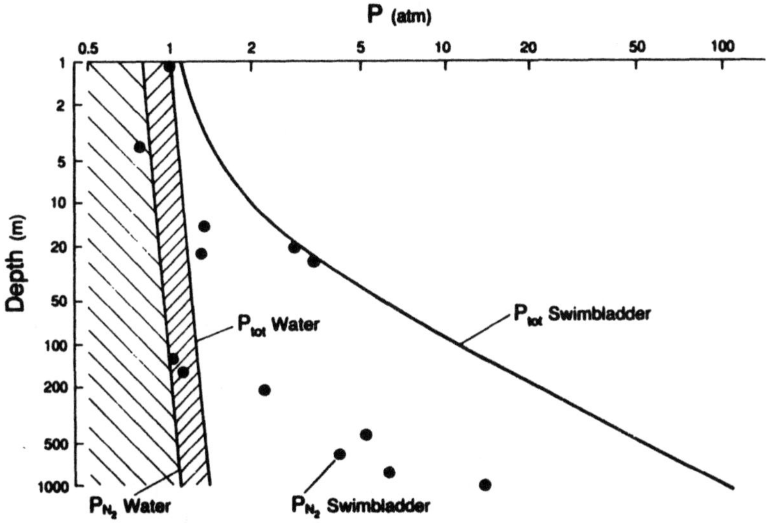

Fig. 3. Nitrogen partial pressure (•) measured in the swimbladder of various fish caught at different water depths; each *symbol* represents a different species. Also given are the depth dependencies for nitrogen partial pressure and for the total gas pressure in the water as well as for the total gas pressure in the swimbladder. The difference between the nitrogen partial pressure and the total gas pressure in the swimbladder is almost exculsively made up by oxygen. Note logarithmic coordinate scales

dissolved gases hardly increase with depth in the oceans. At a depth of 2000 m, Po_2 in the swimbladder may thus be close to 180 atm and P_{N_2} around 20 atm, while in arterial blood, Po_2 is below 0.2 atm and P_{N_2}, 0.8 atm.

3.2 What Prevents Gas from Leaving the Swimbladder?

Irrespective of how the gases entered the swimbladder in the first place, there is tremendous pressure head for them to leave the swimbladder and enter the blood or the surrounding tissues. What keeps gases inside the swimbladder? We mention three factors of importance.

1. The entrance to the pneumatic duct, which constitutes the resorbing part (see below), may be closed off by muscles in the wall.
2. The swimbladder wall is covered by a silvery layer that is impregnated with crystals of guanine and some hypoxanthine, constituting an impermeable coat. Removal of this silvery layer has been shown to increase the gas permeability of the swimbladder wall some 100-fold (Denton et al. 1972; Lapennas and Schmidt-Nielsen 1977).
3. The rete countercurrent system acts as a barrier in that gas is shunted from the rete venous capillaries to the rete arterial capillaries. This is analogous to the

diffusion shunt that has been assumed for muscle tissue in mammals (Piiper et al. 1984; Piiper 1987; Kobayashi et al. 1989a). The rete, thus, not only plays a key role in gas deposition (see below), but also in its maintenance. It should, however, be noted that complete cessation of blood flow to the swimbladder epithelium would be an even more effective way of preventing gas loss.

3.3 How Can Gas Be Released from the Swimbladder?

When fish decide to ascend to and float at a shallower depth, the gas volume of the swimbladder must be reduced. In physostome fish, gas can exit through the gut, and bubbles can indeed be observed leaving from the fish's mouth. In physoclist, where there is no such connection, gas is resorbed in structures like the pneumatic duct of the eel, or in resorptive parts of the swimbladder wall (see Fig. 1B) normally separated from the main swimbladder by muscular activity. These resorptive epithelia are well vascularized, and the resorptive process apparently employs the same mechanisms as, for example, in the alveolar lung (Denton 1961).

4 Mechanisms for Deposition of Gas: The Classical Model

The most interesting and least completely understood mechanism is the deposition of gas into the swimbladder. This was earlier thought to occur as an active process of gas secretion. Although it is now generally accepted that passive diffusion governs gas deposition, the term gas secretion has been retained (Fänge 1983).

For gases to be deposited by passive diffusion from the blood in the capillaries of the swimbladder into its gas phase, high partial pressures must be created in this blood, at least as high as in the gas phase. Two independent mechanisms are thought to be responsible for the formation of high partial pressures in the swimbladder blood: (1) reduction of effective gas solubility in blood by the metabolic action of the swimbladder epithelium; (2) countercurrent enhancement of gas partial pressure in the rete mirabile.

4.1 Reduction of Physical Gas Solubility and Gas Binding

The swimbladder epithelium is known to form lactic acid, even at high O_2 levels that, in other tissues, would allow oxidative ATP formation. Lactic acid is released into the capillary blood, lowering its pH and increasing its salt (lactate) concentration. This acts on the effective solubility of inert gases as well as of O_2 and CO_2.

4.1.1 Salting-Out Effect

All gases are known to display a salting-out effect, i.e. a reduction of their physical solubility when the solute (salt) concentration is increased (Enns et al. 1967; Gerth and Hemmingsen 1982). Many studies have been performed in water as the solvent, and the salting-out effect of lactate is well established, although the Setchenow coefficient, a quantitative measure of the salting-out effect, is quite small (Gerth and Hemmingsen 1982). Measurements in blood of the European eel revealed that, across the range of lactate concentrations found in swimbladder blood, a solubility change of no more than about 1% is expected to occur due to the salting-out effect (Pelster et al. 1988, 1990). However, even this small change in solubility may suffice to lead to a partial pressure enhancement in the countercurrent system of the rete (see below).

The experiments of Pelster et al. (1988) disproved the observation of Steen (1963a) who reported a sharp decline in Ar and N_2 solubilities when pH was reduced from about 7.8 to 7.6.

4.1.2 Root Effect

Reduction in pH leads to reduced O_2 affinity of hemoglobin (Hb) via the Bohr effect, and thus to increased Po_2 in blood (if the total amount of O_2 in blood is assumed for the moment to remain constant). Some fish display in addition to the Bohr effect, a Root effect, i.e. a reduction in O_2 capacity when pH is reduced (Root 1931; Bridges et al. 1983). At low pH the Root hemoglobin is fixed in the deoxygenated state and the oxygen-binding curves of blood containing Root hemoglobin become nearly asymptotic with respect to the abscissa before saturation of all heme molecules is complete (Brittain 1987; Riggs 1988; Pelster and Weber 1991). This leads to a very large increase in Po_2 with acidification, particularly when the oxygen saturation of the blood was high before acidification. Assuming full oxygenation of the hemoglobin in arterial blood in the eel swimbladder, a reduction in the oxygen-carrying capacity of about 40% can be expected with the pH change encountered in the swimbladder capillaries (Pelster and Weber 1991). Most fish equipped with a swimbladder display a Root effect (Farmer et al. 1979; Pelster and Weber 1991).

4.1.3 CO_2 Formation from HCO_3^-

The release of lactic acid is also responsible for the increase in blood Pco_2 during the passage of the swimbladder capillaries. Assuming the vessels of the swimbladder represent a closed system, the proton release converts HCO_3^- into CO_2. This reaction, however, will not change the total CO_2 content (Cco_2) of the blood, analogous to the Root and salting-out effects which also do not change the gas concentration.

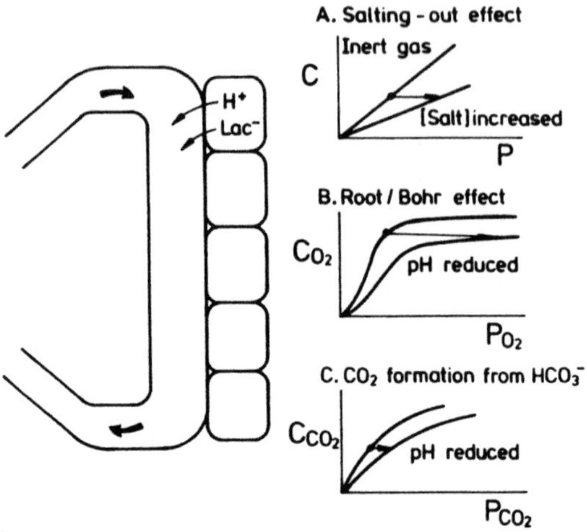

Fig. 4. Mechanisms that reduce the effective gas solubility in blood during passage through the swimbladder, leading to an increase in gas partial pressure. Lactic acid released from the gas gland cells leads to an increased inert gas partial pressure via the salting-out effect (**A**), to an increased Po_2 via the Root effect (**B**), and to an increased Pco_2 via conversion of HCO_3^- to CO_2 (**C**). Gas deposition, which would decrease the gas concentration in the swimbladder vessels, is neglected

In Fig. 4, these three mechanisms for increasing gas partial pressures in swimbladder blood are summarized. It should be noted that the increases in partial pressures will be smaller when gases are secreted into the swimbladder lumen, which then would lead to a reduction in the total gas concentrations.

4.2 Countercurrent Enhancement of Gas Partial Pressures in the Rete Mirabile

When blood with increased partial pressure enters the efferent limb of the rete capillaries (vi in Fig. 2), a partial pressure head from the venous (=efferent) to the arterial (=afferent) capillaries exists along which gases will diffuse back. This will reduce the partial pressure in the efferent capillaries, but will enhance them in the blood of the afferent capillaries, on its way towards the swimbladder (Fig. 5).

It was mainly Kuhn and his colleagues who worked out the details of how this hairpin countercurrent concentration operates in the swimbladder rete as well as in the hairpin vessels of the renal medulla (Kuhn et al. 1963; cf. Fänge 1983).

Figure 6 shows a theoretical plot of the concentrating ability of the rete, for the case of an inert gas (i.e. without chemical binding of gas). It appears that two factors are decisive for quantifying the enhancement in gas partial pressure along the afferent rete capillary (Pae/Pai; logarithmic ordinate scale):

1. The conductance ratio, $D/(\dot{Q}\alpha_a)$, where D is the diffusing capacity of the barrier between venous and arterial capillaries in the rete; $\dot{Q}$, blood perfusion of the rete; α_a, inert gas solubility in arterial blood.
2. The salting-out effect, expressed as the solubility ratio in venous and arterial blood, α_v/α_a. It is assumed here that α_v is reduced relative to α_a (salting-out effect) by the salt (lactate) released by the swimbladder epithelium; and that this salt does not diffuse back in to the rete.

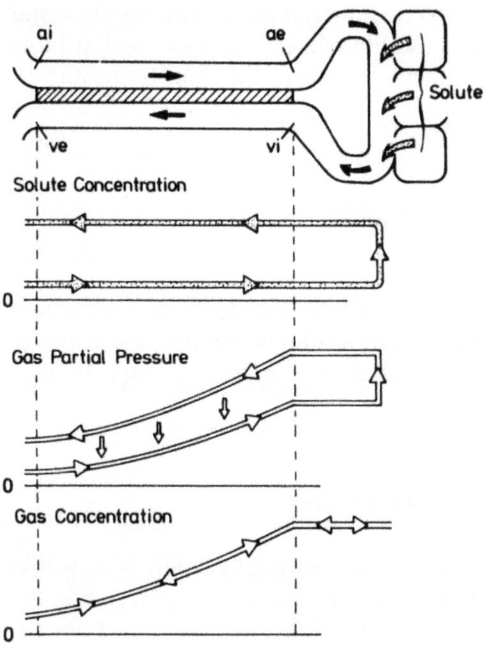

Fig. 5. Schema of countercurrent enhancement of gas in the rete according to the classical concept. Note that it is assumed that salt does not diffuse back into the rete

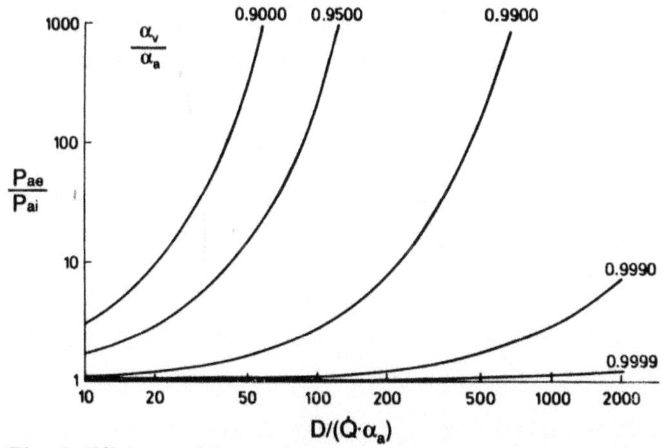

Fig. 6. Efficiency of the rete in enhancing inert gas partial pressure, calculated as the ratio of partial pressures in the arterial efflux and influx, Pae/Pai (cf Fig. 2B). This efficiency is given by the conductance ratio $D/\dot{Q}\alpha_a$ (D, diffusing capacity of the rete; $\dot{Q}$, blood perfusion; α_a, physical solubility in rete afferent blood) and by the salting-out effect, expressed as the solubility ratio, α_v/α_a. The calculation assumes that gas deposition is zero. See text for further explanations

It is unfortunate that values cannot yet be assigned to the ratio $D/\dot{Q}\alpha_a$. The morphometric values available do not allow the determination of total area and diffusion distance with enough certainty, nor are physiological values available. It is, however, evident that with large conductance ratios only small values for the salting-out effect would suffice to produce substantial enhancements of partial pressures in the swimbladder capillaries (Kobayashi et al. 1989c). Similar plots apply to O_2 and CO_2 as well, although their shape may differ due to the alinear binding characteristics in blood.

5 Recent Additions to the Classical Model

We have recently resumed experimental and theoretical work on the mechanisms of gas deposition in the eel swimbladder. The results show that the classical concepts remain valid for the most part, but that some additions appear to be of quantitative significance.

5.1 Flow Must Be Balanced in the Rete Countercurrent

Loss of fluid from the vascular bed in the swimbladder, e.g. by lymph flow, would yield a lower flow rate in the venous compared with the arterial rete capillaries. Niesel and Röskenbleck (1963) showed that such a flow imbalance in a hairpin countercurrent system, the fish rete as well as the kidney hairpin vessels, would greatly compromise its concentrating ability. In fact, with perfect equality of concentration of substance x in afferent and efferent blood, but with higher efferent than afferent blood flow, more substance would be transported away

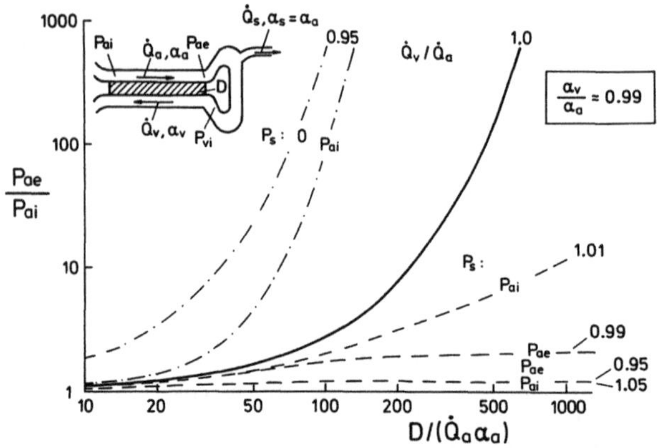

Fig. 7. Influence of blood flow mismatch ($\dot{Q}_v/\dot{Q}_a$) on the efficiency of gas partial pressure enhancement (Pae/Pai), calculated for a salting-out effect that leads to a 1% reduction in solubility ($\alpha_v\alpha_a$ = 0.99; cf. Fig. 6). Curves were calculated for various values of flow mismatch ($\dot{Q}_v/\dot{Q}_a$). For flow mismatch, there is an increase in efficiency only with $\dot{Q}_v/\dot{Q}_a < 1$ and, simultaneously, with gas partial pressure in shunt flow (Ps) equal to, or below, systemic partial pressure (Pai). In all other cases, efficiency diminishes with flow mismatch. (Kobayashi et al. 1989b)

from the tissue served by the hairpin than would be transported into it, leading to loss of x from the tissue.

We have assessed the water shift from eel capillaries to the swimbladder epithelium, and have found this to be small indeed, at best 5% of the total flow (Kobayashi et al. 1989b). This would indicate virtually matched rete afferent with efferent blood flow.

We have theoretically investigated the problem of flow mismatch. Thus, any flow mismatch, higher or lower in afferent than efferent rete capillaries, virtually destroys the rete concentrating ability (Fig. 7; Kobayashi et al. 1989b). There is, however, one exception to this prediction, and that is for O_2. If plasma water is drained in the swimbladder wall, its O_2 concentration will be much lower than that of whole blood. Under this condition, the concentrating efficiency can be strongly enhanced by the reduction of venous compared with arterial blood flow. This mechanism, however, would not work for other gases, since the draining plasma water would contain gases at about the same concentration as that of whole blood (Kobayashi et al. 1989b). This analysis shows that, except perhaps for O_2, flow match may be of vital importance for the rete function.

5.2 Salt Back-Diffusion in the Rete

Most studies of rete function have concentrated on back-diffusion of gases from the afferent to the efferent rete capillaries as the main mechanism for partial pressure enhancement in the rete. Back-diffusion of salt, on the other hand, has been assumed to exert a negligible effect on the partial pressure-enhancing ability of the rete (Kuhn et al. 1963) or even to be detrimental (Steen 1970). However, since undebatable evidence for lactate back-diffusion from venous to arterial capillaries in the rete exists (Steen 1963b; Kobayashi et al. 1989b), we have theoretically re-investigated its implications (Kobayashi et al. 1989c). The analysis demonstrates that salt back-diffusion significantly enhances the concentrating ability of the rete countercurrent system. This can be intuitively understood, since any salt diffusing back into the afferent capillary will exert its salting-out effect there, thus enhancing the inert gas partial pressure already along the afferent capillary (Fig. 8).

Furthermore, diffusive exchange between the rete capillaries of CO_2, HCO_3^-, or H^+ ions will affect the effective O_2 solubility via the Root effect. Measurements, thus, suggest that most of the enhancement in O_2 partial pressure along the afferent capillary results from this back-diffusion, leading to a decrease in pH and a concomitant decrease in hemoglobin oxygen-carrying capacity in arterial capillaries of the rete mirabile; back-diffusion of O_2 itself was found to be very small in the experimental situation (Kobayashi et al. 1990).

5.3 Metabolism of the Gas Gland Cells: Anoxidative Formation of CO_2

An essential step in the initiation of gas deposition is to decrease the physical gas solubility or the gas-carrying capacity of the blood during its passage through the

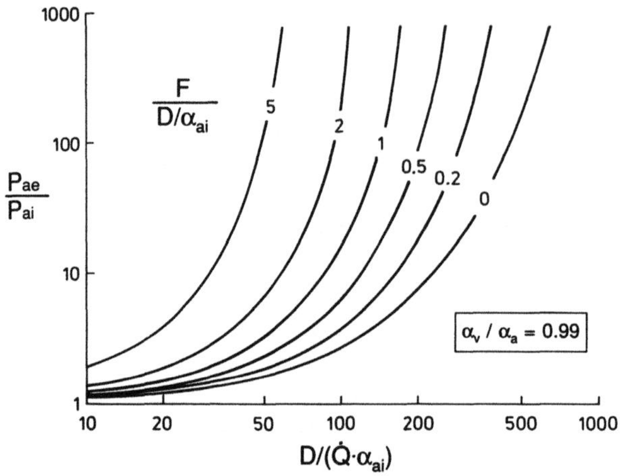

Fig. 8. Effect of solute back-diffusion on the efficiency of partial pressure enhancement in the rete (without gas deposition), calculated for a solubility change of 1%. The efficiency increases with increasing permeability ratio of the rete, expressed as the rate of solute transfer (F) and D/α_{ai}. (Adapted from Kobayashi et al. 1989c)

swimbladder epithelium. This is achieved by metabolites that are produced and secreted from the gas gland cells of the swimbladder epithelium. Blood glucose acts as fuel for the metabolism of these epithelial cells. The gas gland tissue of the scup, *Stenotomus chrysops,* has been shown to convert in vitro 70–90% of the glucose to lactic acid (Ball et al. 1955), and gas gland cells of the vermilion rockfish, *Sebastodes miniatus,* were found to produce lactate even under an oxygen pressure of 51 atm (D'Aoust 1970). Likewise, in a Ringer-perfused preparation of the eel swimbladder, about 60% of the glucose in the perfusate was converted to lactate (Pelster et al. 1989). The importance of lactate formation has also been demonstrated in vivo where increases in blood lactate concentration during passage of the gas gland cells of 2–7 $mmol \cdot l^{-1}$ were observed. The highest lactate concentration measured in blood sampled from swimbladder vessels in these experiments was about 21 $mmol \cdot l^{-1}$ (Steen 1963b; Enns et al. 1967; Kobayashi et al. 1989b).

Oxidative metabolism appears to be of minor importance, mitochondrial density (Dorn 1961; Morris and Albright 1975) and activities of citric acid cycle enzymes in gas gland tissue are low (Boström et al. 1972; Pelster and Scheid 1991). In the Ringer-perfused swimbladder preparation, oxidative metabolism contributed only 5% to glucose degradation (Pelster et al. 1989).

In this preparation, however, aerobic metabolism and lactate production did not balance glucose uptake and the respiratory exchange ratio was found to be far above unity. Similar results were obtained for the eel swimbladder in situ, where about 80% of the glucose removed from the blood was converted to lactate, while only 1% or even less was oxidized completely (Pelster and Scheid

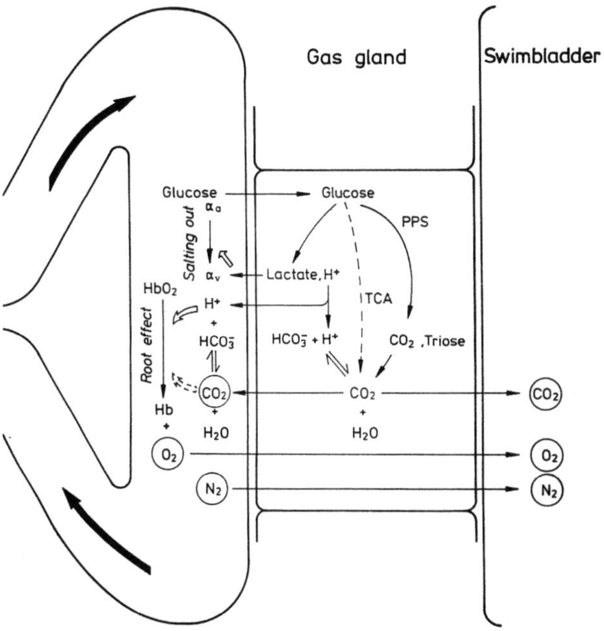

Fig. 9. Present concept of metabolism of the gas gland cells and its influence on the physical solubility of inert gases or the release of gas from a chemical binding site in the blood. *TCA* Tricarboxylic acid cycle; *PPS* pentose phosphate shunt. *Open arrows* indicate influence or possible influence on inert gas solubility and hemoglobin oxygen-binding characteristics. *Filled arrows* indicate movement of substance

1992b, 1993). These data suggest that part of the glucose is metabolized in the pentose phosphate shunt, where CO_2 is formed in an oxidative decarboxylation reaction without involvement of oxygen (Fig. 9). High activities of glucose-6-phosphate dehydrogenase and 6-phosphogluconate dehydrogenase, key enzymes of the pentose phosphate shunt, have indeed been found in cod and eel swimbladder epithelium (Boström et al. 1972; Ewart and Driedzic 1990; Pelster and Scheid 1991). In the pentose phosphate shunt, glucose is specifically decarboxylated in position C_1, not in C_6, while in the aerobic metabolism these two carbon atoms cannot be distinguished. Using differently labelled glucose and measuring the rate $^{14}CO_2$ production, the contribution of the pentose phosphate shunt to the formation of CO_2 was established for isolated swimbladder tissue of the toadfish *Opsanus beta* in vitro (Walsh and Milligan 1993), and also for the blood-perfused swimbladder of the American eel *Anguilla rostrata* (Pelster et al. 1994).

These results demonstrate that the swimbladder epithelium is specialized for the formation of CO_2 in metabolic pathways known as anaerobic pathways, although the tissue is continuously exposed to hyperoxia. As a consequence, the highest P_{CO_2} would be expected to be present in the gas gland cells themselves, from where it diffuses into the swimbladder as well as into the blood, causing an acidification. Indeed, P_{CO_2} in swimbladder gas was observed to exceed that in swimbladder blood (Kobayashi et al. 1990). This is contrary to all other gases, which diffuse from the blood through the swimbladder epithelium into the bladder.

5.4 Kinetics of the Root Effect

Forster and Steen (1969) measured the kinetics of the Root-on and -off effects. Their results showed the off kinetics, i.e. the kinetics of O_2 release from Hb due to acidification in the swimbladder capillaries and in the afferent rete capillaries, to be fast, with a half-time of around 30–40 ms. In contrast, the on kinetics, occurring in the efferent capillary blood, was found to be more than 100 times slower. Back-diffusion of CO_2 in the rete (see above) results in pH changes in blood passing through the rete capillaries; this leads to a Root-off effect in afferent, and a Root-on effect in efferent rete blood. With a fast Root-off effect, as suggested by Forster and Steen (1969), the Po_2 in the afferent vessels would increase, but Po_2 reduction in the efferent blood would be delayed because of the slower Root-on effect. This would reinforce the countercurrent Po_2 enhancement.

Pelster et al. (1992) re-assessed the Root kinetics using thin blood films exposed directly to gas (Heidelberger and Reeves 1990); their results suggest that both the on and off kinetics are essentially identical (Fig. 10), thus lending no support to the kinetic enhancement proposed by Forster and Steen (1969) and Sund (1977).

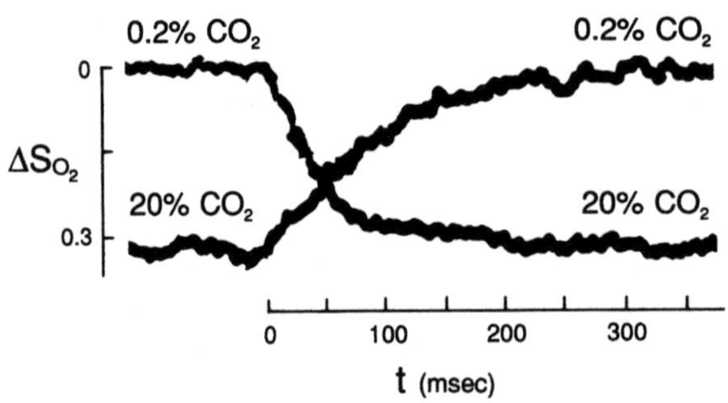

Fig. 10. Kinetics of the Root effect in eel blood measured with a thin blood film preparation. The blood film was equilibrated with 40% O_2, and the Root effect was switched on or off by changing between 0.2 and 20% CO_2. In this particular preparation, reaction half-times for Root-on and -off effects were 86 and 47 ms, respectively. (Pelster et al. 1992)

6 Regulation in the Swimbladder

6.1 How Can Gas Deposition Be Regulated?

When fish descend, the amount of gas in their swimbladder has to increase, and this process must occur rather rapidly. This calls for mechanisms controlling the rate of gas deposition. Although the regulation of gas deposition is not well understood, we can conceive two alternative possibilities where regulation could take place, either the metabolic activity of the gas gland cells in the swimbladder epithelium or the swimbladder perfusion could be regulated. Both processes are expected to impinge on the extent of partial pressure enhancement and, thus, on the amount of gases deposited in the swimbladder. Other parameters that are important for partial pressure enhancement, notably the diffusive conductance (D) of the rete for gases or other substances, the extent of the Root effect or of the salting-out effect (change in solubility for a given change in salt concentration), do not appear to be available for short-term regulation.

6.2 Gas Gland Metabolism

The swimbladder tissue is autonomically innervated, mainly by the splanchnic nerve, which departs from the celiac ganglion, and the vagosympathetic trunk. Cutting the vagosympathetic trunk (vagotomy) has been shown to inhibit gas deposition into the swimbladder (Bohr 1894; Fänge et al. 1976; Lundin and Holmgren 1991). In *Ctenolabrus rupestris*, a similar effect was observed by applying cholinergic blocking agents, like mecamylamine or atropine (Fänge et al. 1976), and histological studies demonstrated the presence of acetylcholine esterase activity in the secretory portion of cod swimbladder. On the other hand, no experimental evidence has been found for the presence of adrenergic nerve endings, suggesting that the gas gland cells are under cholinergic, parasympathetic control of the vagus nerve (McLean and Nilsson 1981).

We have obtained further evidence for the importance of gas gland metabolism in determining the rate of gas deposition into the swimbladder: a good correlation was observed in the eel between the rate of gas deposition and proton release of the gas gland cells (Pelster and Scheid 1992b).

6.3 Blood Perfusion

Blood perfusion has an impact on gas deposition. Figure 6 suggests that the partial pressure enhancement (Pae/Pai) increases with decreasing blood perfusion ($\dot{Q}$), but the amount of gas supplied to the swimbladder epithelium for deposition into the gas phase will likewise decrease with flow. Sund (1977), furthermore, assumed that diffusion along the rete vessels will compromise the countercurrent enhancement when flow is low. Hall (1924) and Fänge (1953) indeed reported vasodilation during a high rate of gas deposition. We have similarly observed a significant positive correlation between blood perfusion of the swimbladder tissue and gas deposition into the bladder under normoxic conditions (Pelster and Scheid 1992b).

Neuronal control of vasomotor activity in the swimbladder tissue appears to be dominated by adrenergic effects, although a cholinergic influence cannot be excluded. Electrical stimulation of the splanchnic nerve or the vagosympathetic trunk elicits vasoconstriction (Nilsson 1972; Wahlquist 1985); both nerves stain for catecholamines (McLean and Nilsson 1981). Perfusing the swimbladder of the eel (*Anguilla vulgaris*) or cod (*Gadus morhua*) with saline-containing catecholamines also causes vasoconstriction; this effect is blocked by phentolamine or phenoxybenzamine, but not by propranolol (Stray-Pedersen 1970; Nilsson 1972; Wahlquist 1985). Presence of α-adrenergic vasoconstriction in swimbladder vessels close to the rete mirabile of the eel has also been verified using a blood-perfused swimbladder preparation (Pelster 1994). Recent experiments indicate that vasoactive substances, like VIP or substance P, under certain conditions also modify gas deposition (Lundin and Holmgren 1991).

Not only the swimbladder vasculature, but also the muscle cells in the swimbladder wall are under the control of the autonomic nervous system. In the eel, stimulation of the vagosympathetic trunk causes alterations in shape and volume of the swimbladder (Stray-Pedersen 1970). Isolated strips of the anterior (secretory) part of the bladder of *Ctenolabrus rupestris* showed α-adrenergic contraction, while β-adrenergic relaxation can be observed in the posterior (resorbing) part (Fänge et al. 1976). Increased adrenergic tone, thus, will result in a forward movement of the diaphragm separating the two parts of the bladder, and gas will be resorbed in the posterior part of the bladder; decreased adrenergic tone is supposed to result in a backward movement of the diaphragm. The diaphragm of *Ctenolabrus rupestris* is homologous with the oval, found, for example, in *Gadus morhua*. Accordingly, an α-adrenergic relaxation of the circular oval muscle, coinciding with a contraction of radial smooth muscles in the oval, has been reported, which causes an opening of the oval and hence favors gas resorption (Nilsson 1971; Ross 1978; Wahlquist 1985).

7 Open Questions

It is quite obvious from this brief review that many questions still remain to be answered. Many pertain to the biology of regulation of swimbladder function, but also the physical mechanisms underlying gas partial pressure enhancement are far from being completely understood. One should not forget that most of the information on the mechanisms of gas deposition has been obtained under laboratory conditions, notably at atmospheric pressure. It would certainly be important to analyze swimbladder function under more natural conditions, e.g. at higher than atmospheric pressure; however, this appears to be technically extremely difficult.

Solubility changes due to the salting-out effect are in the range of only 1%. According to the model of Kobayashi et al. (1989c), we need more information on diffusivities and permeabilities in the rete of gases and substances like lactate, HCO_3^-, or H^+, to decide whether inert gas deposition can be explained on the basis of this solubility change and the enhancement of gas partial pressure by salt back-diffusion.

Information is still lacking on the regulation of the metabolic activity in the gas gland cells and on the mechanisms of ion regulation.

The swimbladder tissue can tolerate extremely high levels of oxygen, and it is not well understood how it prevents oxygen toxicity. More information is, thus, needed about the metabolic pathways involved in protection against oxygen toxicity.

References

Alexander R McN (1966) Physical aspects of swimbladder function. Biol Rev 41: 141–176

Ball EG, Strittmatter CF, Cooper O (1955) Metabolic studies on the gas gland of the swim bladder. Biol Bull 108: 1–17

Bohr C (1894) The influence of section of the vagus nerve on the disengagement of gases in the airbladder of fishes. J Physiol 15: 494–500

Boström SL, Fänge R, Johansson RG (1972) Enzyme activity patterns in gas gland tissue of the swimbladder of the cod (*Gadus morrhua*). Comp Biochem Physiol 43B: 473–478

Bridges CR, Hlastala MP, Riepl G, Scheid P (1983) Root effect induced by CO_2 and by fixed acid in the blood of the eel, *Anguilla anguilla*. Respir Physiol 51: 275–286

Brittain T (1987) The Root effect. Comp Biochem Physiol 86B: 473–481

D'Aoust BG (1970) The role of lactic acid in gas secretion in the teleost swimbladder. Comp Biochem Physiol 32: 637–668

Denton EJ (1961) The buoyancy of fish and cephalopods. Progr Biophys Biophys Chem 11: 178–234

Denton EJ, Liddicoat JD, Taylor DW (1972) The permeability to gases of the swimbladder of the conger eel (*Conger conger*). J Mar Biol Assoc UK 52: 727–746

Dorn E (1961) Über den Feinbau der Schwimmblase von *Anguilla vulgaris L.* Licht-und elektronenmikroskopische Untersuchungen. Z Zellforsch 55: 849–912

Enns T, Douglas E, Scholander PF (1967) Role of swimbladder rete of fish in secretion of inert gas and oxygen. Adv Biol Med Physiol 11: 231–244

Ewart HS, Driedzic WR (1990) Enzyme activity levels underestimate lactate production rates in cod (*Gadus morhua*) gas gland. Can J Zool 68: 193–197

Fänge R (1953) The mechanisms of gas transport in the euphysoclist swimbladder. Acta Physiol Scand 30: 1–133

Fänge R (1966) Physiology of the swimbladder. Physiol Rev 46: 299–322

Fänge R (1983) Gas exchange in the fish swim bladder. Rev Physiol Biochem Pharmacol 97: 111–158

Fänge R, Holmgren S, Nilsson S (1976) Autonomic nerve control of the swimbladder of the goldsinny wrasse, *Ctenolabrus rupestris*. Acta Physiol Scand 97: 292–303

Farmer M, Fyhn HJ, Fyhn UEH, Noble RW (1979) Occurrence of Root effect hemoglobins in Amazonian fishes. Comp Biochem Physiol 62A: 115–124

Forster RE, Steen JB (1969) The rate of the 'Root shift' in eel red cells and eel haemoglobin solutions. J Physiol 204: 259–282
Gerth WA, Hemmingsen EA (1982) Limits of gas secretion by the salting-out effect in the fish swimbladder rete. J Comp Physiol B 146: 129–136
Hall FG (1924) The functions of the swimbladder of fishes. Biol Bull 47: 79–124
Heidelberger E, Reeves RB (1990) O_2 transfer kinetics in a whole blood unicellular thin layer. J Appl Physiol 68: 1854–1864
Kobayashi H, Pelster B, Piiper J, Scheid P (1989a) Significance of the Bohr effect for tissue oxygenation in a model with counter-current blood flow. Respir Physiol 76: 277–288
Kobayashi H, Pelster B, Scheid P (1989b) Water and lactate movement in the swimbladder of the eel, *Anguilla anguilla*. Respir Physiol 78: 45–57
Kobayashi H, Pelster B, Scheid P (1989c) Solute back-diffusion raises the gas concentrating efficiency in counter-current flow. Respir Physiol 78: 59–71
Kobayashi H, Pelster B, Scheid P (1990) CO_2 back-diffusion in the rete aids O_2 secretion in the swimbladder of the eel. Respir Physiol 79: 231–242
Kuhn W, Ramel A, Kuhn H, Marti E (1963) The filling mechanism of the swimbladder. Generation of high gas partial pressures through hairpin countercurrent multiplication. Experientia 19: 497–552
Lapennas GN, Schmidt-Nielsen K (1977) Swimbladder permeability to oxygen. J. Exp Biol 67: 175–196
Lundin K, Holmgren S (1991) An X-ray study of the influence of vasoactive intestinal polypeptide and substance P on the secretion of gas into the swimbladder of a teleost *Gadus morhua*. J Exp Biol 157: 287–298
McLean JR, Nilsson S (1981) A histochemical study of the gas gland innervation in the Atlantic cod, *Gadus morhua*. Acta Zool 62: 187–194
Morris SM, Albright JT (1975) The ultrastructure of the swimbladder of the toadfish, *Opsanus tau L.* Cell Tissue Res 164: 85–104
Niesel W, Röskenbleck H (1963) Die Bedeutung der Stromgeschwindigkeiten in den Gefäßsystemen der Niere und der Schwimmblase für die Aufrechterhaltung von Konzentrationsgradienten. Pflügers Arch 277: 302–315
Nilsson S (1971) Adrenergic innervation and drug responses of the oval sphincter in the swimbladder of the cod (*Gadus morhua*). Acta Physiol Scand 83: 446–453
Nilsson S (1972) Autonomic vasomotor innervation in the gas gland of the swimbladder of a teleost (*Gadus morhua*). Comp Gen Pharmacol 3: 371–375
Pelster B (1994) Adrenergic control of swimbladder perfusion in the European eel *Anguilla anguilla*. J Exp Biol 189: 237–250
Pelster B, Scheid P (1991) Activities of enzymes for glucose catabolism in the swimbladder of the European eel, *Anguilla anguilla*. J Exp Biol 156: 207–213
Pelster B, Scheid P (1992a) Counter-current concentration and gas secretion in the fish swimbladder. Physiol Zool 65: 1–16
Pelster B, Scheid P (1992b) The influence of gas gland metabolism and blood flow on gas deposition into the swimbladder of the European eel *Anguilla anguilla*. J Exp Biol 173: 205–216
Pelster B, Scheid P (1993) Glucose metabolism of the swimbladder tissue of the European eel *Anguilla anguilla*. J Exp Biol 185: 169–178
Pelster B, Weber RE (1991) The physiology of the Root effect. Adv Comp Environ Physiol 8: 51–77
Pelster B, Kobayashi H, Scheid P (1988) Solubility of nitrogen and argon in eel whole blood and its relationship to pH. J Exp Biol 135: 243–252
Pelster B, Kobayashi H, Scheid P (1989) Metabolism of the perfused swimbladder of European eel: oxygen, carbon dioxide, glucose and lactate balance. J Exp Biol 144: 495–506
Pelster B, Kobayashi H, Scheid P (1990) Reduction of gas solubility in the fish swimbladder. In: Piiper J, Goldstick TK, Meyer M (eds). Oxygen transport to tissue, vol XII. Plenum Press, New York, pp 725–733
Pelster B, Scheid P, Reeves RB (1992) Kinetics of the Root effect and of O_2 exchange in whole blood of the eel. Respir Physiol 185: 169–178

Pelster B, Hicks J, Driedzic WR (1994) Contribution of the pentose phosphate shunt to the formation of CO_2 in swimbladder tissue of eel. J Exp Biol 197: 119–128

Piiper J (1987) Role of diffusion shunt in transfer of inert gases and O_2 in muscle. In: Mochizuki M, Honig CR, Koyama T, Goldstick TK, Bruley DF (eds) Oxygen transport to tissue, vol X. Plenum Press, New York, pp 55–61

Piiper J, Meyer M, Scheid P (1984) Dual role of diffusion in tissue gas exchange: Blood-tissue equilibration and diffusion shunt. Respir Physiol 56: 131–144

Riggs A (1988) The Bohr effect. Annu Rev Physiol 50: 181–204

Root RW (1931) The respiratory function of the blood of marine fishes. Biol Bull 61: 427–456

Ross LG (1978) The innervation of the resorptive structures in the swimbladder of a physoclist fish, *Pollachius virens* (*L.*). Comp Biochem Physiol 61C: 385–388

Scheid P, Pelster B, Kobayashi H (1990) Gas exchange in the fish swimbladder. In: Piiper J, Goldstick TK, Meyer M (eds) Oxygen transport to tissue, vol XII. Plenum Press, New York, pp 735–742

Steen B (1963a) The physiology of the swimbladder in the eel *Anguilla vulgaris*. I. The solubility of gases and the buffer capacity of blood. Acta Physiol Scand 59: 124–137

Steen B (1963b) The physiology of the swimbladder in the eel *Anguilla vulgaris*. III. The mechanism of gas secretion. Acta Physiol Scand 59: 221–241

Steen B (1970) The swim bladder as a hydrostatic organ. In: Hoar WS, Randall D (eds) Fish Physiology, vol 4. Academic Press, New York, pp 413–443

Stray-Pedersen S (1970) Vascular responses induced by drugs and by vagal stimulation in the swimbladder of the eel, *Anguilla vulgaris*. Comp Gen Pharmacol 1: 358–364

Stray-Pedersen S, Nicolaysen A (1975) Qualitative and quantitative studies of the capillary structure in the rete mirabile of the eel, *Anguilla vulgaris* L. Acta Physiol Scand 94: 339–357

Sund T (1977) A mathematical model for counter-current multiplication in the swim-bladder. J Physiol 267: 679–696

Wahlqvist I (1985) Physiological evidence for peripheral ganglionic synapses in adrenergic pathways to the swimbladder of the Atlantic cod, *Gadus morhua*. Comp Biochem Physiol 80C: 269–272

Walsh PJ, Milligan CL (1993) Roles of buffering capacity and pentose phosphate pathway activity in the gas gland of the gulf toadfish *Opsanus beta*. J Exp Biol 176: 311–316

Wittenberg B, Schwend M, Wittenberg BA (1964) The secretion of oxygen into the swimbladder of fish. III. The role of carbon dioxide. Gen Physiol 48: 337–355

Chapter 4

Regulation of Respiration in Lower Vertebrates: Role of CO_2/pH Chemoreceptors

W.K. Milsom

Contents

1 Introduction

In the phylogenetic progression from fish to birds and mammals there are a number of progressive changes which reflect the increasing importance of the role of CO_2/pH chemoreceptors in respiratory control. There is the shift from water to air as a respiratory medium with the accompanying changes in the relative solubilities of O_2 and CO_2. Accompanying this are changes in the mean levels of arterial CO_2 partial pressure (Pa_{CO_2}) and pH (pHa) and an increase in the

Department of Zoology, University of British Columbia Vancouver, British Columbia, V6T 1Z4, Canada

Advances in Comparative and Environmental Physiology, Vol. 21

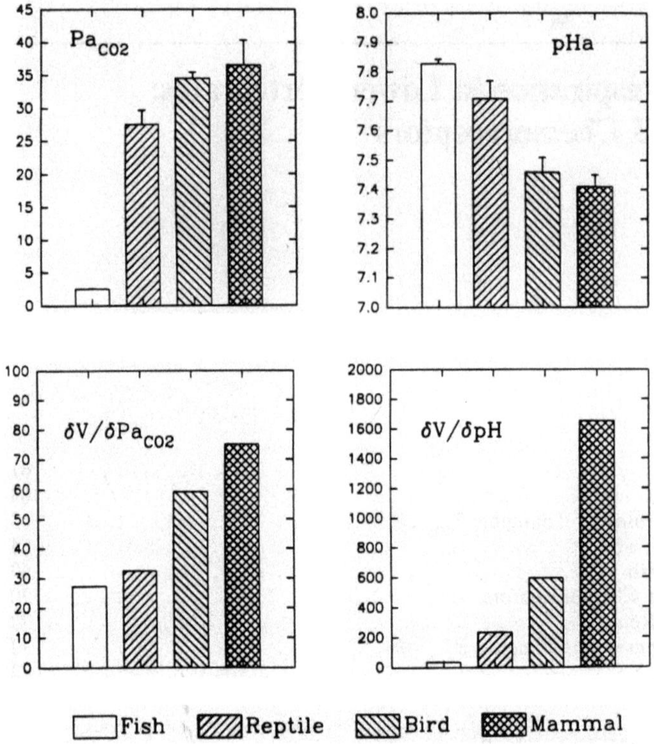

Fig. 1. Comparison of the levels of arterial P_{CO_2} (Pa_{CO_2}, in torr), pH (pHa) and ventilatory sensitivity (expressed as the % change in minute ventilation divided by the change in Pa_{CO_2} or pHa) for a fish (*Raja ocellata*; Graham et al. 1990), reptile (*Chrysemys picta*; Funk and Milsom 1987), bird (*Anas platyrhynchos;* Dodd and Milsom 1987) and a mammal (*Spermophilus lateralis;* McArthur and Milsom 1991)

sensitivity of animals to CO_2/pH as respiratory stimuli (Fig. 1). These data superficially suggest a smooth continuum in the change in respiratory control; from a system primarily driven by the need to obtain O_2 to one primarily driven by the need to excrete CO_2, accompanied by a progressive increase in the role of changes in Pa_{CO_2}/pHa as a modulating influence of respiratory motor output. This view may be misleading, however.

Breathing is continuous and rhythmic in water-breathing fish as well as in air-breathing birds and mammals and this produces relatively stable blood gases and acid-base status. In between these groups are the air-breathing fish, amphibia and reptiles. Many of these vertebrates utilize bimodal gas exchange and almost all exhibit intermittent breathing. These behaviours, which may reflect the evolutionary steps associated with the transition from aquatic to terrestrial life, produce unsteady blood gas and acid-base states (Fig. 2). This runs counter to any suggestion of a smooth progressive change in respiratory control and has raised questions concerning the extent to which trends and similarities seen in

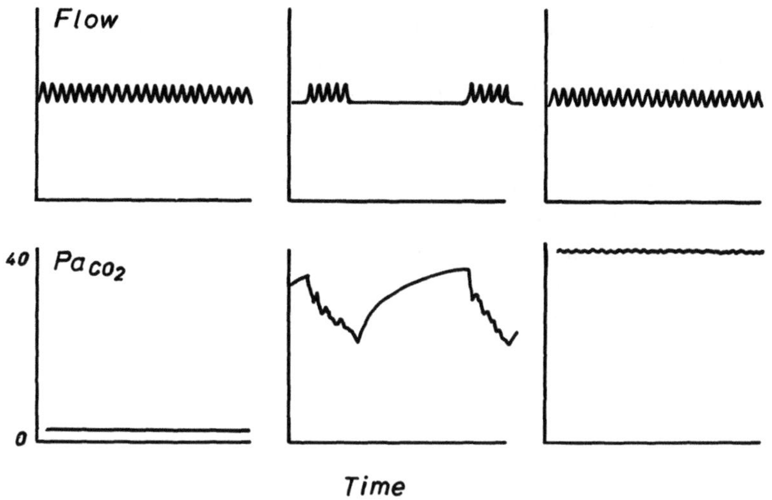

Fig. 2. Schematic diagram of the breathing patterns and associated arterial P_{CO_2} fluctuations in a fish (*left*), amphibian (*middle*) and mammal (*right*)

ventilatory control in different vertebrate groups can reflect continuity in underlying mechanisms (cf. Shelton et al. 1986). Thus, it has been suggested that changes in arterial P_{CO_2}/pH per se have no influence on ventilatory control in fish (Smith and Jones 1982), that the input from CO_2/pH chemoreceptors in air-breathing fish, amphibians and reptiles is an essential component of the mechanism responsible for the genesis of resting ventilation (Shelton et al. 1986; Smatresk 1990b), and that in birds and mammals this input simply modulates a centrally generated respiratory rhythm (von Euler 1986; Feldman et al. 1990).

In addition to the questions concerning the presence, nature and role of CO_2/pH chemoreception in overall ventilatory control in the different vertebrate groups are questions concerning the locations and relative roles of putative chemoreceptor groups involved in CO_2/pH ventilatory responses. Much of the literature pertaining to these questions is also equivocal.

It is in the context of such questions of qualitative and quantitative function that the literature pertaining to the role of CO_2/pH chemoreceptors in the control of breathing in lower vertebrates will be reviewed.

2 Fish

2.1 Ventilatory Responses to Changing P_{CO_2}/pH

2.1.1 Water-Breathing Fish

Due to the low capacitance of water for oxygen, fish must produce a high flow rate of water across their gills to obtain sufficient O_2 to meet metabolic demands.

Because of the high capacitance of water for CO_2, however, these high water flow rates effect rapid CO_2 excretion. The consequence of this difference in water capacitance for O_2 and CO_2 is that the arterial blood of most resting fish usually has an oxygen tension (Pa_{O_2}) of 30 to 100 torr while the CO_2 tension (Pa_{CO_2}) is typically 3–6 torr. The low levels of Pa_{CO_2} found in the blood of most fish are accompanied by equally low [HCO_3^-] and, thus, pH levels which are only slightly alkaline by vertebrate standards (Fig. 1). Regulation of arterial P_{CO_2} and pH at these levels can be achieved by ion-transporting mechanisms that are largely independent of ventilation. Thus, not surprisingly, O_2 rather than CO_2/pH appears to exert the dominant chemical control over gill ventilation.

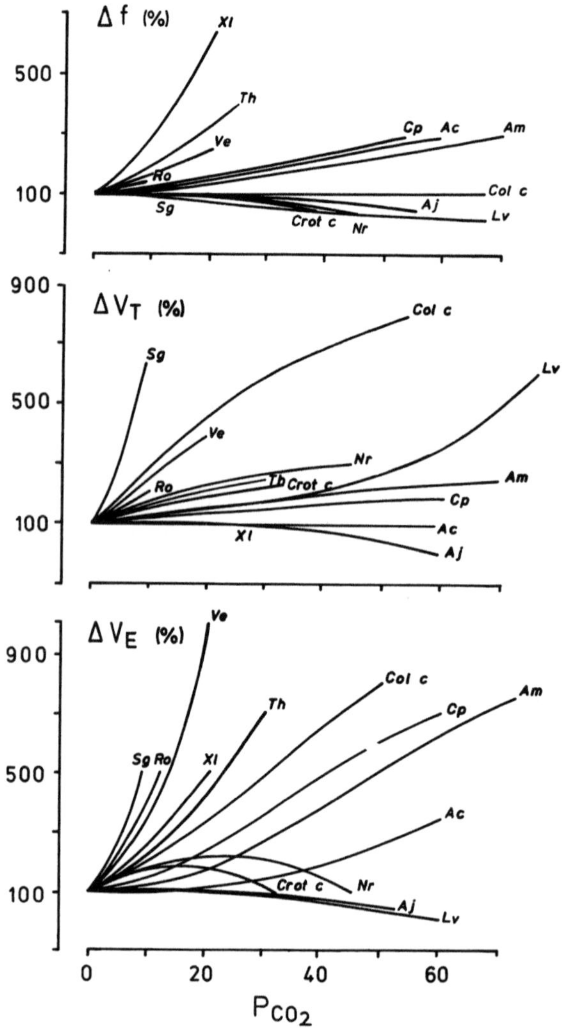

Fig. 3. Relationship between the percent change in minute ventilation ($\triangle\dot{V}$), tidal volume ($\triangle V_T$) and breathing frequency ($\triangle f$) and the partial pressure of CO_2 in inspired air (P_{CO_2}) in various species. Sources of data: fish: *Ro, Raja oscillata* (Wood et al. 1990); *Sg, Salmo gairdneri* (Janssen and Randall 1975). Air-breathing fish: *Ac, Amphipnous cuchia* (Lomholt and Johansen 1974). Amphibia: *Xl, Xenopus laevis* (Jones and Chu 1988). Snakes: *Col c, Coluber constrictor* (Nolan and Frankel 1982); *Aj, Acrochordus javanicus* (Glass and Johansen 1976); *Nr, Natrix rhombibera* (Gratz 1979). Lizards: *Lv, Lacerta viridis* (Nielsen 1962); *Ve, Varanus exanthematicus* (Mitchell et al. 1981); *Crot c, Crotaphytus collaris* (Templeton and Dawson 1963). Crocodilians: *Am, Alligator mississippiensis* (Davies, 1978). Turtles: *Cp, Chrysemys picta* (Funk and Milsom 1987); *Th, Testudo horsfieldii* (Benchetrit and Dejours 1980)

That aside, environmental acidosis, hypercapnia and the arterial acidosis associated with exercise have all been demonstrated to elevate ventilation in fish (Fig. 3; Hall 1931; Randall and Jones 1973; Janssen and Randall 1975; Eddy 1976; Hargis 1976; Dively et al. 1977; Cameron 1978; Thomas and LeRuz 1982; Thomas 1983; Heisler et al. 1988; Aota et al. 1990; Wood et al. 1990; Kinkead and Perry 1991). Because of the interrelationship between P_{CO_2} and pH, as defined by the Henderson-Hasselbach equation, and the close equilibrium between water and blood pH/P_{CO_2}, however, it has been difficult to distinguish both the nature of the specific ventilatory stimulus and its site of action.

Early experiments specifically aimed at separating the effects of $[H^+]$ from those of CO_2 lead to the conclusion that, for salmonids, changes in external or internal $[H^+]$ did not affect ventilation unless accompanied by hypercapnia (Neville 1979; Thomas and LeRuz 1982). The stimulating effects of low levels of hypercapnia on ventilation, however, seem to be transient in most fish. With chronic exposure to hypercapnia, minute ventilation increases initially but starts to decline after a few hours, returning to roughly prehypercapnic levels after about 24 h (Saunders 1962; Janssen and Randall 1975). The time course of this decline in minute ventilation is not identical to that of the blood pH compensation that results from an increase in arterial $[HCO_3^-]$, however (Lloyd and White 1967; Cameron and Randall 1972; Janssen and Randall 1975; Cameron 1978; Thomas and LeRuz 1982). Recent attempts to correlate the changes in ventilation seen in the skate during hypercapnic exposure to changes in arterial P_{CO_2}, arterial pH, cerebrospinal fluid pH and calculated brain intracellular pH (pHi) could find no close correlation with any of these variables and concluded that the best correlation was between gill ventilation and either pHa or pHi (Fig. 4; Wood et al. 1990).

It has been suggested that the ventilatory responses to changes in Pa_{CO_2}/pHa in fish are an indirect effect of the marked Bohr and Root effects that can occur in hypercapnic fish blood. These effects impair blood oxygen transport capability (Black and Irving 1937; Hughes and Shelton 1962; Cameron 1971; Eddy 1971, 1973) and are believed to indirectly stimulate ventilation via receptors sensitive to changes in arterial oxygen content (Smith and Jones 1982; Perry et al. 1989). Thus, blood oxygen saturation and content are reduced in normoxic hypercapnic trout and minute ventilation is elevated (Randall and Jones 1973; Eddy 1976; Smith and Jones 1982; Kinkead and Perry 1991). If water P_{O_2} is increased and blood O_2 content is raised, however, then minute ventilation returns to prehypercapnic levels despite sustained increases in Pa_{CO_2} (Smith and Jones 1982). Furthermore, in chronic hypercapnia, the oxygen saturation of trout blood returns to prehypercapnic levels with a similar time course to the decline in minute ventilation (Eddy and Morgan 1969; Eddy 1976). This argument suggests that in all cases when environmental P_{O_2} is high the effect of hypercapnia should be minimal. This is supported by the observation that gill ventilation in teleosts falls markedly in hyperoxia despite rising Pa_{CO_2} and falling pHa (Fig. 5; Randall and Jones 1973).

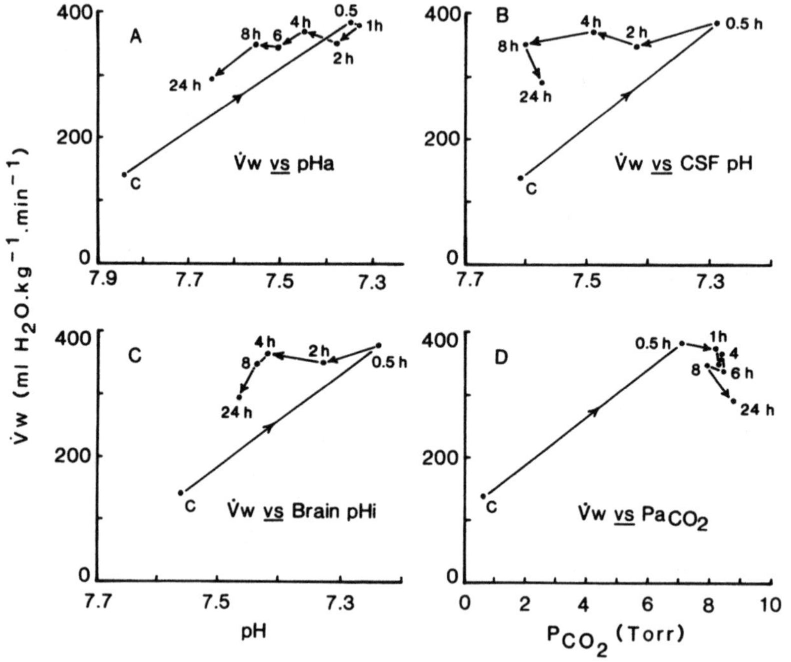

Fig. 4. Relationships between minute ventilation and **A** arterial pH (pHa), **B** cerebrospinal fluid pH (CSF pH), **C** brain intracellular pH (pHi) and **D** arterial CO_2 tension (P_{CO_2}) in *Raja oscellata* during 24h exposure to normoxic hypercapnia. *C* Control; all other values were measured at times indicated. (Wood et al. 1990)

There are some difficulties, however, with the suggestion that the ventilatory response to hypercapnia in teleosts and cyclostomes is simply a reflection of reduced blood O_2-transport capability. Goldfish blood, for example, shows Bohr and Root effects but these fish show no respiratory response to hypercapnia or acid exposure (Dejours 1972, 1973). Furthermore, hypercapnia does affect breathing in elasmobranchs but their blood shows neither Bohr nor Root effects and these species show little or no ventilatory response to hypoxia (Lenfant and Johansen 1966; Butler and Taylor 1971, 1975). As a consequence, another indirect means has been suggested more recently to link changes in Pa_{CO_2}/pHa to increases in ventilation and this is the release of catecholamines (Randall and Taylor 1991). There are studies which show both that acidosis elicits the release of catecholamines and that under many circumstances, levels of circulating catecholamines and ventilation are both elevated (Boutilier et al. 1986; Aota et al. 1990), providing a basis for this suggestion. Unfortunately, there are also studies which show that catecholamines have little or no effect on ventilation and that endogenous catecholamine release is not due to hypercapnia but instead is correlated to low blood O_2 content (Perry et al. 1989; Kinkead et al. 1991; see Perry et al. 1992 for review).

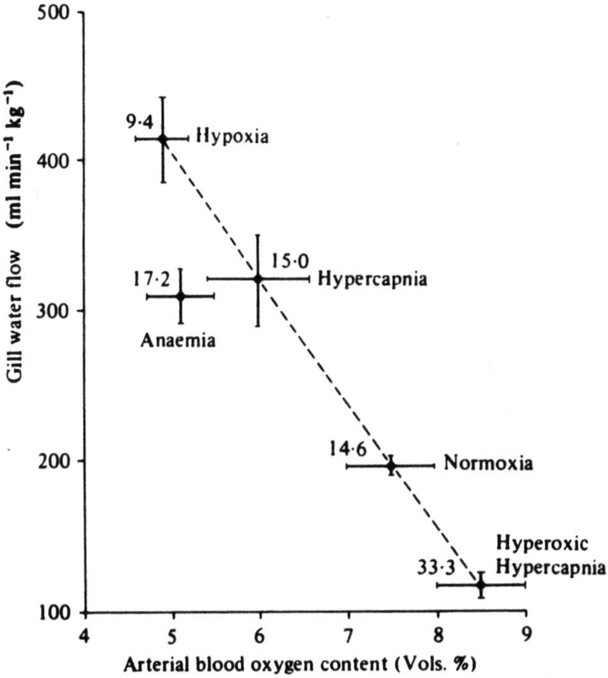

Fig. 5. Relationship between water flow over the gills and arterial O_2 content of dorsal aortic blood in *Salmo gairdneri*. *Horizontal* and *vertical bars* indicate $\pm$SEM. The *number* by each point is the corresponding Pa_{O_2} in kPa. Arterial O_2 content was changed by bubbling N_2, CO_2, and/or O_2 into water containing fish or by making fish anemic by withdrawing red blood cells. (Adapted from Smith and Jones 1982, and Randall 1982)

Finally, several recent studies have demonstrated significant respiratory responses to the acidosis accompanying hyperoxia or exercise independent of changes in arterial O_2 content or catecholamines (Heisler et al. 1988; Wood et al. 1990). In *Scyliorhinus stellaris*, on exposure to hyperoxic conditions, gill ventilation decreases and Pa_{CO_2} rises (Heisler et al. 1988). Initially, the rise in Pa_{CO_2} overrides the bicarbonate-equivalent transfer process from the environmental water and pHa falls. Gill ventilation then increases somewhat restoring plasma pH to control levels. Gill ventilation slowly falls over time as plasma $[HCO_3^-]$ rises with plasma pH being maintained in the range of control values (Heisler et al. 1988). Since Pa_{CO_2} is still rising slowly at this time, this data would suggest that pHa is the adequate stimulus. Manipulation of arterial acid-base status by application of various combinations of inspired CO_2 and intra-arterial infusion of bicarbonate under these conditions produced the correlation between gill ventilation and pHa depicted in Fig. 6 (Heisler, Andersen, Iwama, Boutilier and Claiborne, unpubl., cited in Heisler 1988). This suggests that when O_2 demand is not limiting to these animals, respiratory drive is mainly supplied by pH-sensitive structures in, or closely related to the arterial bloodstream. A similar correlation

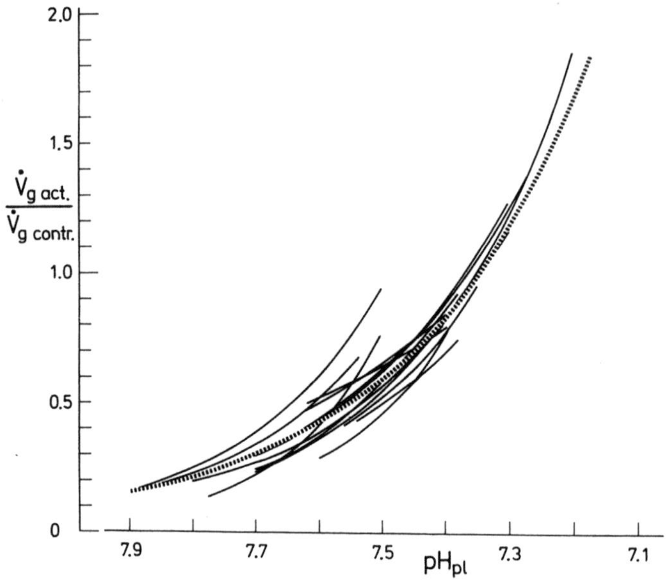

Fig. 6. Gill ventilation in *Scyliorhinus stellaris* during hyperoxia ($\dot{V}_{g\,act}$), in relation to gill ventilation during normoxia ($\dot{V}_{g\,contr}$), as a function of plasma pH (pH_{pl}). Each *solid curve* represents the least-square exponential regression of an individual specimen; the *dashed line* is the general exponential regression of the data pairs of all individuals. (Data from Heisler, Andersen, Iwama, Boutilier and Claiborne, unpubl. cited in Heisler 1988)

has been demonstrated in rainbow trout during post-exercise acidosis and during acid injections, both under normoxic conditions (Wood and Munger, pers comm.). Unfortunately, the study on *Scyliorhinus* also showed that when bicarbonate-equivalent transfer processes became saturated and could no longer increase [HCO_3^-] to match the rising P_{CO_2}, and pH began to fall again, ventilation now remained unchanged (Heisler 1988).

It is difficult to identify any single mechanism which might explain the sum of this data. It is possible that the hypercapnic ventilatory response is a consequence of the integration of the effects of both Pa_{CO_2} and pHa acting at separate receptor sites. Furthermore, the response may also arise, in large part, from changes in intracellular pH (pHi), probably at some specific locus. Changes in pHi should be more sensitive to changes in Pa_{CO_2} than pHa under conditions of constant [HCO_3^-] but more closely related to changes in pHa than Pa_{CO_2} under conditions of changing [HCO_3^-].

To summarize briefly, most fish do respond to environmental acidosis, hypercapnia and arterial acidosis. These responses may be a net effect of both direct and indirect actions. There is a growing body of evidence to suggest that fish must possess CO_2/pH chemoreceptors which directly elicit such changes. There is also good evidence to suggest that indirect effects of changing arterial P_{CO_2}/pH are involved in producing the ventilatory responses seen in many fish.

These appear to include changes in O_2 content, acting via O_2-sensitive chemoreceptors, and possibly changes in the levels of circulating catecholamines. The latter, which might be due to either the changes in P_{CO_2}/pH per se, or due to the changes in O_2 content they can elicit in some species, would appear to act on central respiratory neurons only and only if exceptionally high levels are attained. It is suggested that changes in Pa_{CO_2}/pHa probably act, at least in part, by changing pHi at localized sites. The magnitude and character of the ventilatory response will vary from species to species, reflecting the net balance of these direct and indirect effects as well as the level of other respiratory drives. Hypercapnia/acidosis represents a relatively minor component of the normal respiratory drive and appears to be easily overwhelmed by other inputs.

2.1.2 Air-Breathing Fish

Air-breathing fish exhibit bimodal breathing, utilizing gills and lungs to varying degrees in different species, to exchange gases with both water and air. In well-aerated waters, the branchial ventilation of facultative air-breathing fish, i.e. those species which utilize lung ventilation to supplement aquatic gas exchange, is continuous and rhythmic and air breathing generally consists of solitary air breaths occurring at irregular intervals. In the more advanced or obligate air-breathing fish, the gills are degenerate and branchial ventilation is increasingly weak and irregular. Air breathing now predominates and, in many cases, solitary lung ventilations are replaced by multiple inflation cycles (Johansen 1970; Shelton et al. 1986; Smatresk 1988). These changes seen in obligate air-breathing species are often accompanied by elevated arterial P_{CO_2} since the gills remain the predominant site for CO_2 elimination in most air-breathing fish. The dissociation of sites for O_2 and CO_2 exchange and the discontinuous nature of lung and branchial ventilation can lead to respiratory-related oscillations in blood gases. This is particularly true during emersion in those species which leave the water for any period of time (Shelton et al. 1986; Fishman et al. 1989). Thus, one might reasonably predict that there would be a progressive increase in changes in P_{CO_2}/pH as a source of respiratory drive for air-breathing fish, particularly for the obligate species.

Exposure to aquatic hypercapnia stimulates air breathing in most air-breathing fish, but not all (Fig. 3). In general, gill breathing is also stimulated by low levels of aquatic hypercapnia but the effect is relatively small and progressive hypercapnia generally inhibits branchial ventilation (for reviews, see Johansen 1970; Shelton et al. 1986; Smatresk 1988). The level of aquatic CO_2 required to inhibit gill ventilation ranges from 2 to 10% in different species as does the level required to stimulate air breathing (Johansen et al. 1967, 1968; Johansen and Lenfant 1968; Johansen et al. 1970; Hughes and Singh 1970, 1971; Singh and Hughes 1973; Lomholt and Johansen 1974; Graham 1976; Graham et al. 1977; Burggren 1979; Smatresk and Cameron 1982a). In any given species, air breathing is stimulated before gill ventilation ceases (Jesse et al. 1967; Hughes and Singh

1970; Graham et al. 1977). These thresholds are affected by the level of aquatic P_{O_2} (Jesse et al. 1967; Hughes and Singh 1970).

Aerial hypercapnia has no effect on ventilation in many species but does stimulate ventilation in others (Johansen et al. 1967, 1968; Lomholt and Johansen 1974; Burrgren 1979; Glass et al. 1986). The nature of the response appears to depend on the efficacy of gill ventilation in eliminating CO_2. In species in which branchial CO_2 excretion is rapid, inhalation of CO_2 or injection of CO_2 into the air-breathing organ may not result in any change in arterial P_{CO_2}. All of the CO_2 taken up by the air exchanger is eliminated by the aquatic exchanger. The implication of this data is that changes in gill and lung ventilation only occur if changes occur in P_{CO_2}/pH at the gills or in the arterial circulation.

The extent to which these effects are mediated by indirect means [such as changes in O_2 content or catecholamine concentration] or by changes in P_{CO_2}/pH per se is uncertain. As in the case with water-breathing fish, there is evidence to suggest that the ventilatory responses to hypercapnia and acidosis in air-breathing fish are a consequence of changes in the O_2 content of the blood. Thus, in the gourami, *Trichogaster*, the increased aerial ventilation associated with aquatic hypercapnia was suppressed when the water was also hyperoxic (Burggren 1979). This was also true for the increase in aerial ventilation associated with acid injections in the bowfin, *Amia calva* (Hedrick et al. 1991; McKenzie et al. 1991). Again, however, such an explanation would not seem to account for all of the data. In several species, chronic exposure to high levels of inspired CO_2 (aquatic or aerial) stimulates air breathing only transiently. The time course of this stimulation appears to follow changes in extracellular acid-base balance, suggesting that pH rather than P_{CO_2} may be the adequate stimulus (Graham and Baird 1982; Smatresk and Cameron 1982a). Air breathing is also stimulated in estivating lungfish, a time when Pa_{O_2} and Pa_{CO_2} both increase (Fishman et al. 1989), which also implicates peripheral or central CO_2/pH chemosensitivity. Although O_2 content may still decrease under both of these conditions, as pHa still falls in both, the changes would not appear sufficient to explain the magnitude of the ventilatory responses.

2.2 CO_2/pH-Sensitive Chemoreceptors

The literature reviewed in the previous sections indicates that most fish do respond to changes in P_{CO_2}/pH in their external and/or internal environment. Although it is not at all clear whether the specific stimulus modality is a change in Pa_{CO_2}, pHa or pHi or whether the stimulus acts directly on chemoreceptive tissue or indirectly by some other means, there is some suggestion that specific P_{CO_2}/pH-sensitive chemoreceptors may be involved in these responses. No such receptors, however, have been identified as yet.

Given that systemic arterial chemoreceptors in all air-breathing vertebrates examined respond to changes in both arterial O_2 and CO_2 tensions (Smatresk 1990b), it seems probable that the branchial chemoreceptors of fish might

Table 1. Possible location and innervation of putative CO_2/pH-sensitive receptors in lower vertebrates which influence ventilation

Fish Air-breathing fish Tadpoles	Adult amphibians	Reptiles	Innervation
Nasal passages	Nasal passages	Nasal passages	I
	Lungs:	Lungs:	
	mechanoreceptors	mechanoreceptors	X
		chemoreceptors	X
1st Gill arch	Carotid labyrinth	Carotid Artery or aortic arch	IX, X
Remaining gill arches	Aortic arch	Aortic arch	X
	Pulmocutaneous artery	Pulmocutaneous artery	X
Brain	Brain	Brain	

mediate reflex responses to hypercapnic acidosis as well as to hypoxemia (Table 1). These receptors are located on the gill arches, innervated by the glassopharyngeal and vagus nerves and are believed to be the homologues of the carotid and aortic bodies of higher vertebrates. Although studies of the discharge characteristics (Milson and Brill 1986; Burleson and Milsom 1990) and reflex roles (cf. Burleson and Smatresk 1990; McKenzie et al. 1991) of these receptors document their involvement in hypoxic ventilatory responses, none have examined the possibility that they may be involved in hypercapnic ventilatory responses. Given what is known of the discharge characteristics of these receptors in higher vertebrates, it would be predicted that reflex stimulation of these receptors by hypercapnic acidosis would produce responses which are likely to be weak at the relatively low temperatures and low P_{CO_2} found in blood and water. This is consistent with observations and it could be predicted that at elevated temperature the response to P_{CO_2} and acidosis should be more pronounced. This has also not been investigated.

It is also possible that the respiratory centers of fish may respond directly to acidosis or that there may be distinct central chemoreceptors in fish as there are in terrestrial vertebrates (Table 1). Superfusion of the brain with acidotic solutions has been shown to elicit a modest ventilatory stimulation in lampreys while superfusion with bicarbonate solutions elicited a modest hypoventilation in both lampreys and carp (Hughes and Shelton 1962; Rovainen 1977). In more recent studies, however, no significant effects of intracisternal or extradural fluid acidosis have been found in the skate or bowfin (Graham et al. 1990; Wood et al. 1990; Hedrick et al. 1991).

Aquatic hypercapnia has little effect on arterial P_{CO_2} in lungfish, because of the reduction in gill surface area in these obligate air-breathing fish. Nonetheless, it still stimulates air breathing and inhibits branchial ventilation (Johansen et al. 1967; Johansen and Lenfant 1968). This also occurs in gar where it has been attributed to the stimulation of olfactory defense receptors (Smatresk 1990a; Table 1). Taste or olfactory receptors, sensitive to CO_2 have been identified in carp (Yoshii et al. 1979), trout (Yamashita et al. 1989), gar (Smatresk 1990a) and catfish (Caprio and Smatresk, pers. comm.) but there is no indication that stimulation of these receptors leads to an increase in ventilation. In fact, as will be discussed in a later section, in air-breathing vertebrates stimulation of olfactory receptors by CO_2 is more likely to inhibit ventilation.

In summary, there are no studies to date which unequivocally demonstrate the presence of chemoreceptors in fish which are sensitive to changes in P_{CO_2}/pH per se and which participate in the regulation of resting ventilation or in the ventilatory responses to hypercapnic acidosis. Although this would support the idea that such changes act by other indirect means, it might also simply reflect the dearth of investigation into this specific question.

3 Amphibians and Reptiles

3.1 Ventilatory Responses to Changing P_{CO_2}/pH

3.1.1 Tadpoles

In anuran amphibians, breathing goes through important developmental changes. Initially, the aquatic breathing of the anuran larva (tadpole) is quite similar to that of water-breathing fish. Immediately after hatching, when the gills are still external to the body, gill ventilation frequency is very high; $\simeq$ 110 cycles/min. When the gills become enclosed internally within the paired branchial chambers, gill ventilation falls sharply to about 50 cycles/min. This degree of buccal activity characterizes older larvae and persists even in the adult (Burggren and Doyle 1986). Although the lungs are well developed in older larvae, air breathing is a rare occurrence before metamorphosis ($\simeq$ 4/h at 20 °C). After metamorphosis to the adult, lung ventilation increases to about 1–3 breaths/min (Burggren and Doyle 1986). The emergence of a ventilatory response to hypercapnia is also part of the developmental scenario. The early larvae are insensitive to changes in P_{CO_2} until late premetamorphic stages when the lungs begin to display an active role in gas exchange and breathing patterns take on an adult appearance (Infantino 1990). Correspondingly, it has been shown that an in vitro brainstem-spinal cord preparation of the tadpole is insensitive to changes in pH over a range from 7.2 to 7.8 produced by changing the CO_2 concentration in the superfusion solution, suggesting that no active central CO_2 chemoreceptors are present in this preparation (Walker et al. 1990).

3.1.2 Adult Amphibians

As in air-breathing fish, oxygen appears to play a dominant role in ventilatory control of amphibians. Several species tolerate wide swings in blood P_{CO_2} and pH and show little or no ventilatory response. For other species, however, responses to hypoxia are weak while the addition of CO_2 to the water or to inspired air causes marked increases in ventilation (MacIntyre and Toews 1976; Toews and MacIntyre 1978; Boutilier et al. 1979; Jackson and Braun 1979; Boutilier et al. 1980; Boutilier and Toews 1981; Heisler et al. 1982; Jones 1982; West et al. 1987; Boutilier 1988; Jones and Chu 1988).

Urodele Amphibians. Urodeles spend most of their lives in water. Two species have been studied in most detail and have yielded contrasting results. Loading CO_2 via the skin in the urodele, *Amphiuma tridactylum*, had no effect on air breathing even though Pa_{CO_2} increased 2.5 times to >40 mm Hg (Toews 1971). This was not the case in the hellbender, *Cryptobranchus alleganiensis*. In this species, aquatic hypercapnia (2.5% CO_2) led to an increase in lung ventilation, aquatic buccal pumping and rocking motions that ventilated the skin (Boutilier and Toews 1981). More than 90% of CO_2 elimination normally occurs across the skin in this species (Guimond and Hutchison 1976). Even though the ventilation frequency was quadrupled, Pa_{CO_2} increased from 6.2 to over 20 torr (Fig. 7; Boutilier and Toews 1981). As with other aquatic species such as the salamanders *Amphiuma* and *Siren*, there was virtually no compensation of extracellular pH during sustained hypercapnia (Boutilier et al. 1980; Boutilier and Toews 1981; Heisler et al 1982). There was, however, an increase in Pa_{O_2} which just restored oxyhemoglobin saturation back to control levels (Fig. 7). These data have been cited to support the suggestion that increased ventilatory activities in aquatic vertebrates are a response to the effects of acidosis on arterial oxygen content, as in fish (i.e. CO_2-Bohr effect; Boutilier and Toews 1981). Since Pa_{O_2} is increased above resting levels and O_2 content is restored to resting levels, however, this cannot be the stimulus for the sustained elevation in ventilation, although it does demonstrate a consequence of the effort.

Anuran and Apodan Amphibians. These two groups of amphibians show greater degrees of terrestriality than the urodeles and all species studied to date show an increase in ventilation on exposure to hypercapnic conditions. In most species, there appears to be a threshold for stimulation beyond which minute ventilation increases with increasing levels of hypercapnia or acidosis (anurans: *Rana*: Smyth 1939; *Bufo*: MacIntyre and Toews 1976; West et al. 1987; Branco et al. 1992; *Xenopus*: Jones and Chu 1988; apodans: *Typhlonectes:* MacIntyre and Toews 1976). In some species, sensitivity to changing Pa_{CO_2}/pHa is extremely high; in *Rana catesbeiana*, ventilation volume doubles following a 2 mm Hg rise in Pa_{CO_2} (Jackson and Braun 1979).

In aquatic anurans, the buccal pump is used solely for lung ventilation. Often, breathing consists of episodes of several lung ventilations. In more terrestrial anurans, rhythmic buccal oscillations occur almost continuously between lung ventilations. Furthermore, in many of the terrestrial anurans, episodes of lung ventilation may consist of several lung inflations in succession without an expiration, inflating the lungs to greater than normal pressures (lung inflation

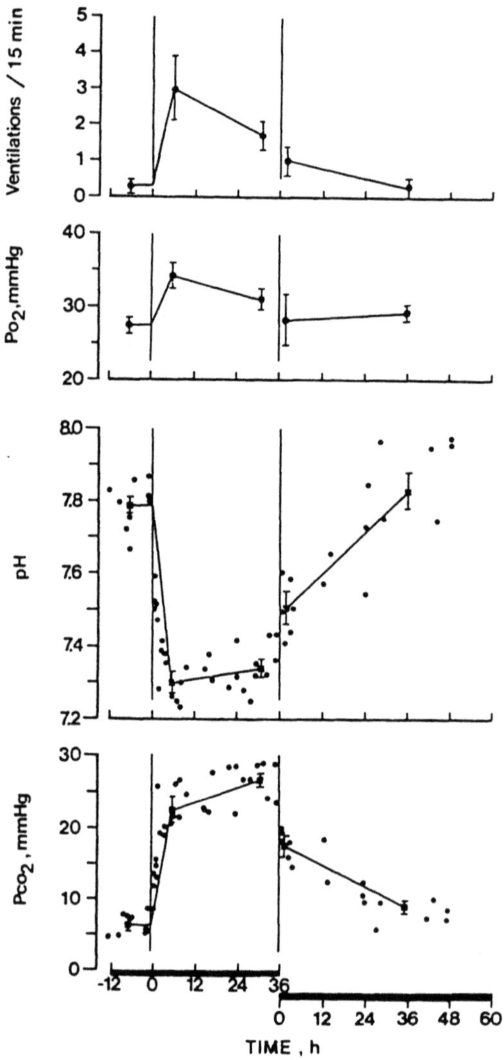

Fig. 7. Changes in lung ventilation frequency and arterial P_{O_2}, pH and P_{CO_2} during 36h exposure to hypercapnia in *Cryptobranchus alleganiensis*. Mean values ($\pm$SEM) are shown for all relationships along with the individual data points for the changes in pH and P_{CO_2}. Data are also shown for up to 12 h before and 48 h following exposure. (Boutilier and Toews 1981)

cycle; Fig. 8; see Shelton et al. 1986, for review). In species exhibiting episodes of lung ventilation, the net effect of hypercapnic stimulation is primarily a decrease in the non-ventilatory period between cycles and an increase in the number of ventilations in each cycle (Fig. 8; West et al. 1987; Smatresk and Smits 1991; Branco et al. 1992). In *Xenopus laevis*, the changes in minute ventilation in response to environmental hypercarbia are almost exclusively due to increases in breathing frequency (Jones and Chu 1988) while, in *Bufo marinus* and *Bufo paracnemis*, changes in tidal volume are the dominant component of the response (Smatresk and Smits 1991; Branco et al. 1992). To the extent that they have been studied, buccal oscillations appear to be progressively reduced by increasing CO_2

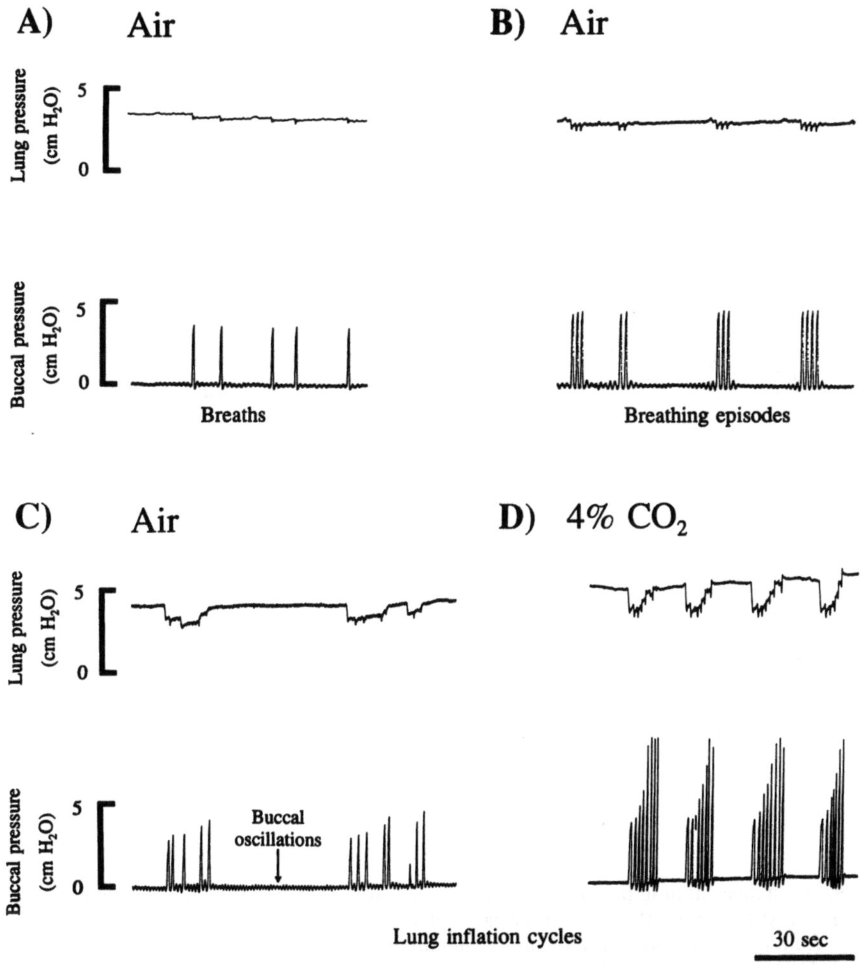

Fig. 8. Traces illustrating breathing in the bullfrog, *Rana catesbeiana. Upper trace* is lung pressure and *lower trace* is buccal pressure in each pair of traces. **A** depicts single breaths (animal breathing air); **B** depicts episodes of breathing (animal breathing air); and **C, D** depict lung inflation cycles (animal breathing Air and 4% CO_2 in air). Note the buccal oscillations (small deflections in the buccal pressure trace) in **A, B** and **C** but not **D**. (Kinkead and Milsom, unpubl.)

levels (Fig. 8; McIntyre and Toews 1976) just as branchial ventilation is in air-breathing fish (see Sect. 2.1.2).

There is recent evidence which would suggest that while the hypercapnic ventilatory responses measured for most anuran amphibians, at least, are probably accurate reflections of the responses of these species to environmental hypercarbia, they underestimate the response to similar changes in Pa_{CO_2}/pHa produced by other means. In *Rana catesbeiana*, a further significant increase in ventilation is seen immediately following the removal of CO_2 from inspired gas (post-hypercapnic hyperpnea; Kinkead and Milsom, unpubl; Fig. 9). At this time,

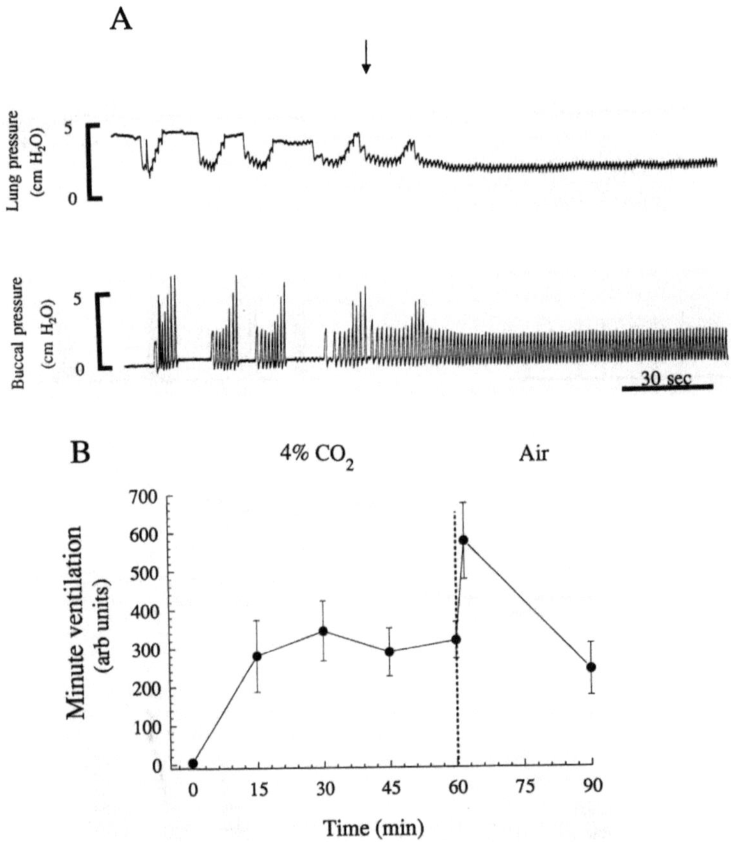

Fig. 9. A Changes in lung pressure and buccal pressure associated with lung ventilations in a bullfrog. At the *arrow*, the respiratory gas is changed from 4% CO_2 in air to air. Note the switch from lung inflation cycles to simple lung inflations. **B** Quantification of the changes in minute lung ventilation [breathing frequency x buccal pressure amplitude (arbitrary units)] for six bullfrogs during a 1-hour exposure to 4% CO_2 and during the first 30 min following this exposure. Note the dramatic increase in minute ventilation associated with the switch from the hypercapnic gas mixture back to air. (Kinkead and Milsom, unpubl.)

arterial P_{CO_2} will still be elevated but the level of P_{CO_2} in the airways will no longer be elevated throughout the ventilation cycle; only phasically during expiration. There is evidence to suggest that there are upper airway receptors, primarily in the olfactory epithelium, which are sensitive to CO_2 and which reflexly inhibit ventilation if stimulated tonically throughout the ventilation cycle (Sakakibara 1978; Coates and Ballam 1990). The actions of these receptors have been studied more intensively in reptiles and will be discussed in more detail in following sections (Sects. 3.1.3 and 3.3.1). The data suggest, however, that while increasing levels of inspired CO_2 are a useful means to study the ventilatory responses of these animals to environmental hypercarbia, they also elicit an inhibitory reflex which will partially mask the full response of these animals to increases in Pa_{CO_2}.

As a result, this is not an appropriate way to examine ventilatory responses to increases in Pa_{CO_2} arising from the metabolic production of CO_2/H^+.

In most cases, if hypercapnia is prolonged, there is some pH compensation which occurs over time as $[HCO_3^-]$ increases. In some aquatic animals, such as the apodan amphibian, *Typhlonectes compressicauda*, blood pH compensation is almost complete with chronic hypercapnic exposure (Toews and MacIntyre 1978), whereas in the terrestrial toad, *Bufo marinus*, blood pH is only compensated by 20–30% (Boutilier et al. 1979; Toews and Heisler 1982). Breathing is reduced in all these species in chronic hypercapnia but it is not at all clear to what extent minute ventilation correlates with the changes in Pa_{CO_2}/pHa. It would appear that in amphibians, as in fish, there is no absolute correlation between minute ventilation and any of Pa_{CO_2}, pHa or O_2 content (Fig. 7), suggesting again that it is a combination of these variables or their effect on some other variable such as pHi, perhaps at a localized site, which is the specific stimulus for the response.

3.1.3 Reptiles

There is a rich literature documenting the ventilatory responses of reptiles to hypercapnia. In general, the hypercapnic ventilatory response is well developed and changes in Pa_{CO_2} rather than Pa_{O_2} now appear to provide the dominant chemical drive to breathe (see Milsom 1990 for review). In most species, the relationship between ventilation and arterial Pa_{CO_2} is roughly linear over the physiological range but there are exceptions to this trend as well as tremendous variability in the slope of this relationship (Fig. 3). The relative contributions of changes in V_T and f to the changes seen in $\dot{V}_E$ are equally varied (Fig. 3). In general, most species increase V_T equally with, or proportionately more than f and the species showing the briskest responses to hypercapnic stimuli exhibit proportionately larger increases in both variables.

Chelonian Reptiles. In chelonians, CO_2 inhalation invariably leads to an increase in minute ventilation. Often there may be a pause in breathing after the first inhalation of CO_2-enriched gas (Randall et al. 1944; Jackson et al. 1979). This appears to be an inhibitory reflex due to stimulation of upper airway chemoreceptors (see Sect. 3.3.1) which is usually followed by a rise in V_T and breathing frequency, the latter being due primarily to shortening of the periods of breath holding (Frankel et al. 1969; Glass et al. 1978; Jackson et al. 1979; Benchetrit and Dejours 1980; Milsom and Jones 1980; West et al. 1989). Thus, inhalation of 4–5% CO_2 at 20–25 °C increases minute ventilation $\simeq$3 to 5 times above resting levels in most species (Fig. 3; *Chrysemys*, Milsom and Jones 1980; Funk and Milsom 1987; *Pseudemys*, Jackson et al. 1974; *Testudo*, Glass et al. 1978; *Chelonia*, Jackson et al. 1974; *Pelomedusa*, Glass et al. 1978; *Chelydra*, West et al. 1989). In all these cases, minute ventilation is increased by changes in both frequency of breathing (2–4 times resting levels) and V_T (1.5–3 times resting levels).

There are two distinct breathing patterns in chelonians. Many have a breath-

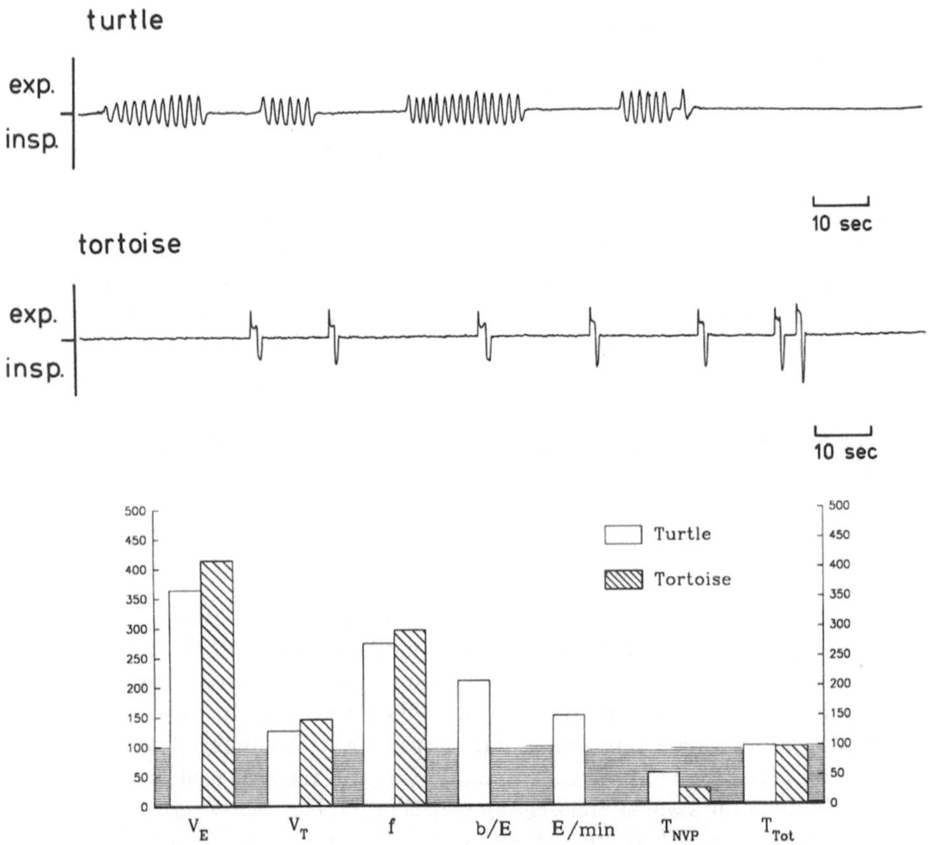

Fig. 10. *Above* Air flow traces depicting the breathing patterns of tortoise and a turtle. *Below* Changes in minute ventilation ($\dot{V}_E$), tidal volume (V_T), breathing frequency (f), the number of breaths in each breathing episode (b/E), the number of breathing episodes per minute (E/min), the length of an individual breath (T_{Tot}; period of air flow) and the length of the non-ventilatory period between episodes (T_{NVP}) in a tortoise and a turtle breathing 5% CO_2. (Data from Benchetrit and Dejours 1980; Milsom and Jones 1980)

ing pattern which consists of single breaths relatively evenly separated by short periods of breath holding, while others have a pattern which consists of episodes of almost continuous breathing separated by much longer periods of breath holding (Fig. 10; see Milsom 1990 for review). The increases in breathing frequency in both groups are due primarily to a shortening of the periods of breath holding rather than to any change in the periods of time actually spent actively inspiring or expiring. In species which exhibit episodic breathing, there is also an increase in the number of breaths in each breathing episode (Fig. 10; Frankel et al. 1969; Glass et al. 1978; Milsom and Jones 1980; Funk and Milsom 1987; West et al. 1989). There is no apparent relationship between breathing pattern or CO_2 sensitivity and either well-adapted diving or strictly terrestrial life-styles in these animals.

Crocodilian Reptiles. In crocodilians, as in chelonians, inhalation of CO_2 causes minute ventilation to rise, again due both to increases in V_T and decreases in the period of breath holding (Randall et al. 1944; Davies and Kopetzky 1976; Glass and Johansen 1979; Davies et al. 1982; Douse and Mitchell 1992a). The latter seems to make the main contribution to ventilatory changes in *Crocodylus* (Glass and Johansen 1979) but in *Alligator* roughly equivalent changes occur in V_T and breathing frequency (Davies et al. 1982; Douse and Mitchell 1992a). All crocodilians studied to date normally breathe episodically and, as in chelonians, the increase in breathing frequency is also accompanied by an increase in the number of breaths in each breathing episode (Douse and Mitchell 1992b). The sensitivity of crocodilians to CO_2 is similar to that of turtles (Fig. 3).

Snakes and Lizards. Inhalation of CO_2 in snakes and lizards produces a variety of responses. Low levels of inspired CO_2 invariably cause minute ventilation to increase, but it can be depresssed by higher levels (>3% CO_2) in some lizards (*Crotaphytus*, Templeton and Dawson 1963; *Lacerta* sp., Nielsen 1961) and snakes (*Acrochordus*, Glass and Johansen 1976; *Natrix* Gratz 1979). In general, CO_2 causes V_T to increase in all these animals, but the higher concentrations usually result in a substantial fall in breathing frequency. In other snakes and lizards, however minute ventilation continues to increase when gas mixtures containing high levels of CO_2 (2–10%) are inspired. The black racer snake, *Coluber constrictor*, responds to the inhalation of 4–7% CO_2 by increasing its minute ventilation between three and eight times (Nolan and Frankel 1982), depending on temperature; the change is caused entirely by increasing V_T. The same general levels of stimulation cause increased ventilation in other snakes (e.g. in *Drymarchon*, Randall et al. 1944), but through pronounced changes in breathing rate with V_T falling slightly. Only in the monitor lizard, *Varanus exanthematicus* (Glass and Wood 1983), do V_T and breathing frequency increase together, though even here the relationship is very variable.

The large variability in the responses of snakes and lizards to CO_2 inhalation is difficult to explain at present. It appears to be the consequence of differences in the importance and relative sensitivities of the various chemoreceptor groups found in these animals (see Sect. 3.3). In *Crotaphytus, Acrochordus, Lacerta* and *Coluber*, all of which show no increase in breathing frequency in response to inhaled CO_2, an immediate hyperpnea is seen when CO_2 is removed from the gas mixture (post-hypercapnic hyperpnea) (Nielsen 1961; Templeton and Dawson 1963; Glass and Johansen 1976; Nolan and Frankel 1982). This is similar to the response already described in anuran amphibians (Fig. 9; Sect. 3.1.2) which was interpreted to suggest that the stimulating effect of systemic hypercapnia is masked by an inhibitory effect of tonically elevated airway CO_2. In the case of the reptiles, there is evidence to suggest that intrapulmonary chemoreceptors as well as upper airway receptors may inhibit ventilation during CO_2 inhalation (see Sects. 3.3.1 and 3.3.3). As pointed out earlier, these observations illustrate that the effects of hypercapnia with and without hypercarbia can be quite different in some species. While increasing levels of inspired CO_2 may be suitable for studies

Table 2. Differences between changing Pa_{CO_2}/pHa by increasing inspired CO_2 versus increasing metabolic CO_2 production (or by venous CO_2 loading) on the discharge and reflex responses of various chemoreceptor groups in reptiles

A. Effects of increasing inspired CO_2

		Reflex Effects	
Receptor	Discharge	$\dot{V}_E$	f/V_T
Upper airway	Increased throughout cycle	Decrease	Decrease f primarily
Intrapulmonary	Decreased tonic Decreased phasic	Decrease	Decrease f Increase V_T
Pulmonary stretch	Decreased	Increase	Increase f Increase V_T
Arterial	Increased	Increase	Increase f Increase V_T
Central	Increased	Increase	Increase f primarily
Net effect		Increase, decrease or no change	Increase/decrease or no change in f Increase V_T

B. Effects of venous CO_2 loading or increased metabolic CO_2 production

		Reflex Effects	
Receptor	Discharge	$\dot{V}_E$	f/V_T
Upper airway	No effect or slight increase during expiration	No effect	
Intrapulmonary	No effect or slight increased phasic	No effect	
Pulmonary stretch	No effect or slight decrease during expiration	No effect or slight increase	Increase f Increase V_T
Arterial	Increased	Increase	Increase f Increase V_T
Central	Increased	Increase	Increase f primarily
Net effect		Increase	Increase f Increase V_T

of the effects of environmental hypercarbia on ventilation in these species, this technique is not suitable for studies of the effects of increases in Pa_{CO_2} arising from increased metabolic production of CO_2/H^+ (Table 2).

Ventilatory responses to metabolic acidosis/alkalosis in reptiles have been less well studied. There is evidence to suggest that central $CO_2/[H^+]$ chemosensors are dominant over peripheral inputs in eliciting the ventilatory response to hypercapnia (see Sect. 3.3.6) and there are several studies which have analyzed the effects of CSF-fixed acid (produced by anoxia) vs CSF-volatile acid (CO_2-

induced) on ventilation in turtles (Hitzig and Jackson 1978; Hitzig 1982; Hitzig and Nattie 1982; Hitzig et al. 1985; Davies and Sexton 1987). The data suggest that changes in breathing correlate better with changes in CSF P_{CO_2} than CSF pH. Changes in CSF pH alone do stimulate ventilation but to a lesser degree than similar changes in CSF pH induced by increasing Pa_{CO_2} (Hitzig and Nattie 1982).

If hypercapnia is prolonged in turtles, very little pH compensation occurs. Because the normally high arterial $[HCO_3^-]$ found in turtles ($\simeq$40 mmol/l) is close to the apparent upper $[HCO_3^-]$ which can be attained in this species (50 mmol/l) (Silver and Jackson 1985; Heisler 1986), arterial $[H^+]$, P_{CO_2}, $[HCO_3^-]$ and minute ventilation all remain elevated and relatively stable (Silver and Jackson 1985). As a consequence, we are offered no insights into the relative roles of changes in P_{CO_2} and/or pH in eliciting the ventilatory response. Insufficient data exist to allow such an analysis, with or without the possible confounding effects of environmental hypercarbia, in any species.

3.2 Role of CO_2/pH in Control of the Breathing Pattern

As mentioned earlier, many amphibians and reptiles ventilate their lungs in an episodic fashion (Figs. 8, 10). In these species, hypercapnia results in a decrease in the pause between episodes of breathing and an increase in the number of breaths in each episode (Frankel et al. 1969; Glass et al. 1978; Milsom and Jones 1980; Funk and Milsom 1987; West et al. 1987, 1989; Smatresk and Smits 1991; Branco et al. 1992; Douse and Mitchell 1992b). This has led to the suggestion that the non-ventilatory period must be regulated, at least in part, by negative feedback from various chemoreceptor groups in response to oscillations in blood gas concentrations and/or pH (Lenfant and Johansen 1968; Johansen et al. 1970b; Emilio and Shelton 1974; Toews et al. 1978; Ackerman and White 1979; Burggren and Shelton 1979; Shelton and Boutilier 1982; Smatresk and Cameron 1982, Shelton et al. 1986). Recent evidence, however, suggests that although changes in blood gases and pH do indeed alter the length of the non-ventilatory period, they act in an additive fashion with other inputs and do not turn breathing on and off per se.

The evidence to support this line of reasoning has been growing for some years. There are observations demonstrating that both hyperoxia and artificial, unidirectional ventilation lead to a cessation of ventilation in many species (Kinney and White 1977; Toews and Kirby 1985; West et al. 1987). More recently, it has been shown that if Pa_{CO_2} is increased or pHa or cerebrospinal pH decreased under these conditions, episodic breathing will be reinitiated (West et al. 1987; Smatresk and Smits 1991; Branco et al. 1992) despite the fact that under these conditions pulmonary gas tensions (Douse and Mitchell 1992b), and arterial blood gases and pH (Kinkead and Milsom, 1994; Fig. 11), remain constant. Furthermore, in the in vitro turtle brainstem-spinal cord preparation, the abbreviated episodic pattern exhibited by vagotomized animals (two and sometimes three breaths in an episode) remains (Douse and Mitchell 1990). This is

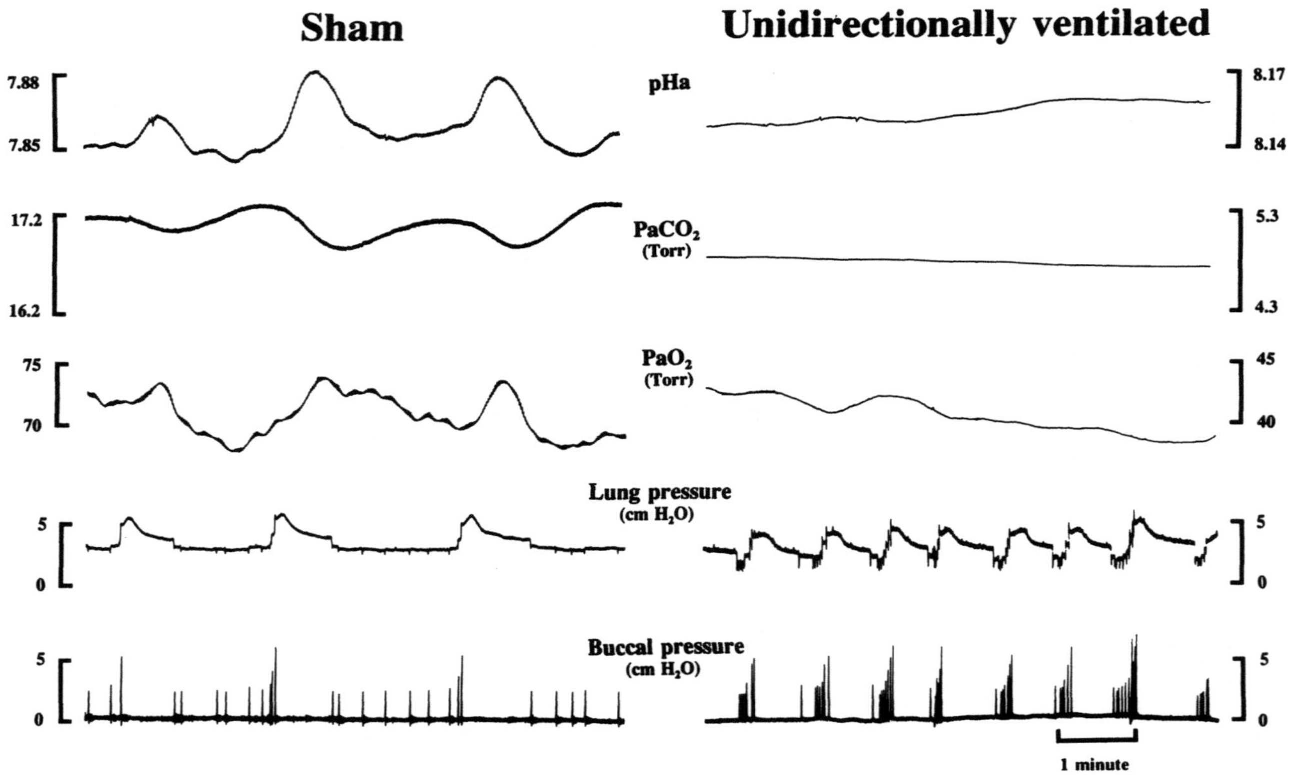

Fig. 11. Changes in arterial pH (pHa) and blood gases (Pa_{CO_2} and Pa_{O_2}) associated with lung inflation cycles (depicted by changes in lung pressure and buccal pressure) in the bullfrog, *Rana catesbeiana*, during spontaneous ventilation (*sham*) and unidirectional artificial ventilation (*unidirectionally ventilated*). Note that breathing remains episodic during artificial ventilation even though there are no accompanying changes in blood gases or pH. (Kinkead and Milsom, 1994.)

a condition under which all peripheral feedback has been removed and in which CO_2 and O_2 levels are held constant. These data suggest that the essential mechanisms generating the episodic breathing patterns seen in these species are intrinsic to the central neural networks associated with ventilatory control (central pattern generator). Negative feedback from various afferent inputs, either tonic or phasic, is required to initiate ventilation and plays an important role in modulating the level of respiratory drive, but it does not appear to be essential for the genesis of the episodic pattern. In light of this new evidence, it is not surprising that previous attempts to explain episodic breathing as a result of oscillations in arterial blood gases and/or pH, and to define clear-cut thresholds in the levels of these variables for initiating and terminating episodes of breathing, have been unsuccessful (Lenfant and Johansen 1968; Johansen et al. 1970; Emilio and Shelton 1974; Toews et al. 1978; Ackerman and White 1979; Burggren and Shelton 1979; Shelton and Boutilier 1982; Smatresk and Cameron 1982b; Boutilier and Shelton 1986; Shelton et al. 1986).

3.3 CO_2/pH-Sensitive Chemoreceptors

The specific characteristics of tissue associated with CO_2/pH chemoreception have not been well defined in any vertebrate. It is certain, however, that the location, innervation, stimulus specificity and reflexogenic roles of CO_2/pH chemoreceptive tissue in amphibians and reptiles are far from homogeneous.

3.3.1 Olfactory, Gustatory and Upper Airway Chemoreceptors

As noted in earlier sections, several species of amphibians and reptiles actually depress ventilation when the concentration of inspired CO_2 is elevated (Fig. 3; Smyth 1939; Nielsen 1961; Templeton and Dawson 1963; Glass and Johansen 1976; Sakikabara 1978; Gratz 1979) and show a significant hyperpnea immediately upon return to breathing room air (Fig. 9; Nielsen 1961; Templeton and Dawson 1963; Glass and Johansen 1976; Nolan and Frankel 1982; Kinkead and Milsom, unpubl.). These responses have been attributed to the actions of inspired CO_2 on upper airway chemoreceptors in both amphibians and reptiles as well as on intrapulmonary chemoreceptors in some reptiles (see Sect. 3.3.3).

A series of denervation studies have isolated the upper airway receptors responsible for reflex inhibition of ventilation during CO_2 inhalation to the nasal sensory epithelium innervated by the olfactory nerve in the bullfrog (*Rana catesbeiana*, Sakikabara 1978; see also Smyth 1939), the Tegu lizard (*Tupinambis nigropunctatus*, Ballam 1984, 1985) and the garter snake (*Thamnophis sirtalis*, Coates and Ballam 1987). The vomeronasal organ, innervated by the trigeminal nerve was also believed at one point to play a possible role in this response (Ballam 1985; Coates and Ballam 1987), but it now appears that CO_2-sensitive receptors in this location actually inhibit the olfactory chemoreceptor-mediated

CO_2 response (Coates and Ballam 1989). The only electrophysiological recordings of the effects of CO_2 on the activity of olfactory chemoreceptors have been performed in tiger salamanders (*Ambystoma trigrinum*, Getchell and Shepherd 1978) and bullfrogs (Coates and Ballam 1990). The latter study indicates that these cells are responsive to levels of CO_2 below normal end-expiratory CO_2 concentrations and supports the proposal that the ventilatory depression observed in response to upper airway CO_2 in some reptiles and amphibians is mediated by olfactory cells (Table 1, 2).

The post-hypercapnic hyperpnea is probably also mediated by the removal of the tonic influence of CO_2 on these cells in amphibians as well as on the intrapulmonary chemoreceptors in some reptiles. Transiently, during the period immediately following the return from CO_2 inhalation to breathing air, the P_{CO_2} of arterial blood will still be elevated, and possibly that of end-expired gas. Inspired gas, however, will now be void of CO_2 and the olfactory cells will receive phasic stimulation rather than tonic stimulation. It has been shown that CO_2 delivered to these receptors during only the inspiratory or the expiratory phase of the ventilation cycle does not inhibit ventilation; ventilation is only depressed if the elevated levels of CO_2 are sustained (Fig. 12; Table 2; Ballam and Coates 1989; Coates et al. 1991). Thus, when CO_2 is removed from inspired gas, the depression of ventilation due to the tonic stimulation of the olfactory cells is removed and the true effect of arterial hypercapnia on ventilation is unmasked transiently until the elevated ventilation returns blood gases to normal levels.

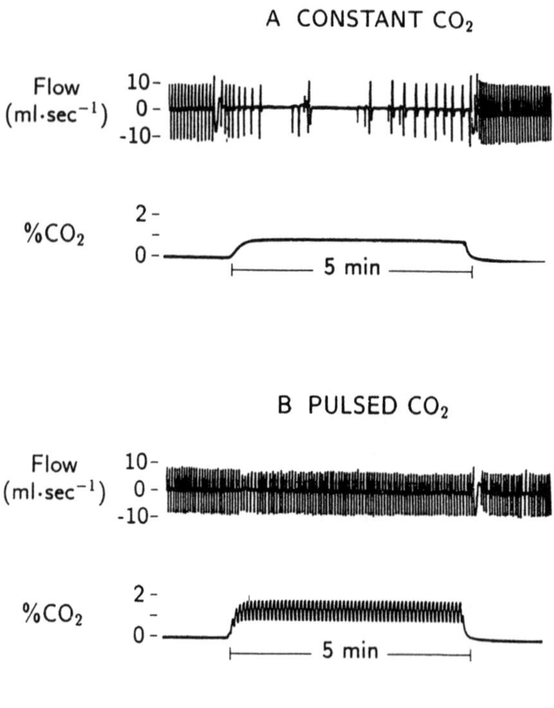

Fig. 12. Ventilatory flow (*upper trace*) and CO_2 composition of gas entering the external nares (*lower trace*) in the tegu lizard illustrating the effects on breathing when CO_2 is **A** held constant or **B** pulsed at the same mean concentration through the upper airways. (Ballam and Coates 1989)

It should be noted that the actual role of the inhibitory reflex arising from the stimulation of the olfactory cells remains unknown. It has been postulated that this response might function to inhibit breathing when the ambient P_{CO_2} is higher than the levels in the systemic blood, reducing CO_2 uptake while the animals seek fresh air (Coates and Ballam 1987). These receptors have also been postulated to be involved in the detection of CO_2 originating from prey or predators in burrows (Ballam 1985), but why this should inhibit ventilation remains an open question.

3.3.2 Pulmonary Stretch Receptors

Although the primary stimulus modalities of these receptors are changes in lung volume, pressure or wall tension, increasing pulmonary P_{CO_2} has been shown to inhibit receptor discharge by varying degrees in amphibians and reptiles (Table 1; frogs: Milsom and Jones 1977; Kuhlmann and Fedde 1979; turtles: Milsom and Jones 1976; Jones and Milsom 1979; McLean et al. 1989; alligators: Powell et al. 1988; Douse et al. 1989; lizards: Fedde et al. 1977; snakes: Furilla and Bartlett 1988). This inhibition is largely independent of the effects of the CO_2 on pulmonary smooth muscle tone (Jones and Milsom 1979), but it is not known whether these effects are due to the changes in pH or the direct effects of CO_2. The reflex effects of CO_2 modulation of pulmonary stretch receptor (PSR) discharge have been best studied in turtles. Increasing intrapulmonary P_{CO_2} in a vascularly isolated lung decreased both inspiratory and expiratory discharge of pulmonary stretch receptors (Table 2; Jones and Milsom 1979) and resulted in increases in tidal volume and breathing frequency (Milsom and Jones 1979). These increases were only 19% of those which resulted when arterial hypercapnia also developed. Thus, the net reflex effect of CO_2 acting on these receptors was small but the data do suggest that CO_2 modulation of PSR discharge can have a physiological effect. This may not only be the case for exposure to environmental hypercarbia but may also be the case in situations in which metabolic CO_2 production is increased. Depending on the net effect of the increase in metabolic rate on ventilation, intrapulmonary CO_2 could increase during the non-ventilatory periods typical of the breathing patterns in amphibians and reptiles and reduce receptor discharge, producing a similar result.

3.3.3 Intrapulmonary Chemoreceptors

Intrapulmonary chemoreceptors (IPC) have now been described in the lungs of several species of reptiles which appear identical to those which have been well documented in birds. They are located within the lung, innervated by the vagus nerve and have a discharge which is inversely proportional to P_{CO_2} (Fig. 13; Table 1). They are rate-sensitive, demonstrating an overshoot and undershoot, respectively, in the change in discharge accompanying the sudden removal or addition of CO_2 to the inspired gas. They are insensitive to changing P_{O_2} and

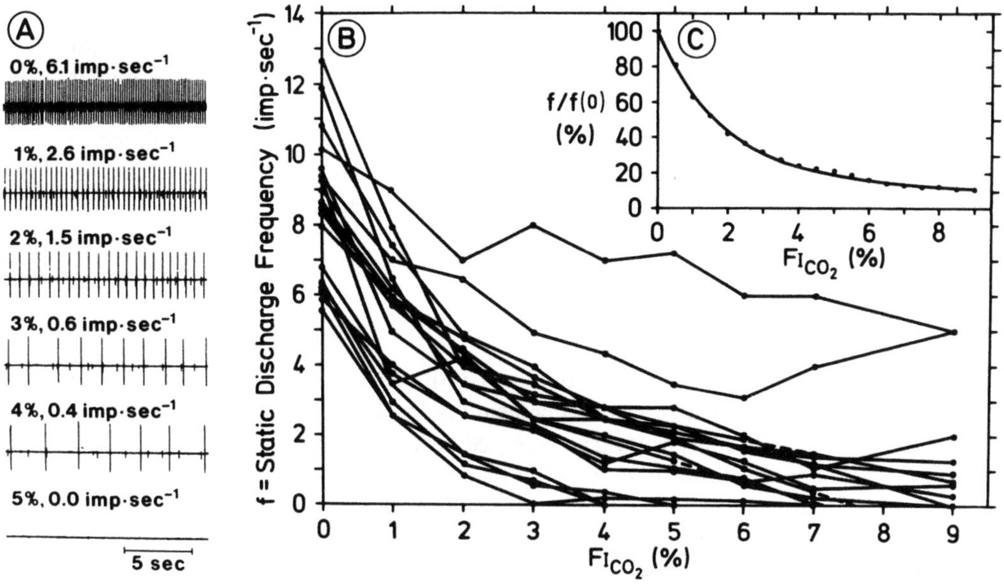

Fig. 13. Response characteristics of intrapulmonary chemoreceptors from the tegu lizard. **A** Static discharge frequency, in impulses/sec, from one receptor at levels of CO_2 from 0 to 5%. **B** The relationship between static discharge frequency and the fractional composition of inspired CO_2 for 18 intrapulmonary chemoreceptors. **C** Mean of the relationships for the 18 receptors

their responses to changes in lung volume or pressure are not significant within physiological limits. The location, innervation and functional characteristics of avian IPC have been more extensively studied and an excellent review of this literature is given by Scheid and Piiper (1986). It is currently believed that both CO_2 and H^+ exert independent effects on avian IPC activity with the sensitivity to CO_2 being somewhat greater (Burger et al. 1974; Powell et al. 1978; Scheid et al. 1978). This has not been examined in reptiles.

The phylogenetic distribution of IPC would suggest this receptor group evolved after mammals split from the reptilian stock (Fig. 14). They have not been found in any amphibian (*Rana catesbeiana.* Milsom and Jones 1977; Kuhlmann and Fedde 1979; *Bufo marinus, Ambystoma trigrinum:* Furilla and Bartlett 1988) nor in mammals (Kunz et al. 1976; Bystrzycka and Nail 1980). They are present in birds (see Scheid and Piiper 1986 for review) and all diapsid reptiles examined so far (lizards: *Tupinambis nigropunctatus*, Fedde et al. 1977; *T. teguixans,* Gatz et al. 1975; snakes: *Thamnophis sirtalis,* Furilla and Bartlett 1987; alligators: *Alligator mississippiensis,* Douse and Mitchell 1987; Powell et al. 1988; Douse et al. 1989). Their distribution in the more ancient group of reptiles, the *Chelonia*, is puzzling. They appear to be present in one species of tortoise (*Testudo hermanii*, Ishii et al. 1986) but not in two species of fresh water turtles (*Chrysemys picta, Pseudemys*

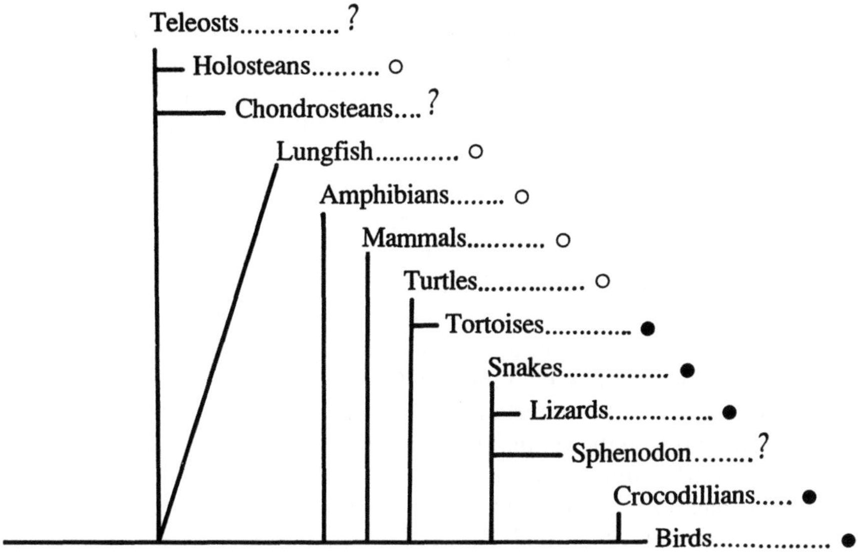

Fig. 14. Cladogram illustrating the phylogenetic relationship between various vertebrate groups. *Open symbols* indicate an absence, and *filled symbols* the presence, of intrapulmonary chemoreceptors for each group

scripta, Jones and Milsom 1979; McLean et al. 1989). The phylogenetic significance of this remains unclear.

The reflex role of this receptor group remains somewhat controversial. In birds, in which this has been most extensively studied, there is good evidence that during quiet ventilation, this receptor group plays a similar role to that performed by pulmonary stretch receptors in other vertebrate groups. As such, these receptors appear to monitor the rate and extent of CO_2 washout/accumulation during inspiration/expiration, reflecting the rate of lung inflation/deflation, and contribute to both inspiratory inhibitory (Burger et al. 1974; Fedde et al. 1974a; Molony 1974; Banzett and Burger 1977; Burger et al. 1978; Milsom 1990) and inspiratory facilitatory reflexes (Berger and Tallman 1982). This also appears to be the case in the garter snake (Furilla and Bartlett 1989; Furilla et al. 1991).

Both birds and reptiles possess pulmonary stretch receptors as well as intrapulmonary chemoreceptors, however, suggesting that these receptors must have some other, unique role. Furthermore, because of the intermittent nature of the breathing patterns found in reptiles, the dynamics of changes in lung volume and intrapulmonary P_{CO_2} over episodes of breathing and breath holding can be significantly different and, thus, the discharge profiles from stretch receptors and intrapulmonary chemoreceptors may also be significantly different (Powell et al. 1988). Because both groups of receptors are innervated by the vagus nerve, however, it has not been possible to selectively denervate one group of receptors independently of the other, making studies of reflex roles difficult. Whatever their role, it has been shown that changing temperature has a much more profound effect on the discharge of IPC than on that of pulmonary stretch receptors and, thus, the relative roles of these two receptor groups to overall ventilatory control should change with temperature (Douse and Mitchell 1987; Furilla and Bartlett 1987; Douse et al. 1989).

It has been suggested that IPC may contribute to the hypercapnic ventilatory response and, in particular, that these receptors might be able to monitor CO_2 flux from venous blood into the pulmonary space and act as a venous CO_2 receptor (see Scheid and Piiper 1986, for review). As such, they could act to minimize or even eliminate any increase in Pa_{CO_2} arising from increased metabolic CO_2 production. This hypothesis has been most extensively examined in birds. In studies employing venous CO_2 loading, the data suggest that although IPC appear to be important in adjusting the pattern of breathing, they play no apparent role in the ventilatory response to changes in Pa_{CO_2}/pHa per se (Milsom et al. 1981; Tallman and Grodins 1982a; Jones et al. 1986). During CO_2 inhalation, it has been argued that IPC alone can evoke maximal ventilatory responses (Fedde et al. 1982) and it has been shown that similar responses can be elicited in their absence (Milsom et al. 1981). The extent to which IPC may perform a unique function may be in eliciting ventilatory responses to hypocapnia (Osborne et al. 1977). The small amount of data available from studies of reptiles sheds no further light on this question. There is one study employing venous loading of CO_2 across the cutaneous circulation in snakes in which significant increases in ventilation were achieved with no increase in Pa_{CO_2} (Furilla et al. 1991). Because of the increase in ventilation and the reduction in the non-ventilatory period, there was also no increase in either the peak concentration of CO_2 in the intrapulmonary space or in mean expired levels of CO_2. Consequently, the magnitude of the phasic swing in IPC discharge during each breath would have been unaltered by this manipulation as has been shown in birds (Tallman and Grodins 1982b; Table 2). The rate of increase in discharge would have changed under these conditions and this, independently of any changes in absolute discharge levels, may have acted to increase ventilation by shortening the non-ventilatory period (Furilla, pers. comm.). Consideration of the nature of reptilian breathing patterns and IPC response profiles would suggest that if this were the case, these receptors may convey information about changes in lung CO_2 stores during breath holding (Powell et al. 1988). It is also possible, however, that the increase in ventilation during the cutaneous CO_2 loading study was due to a reduction in pHa as will be discussed shortly.

It should be noted that, to date, there is no evidence to suggest that IPC bestow any unique responses to species which possess them. Whatever the reflex roles of this receptor group, there is no doubt that, in several species of reptiles, they do account for the differences seen in the ventilatory responses to increasing Pa_{CO_2} by CO_2 inhalation versus venous CO_2 loading. As documented in Section 3.1.3, environmental hypercarbia has variable effects on breathing frequency in reptiles (especially snakes and lizards), and hence $\dot{V}_E$, while hypercapnia produced by other means consistently produces a brisk ventilatory response with significant increases in breathing frequency (Gatz et al. 1975). In the garter snake study mentioned earlier, venous loading of CO_2 across the cutaneous circulation did lead to a fall in pHa. Ventilation increased 3 times for only a 0.07 unit fall in pHa during venous CO_2 loading, whereas it increased only 2 to 2.5 times for a 0.08 to 0.19 unit fall in pHa during CO_2 inhalation (Furilla et al. 1991). Very similar results have been obtained from venous CO_2 loading experiments in birds (cf. Tallman and Grodins 1982b). Although vagotomy greatly altered the breathing pattern in these animals, it also removed the difference between the ventilatory response to venous CO_2 loading versus CO_2 inhalation. In the tegu lizard, ventilation doubled when CO_2 was loaded through the visceral circulation, whereas ventilation decreased when CO_2 was added to inspired air (Table 2; Ballam and Donaldson 1988). As noted earlier, part of this latter inhibition in the tegu lizard is due to the stimulation of upper airway receptors (Sect. 3.3.1). It has also been shown, however, that part of the inhibition to CO_2 inhalation arises from receptors in the lungs and that the upper airway reflex is not elicited by a phasic increase in CO_2 during the expiratory phase of the ventilation cycle, such as occurs with venous CO_2 loading (Ballam 1985; Ballam and Coates 1989; Coates et al. 1991). Finally, in many of these same species, there is a post-hypercapnic hyperpnea which can also be eliminated by vagotomy (Nielsen 1961; Templeton and Dawson 1963; Gatz et al. 1975; Glass and Johansen 1976; Nolan and Frankel 1982). All of these studies indicate that decreasing or reversing the phasic discharge profile of IPC with CO_2 inhalation as opposed to increasing the phasic discharge profile with venous CO_2 loading causes profound differences in the ventilatory response (Table 2). Similar results have been obtained in birds (see Milsom 1990 for review).

Thus, as with the upper airway chemoreceptors, the role of this receptor group remains unknown but the results indicate that, in these animals, environmental hypercarbia and arterial hypercapnia/acidosis produced by metabolic means have very different effects on several receptor groups and elicit different net effects on ventilation (Table 2).

3.3.4 Pulmonary (Mixed Venous) Chemoreceptors

The presence of chemoreceptors within the lungs of mammals responsive to changes in P_{CO_2} or CO_2 content of pulmonary arterial (mixed venous) blood has been debated for many years. In *Chelonia* (turtles and tortoises), "Heymans-type" chemoreceptors have been reported to occur in the truncal region, including the pulmonary artery. These receptors are innervated by the inferior truncal branch of the vagus nerve and are stimulated by changes in both P_{O_2} and P_{CO_2} (Ishii et al. 1985, 1986; Table 1). They appear to produce significant reflex ventilatory responses to stimulation by NaCN (Benchetrit et al. 1977), although it is unknown whether they produce reflex ventilatory responses to changes in P_{CO_2}. Injections of sodium cyanide (Lillo 1980; Hoffman and Cordeiro de Sousa 1982; Van Vliet and West 1986) and lobeline (Van Vliet and West 1986) into the pulmocutaneous arch of bullfrogs also produced increases in ventilation, suggesting that similar receptors might exist in the pulmocutaneous artery of amphibians innervated by the recurrent laryngeal nerve. The role of these receptors or "putative" receptors in the hypercapnic ventilatory response, however, is also undetermined. Bilateral vagotomy, which will eliminate the input from these receptors, also eliminates the influences of CO_2-sensitive PSR, and aortic chemoreceptors and intrapulmonary chemoreceptors in species which possess them (Table 1); no specific denervation studies have been performed.

3.3.5 Arterial Chemoreceptors

Studies of arterial chemoreceptors in amphibians and reptiles have focused primarily on O_2 sensitivity with very little or no emphasis on CO_2 sensitivity. There are strong homologies between the various chemoreceptor groups and the carotid and aortic bodies of birds and mammals, suggesting that they should contribute to the hypercapnic ventilatory response but there is very little evidence to support this.

The carotid labyrinth of anuran amphibians is chemoreceptive and a possible homologue of the mammalian carotid sinus (Table 1; Chowdhary 1951; Adams 1958). It has afferent fibers running in the carotid nerve, a branch of the glassopharyngeal nerve, and receives efferent sympathetic innervation mainly from the fourth spinal root, which joins the carotid nerve (Ishii and Ishii 1970, 1973). This efferent innervation appears to modulate chemoreceptor sensitivity as well as, but independently of, carotid blood flow (Ishii and Ishii 1970, 1973). These chemoreceptors are responsive to changes in Pa_{CO_2} (West and Van Vliet 1992), but denervation of this chemoreceptor group appears to have no effect on the hypercapnic ventilatory responses of either *Xenopus laevis* or *Bufo marinus* (West et al. 1987; Jones and Chu 1988). It should be noted, however, that denervation of the carotid labyrinth does not influence the ventilatory response to hypoxia or NaCN injections in these species either (Evans and Shelton 1984; Van Vliet and West 1986; West et al. 1987; Jones and Chu 1988) but it does inhibit the ventilatory response to hypoxia in *Rana catesbeiana* (Smyth 1939). The effect of denervation on the hypercapnic response of this latter species, however, was still only slight at best (Smyth 1939).

Aortic chemoreceptors have also been described in toads innervated by pharyngeal branches of the vagus nerve (Ishii et al. 1985). It is believed that this branch of the vagus in the toad corresponds to the aortic nerve in mammals. The majority of receptors examined were responsive to both hypoxia and hypercapnia but their role in hypercapnic ventilatory responses is unknown.

In the majority of lizards, there is also an aggregation of chemoreceptive tissue at the bifurcation of the internal and common carotid arteries (Table 1; Adams 1958; Rogers 1967; Kobayashi 1971). This region is innervated by the superior laryngeal branch of the vagus nerve (Rogers 1967) and perhaps also by the glassopharyngeal branch of the vagus nerve (Terni 1931; Adams 1958). There are no electrophysiological data to corroborate receptors in this region, innervated by these nerves and, although lizards do exhibit a ventilatory response to the injection of hypoxic blood into the junction of the carotid arch, which is eliminated by denervation of this region (Courtice 1980), their role in the hypercapnic ventilatory response has not been examined.

In turtles, chemoreceptive tissue is not present at the extant carotid bifurcation. During embryological development, the aortic arches retreat backwards and the carotid bifurcation remains close to the heart. The external carotid arteries then atrophy and the internal carotid arteries (now called common carotid arteries) divide, secondarily, in the head region (West et al. 1981). Chemoreceptive tissue has been described on the aortic arch, innervated by both the inferior and superior truncal nerves (branches of the vagus) (Ishii et al. 1985; Table 1). The former is believed to correspond to the aortic nerve of mammals, while the latter arises from the ganglion trunci of the vagus. Those receptors innervated by the inferior truncal nerve have been shown to respond to changes in Pa_{CO_2} (Ishii et al. 1986). Their contribution to hypercapnic ventilatory responses is also undetermined.

A number of indirect attempts have been made to assign relative significance to this group of chemoreceptors in amphibians and reptiles. Most such studies assess the net effects of stimulating most or all peripheral chemoreceptors and thus include the effects of both inhibitory and excitatory reflexes arising from upper airway, intrapulmonary, mixed venous and arterial chemoreceptors (Table 2). Such attempts must be viewed with caution and are discussed in the next section and in the conclusions.

3.3.6 Central Chemoreceptors

Ventilatory responses to changes in the P_{CO_2}/pH of solutions perfusing either the ventriculocisternal spaces or the surface of the medulla have now been described in frogs, toads and turtles (De Marneffe-Foulon 1962; Hitzig and Jackson 1978; Hitzig 1982; Hitzig and Nattie 1982; Hitzig et al. 1985; Smatresk and Smits 1991; Branco et al. 1992; Table 1). In amphibians, responses have been obtained from perfusion of the fourth ventrical (Branco et al. 1992) and superfusion of the ventral medullary surface (De Marneffe-Foulon 1962; Smatresk and Smits 1991). In turtles, they have been obtained with perfusion of the lateral and fourth ventricles (Hitzig and Jackson 1978; Hitzig 1982; Hitzig et al. 1985). Specific receptors have not been identified and, thus, the exact locations and discharge characteristics of these "putative" central chemoreceptors remain unknown. This is also true for birds and mammals (see Loeschcke 1982; Nattie 1990 for reviews). In mammals, where central CO_2 chemosensitivity has been more extensively studied, the receptors appear to lie within 200 μm of the ventrolateral surface of the medulla between the origins of the VIIth and XIIth cranial nerves.

Attempts have been made to determine whether central chemoreceptors respond primarily to changes in P_{CO_2} or $[H^+]$. All such data are equivocal. It has generally been stated that these receptors in mammals respond primarily to changes in the pH of the cerebral interstitial fluid (ISF) which surrounds them and that the effects of changes in P_{CO_2} are mediated by virtue of their acidity (see Loeschke 1982 for review). That aside, it has been shown in amphibians (Smatresk and Smits 1991), reptiles (Hitzig and Nattie 1982) and mammals (Eldridge et al. 1985; Burton et al. 1989) that hypercapnic acidosis stimulated ventilation more than an equivalent or more severe fixed acidosis (increase in metabolic acids produced by anoxia or an isocapnic acidotic perfusate). It is most likely that the true stimulus is a change in pHi of the chemoreceptive cells which might be more sensitive to changes in P_{CO_2} than ISF pH (Neubauer 1990).

Whatever the exact nature of the primary stimulus, due to the characteristics of the blood brain barrier, cerebral interstitial fluid pH reflects changes in Pa_{CO_2} (CO_2 crosses the barrier easily) more closely than in $[H^+]_a$ (which does not). In stable non-respiratory (metabolic) acidosis and alkalosis, effective mechanisms minimize changes in cerebrospinal fluid (CSF) pH (and presumably ISF pH), while in stable respiratory acidosis and alkalosis, CSF pH changes are maintained (Hitzig and Nattie 1982; Fencl 1986; Davies and Sexton 1987). As a consequence, the relative roles of peripheral and central chemoreceptors to the net ventilatory response should differ under these conditions.

Partitioning relative roles to the central versus peripheral chemoreceptors in producing ventilatory responses is not easy. There are two sorts of studies which offer data which entice people to do so, but such attempts must be viewed with caution. In one group of studies, the brain is perfused with mock cerebrospinal fluid in an attempt to hold the pH/P_{CO_2} of the central chemoreceptors constant while the blood gases and pH at the peripheral chemoreceptors are manipulated and vice versa. Because CO_2 crosses membranes so effectively, however, there is no assurance that such manipulations do indeed maintain any constancy at central chemoreceptive sites. The "peripheral stimulus" in these studies may still have a central component. The second group of studies compares ventilatory responses before and after denervation of peripheral chemoreceptor groups. The only drawback to this approach is that it assumes there is an additive interaction between the chemoreceptor groups. These concerns have been addressed best in studies on mammals. Here, the evidence suggests that the contribution of the arterial chemoreceptors varies both over the time course of the response and as a function of the magnitude of the stimulus. Many studies suggest that the carotid bodies in mammals contribute more to the initial phase of the hypercapnic ventilatory response than to the steady state changes. This corresponds to the much larger dynamic component (overshoot) of the receptor response to a step change in Pa_{CO_2} (Black et al. 1971). It also appears that the carotid bodies only augment the drive of central respiratory neurons at moderate levels of hypercapnia (Heymans and Neil 1958; also see Fidone and Gonzalez 1986; Fitzgerald and Lahiri 1986 for reviews). With higher levels of hypercapnia, the central actions of $CO_2/[H^+]$ are maximal and the carotid bodies can contribute little more. This may also be the case with the ventilatory response to changes in pHa due to metabolic acidosis. The carotid bodies may contribute to the response only until the changes in pH become severe enough to effect CSF pH (Leusen 1954; Hornbein and Roos 1963; Mitchell 1965; Pappenheimer et al. 1965; Fencl et al. 1966; Berger et al. 1977).

Given this caveat, it has recently been shown that the effects of hypercapnia are significantly reduced, but not eliminated, when the brains of toads are perfused with alkaline cerebrospinal fluid (CSF). This indicates that peripheral chemoreceptors, which might include the carotid labyrinth and aortic chemoreceptors, are responsible for at least part of the hypercapnic ventilatory response. In one study on *Bufo paracnemis*, ventilation fell dramatically at all levels of inspired CO_2, when the fourth ventricle was perfused with a mock CSF of pH 7.7 (Branco et al. 1992). The recorded values suggest that the relative role of the peripheral chemoreceptors to the hypercapnic ventilatory response was only about 20% (Branco et al. 1992). It is more difficult to assign relative roles to peripheral and central chemoreceptors based on the data presented in the other study on *Bufo marinus*, but it would appear that the contribution of the peripheral chemoreceptors to the hypercapnic response over the physiolgical range normally encountered by this species could be as high as 50% (Smatresk and Smits 1991). Of note from these two studies is the observation that central chemoreceptors appear to elicit increases in both tidal volume and breathing frequency, the latter due to an increase in the number of breaths in each breathing episode and a reduction in T_{NVP} (Fig. 15; Smatresk and Smits 1991; Branco et al. 1992).

In turtles, bilateral vagotomy, which will eliminate the influences of most peripheral receptors (arterial chemoreceptors, CO_2-sensitive pulmonary stretch receptors and intrapulmonary chemoreceptors in species which have them; Table 1), can reduce the hypercapnic ventilatory response by up to 50% (Benchetrit and Dejours 1980; Milsom and Jones 1980). Studies which manipulate the acid-base status of the central CO_2 chemoreceptors independently of the peripheral chemoreceptors would suggest that more than 50% of the hypercapnic ventilatory response may be attributable to peripheral chemoreceptors (Hitzig et al. 1985). In turtles, all of the increase in tidal volume due to hypercapnia can be ascribed to the stimulation of peripheral receptors, while central chemoreceptors appear to be involved exclusively in changing breathing frequency (Fig. 15; Hitzig et al. 1985).

4 Conclusions

In reviewing the literature on the role of CO_2/pH chemoreceptors in lower vertebrates, the one thing that becomes quite clear is that general conclusions can only be drawn with extreme caution. The literature would suggest that at least some members of all classes of lower vertebrates do exhibit ventilatory responses

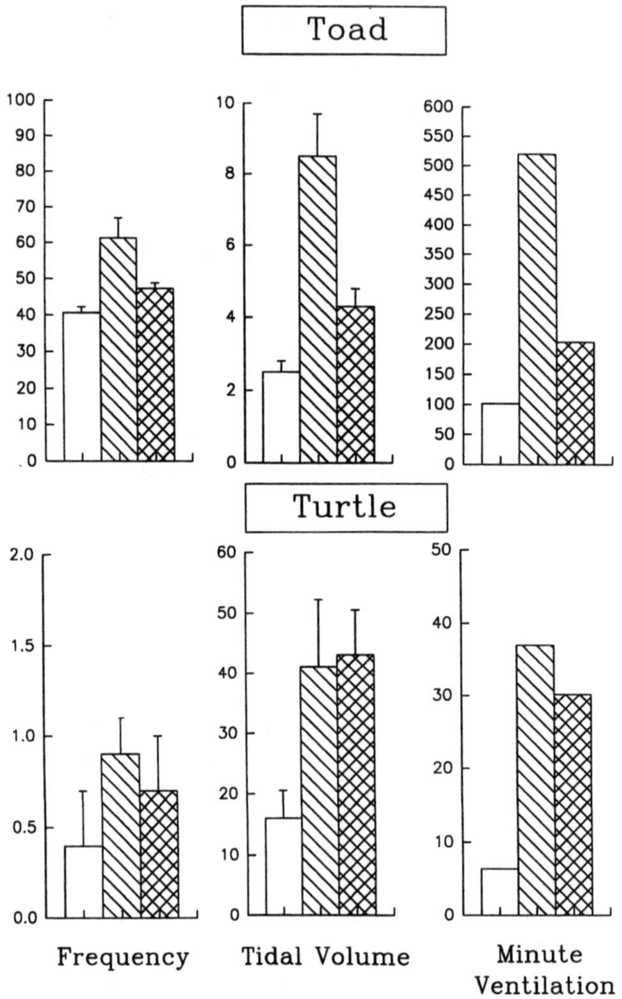

Fig. 15. The changes which occurred in minute ventilation ($\dot{V}_E$; ml/min), tidal volume (V_T; ml) and breathing frequency (f; breaths/min) in the marine toad (*Bufo marinus*) and a turtle (*Pseudemys scripta*) when the cerebral ventricles were perfused with different solutions while the animals inspired gases of different compositions. In the toad, the three conditions represent a control (0% CO_2 inspired, ventricles unperfused, *open bar*), stimulation of both peripheral and central chemoreceptors (6%, unperfused, *diagonal hatching*) and stimulation of the peripheral chemoreceptors alone (6%, perfused, *cross-hatching*). (Data from Branco et al. 1992). In the turtle, the three conditions represent a control [0% CO_2 inspired, ventricles perfused with a solution of strong ion difference (SID) = 32 mmol, *open bar*)], stimulation of both peripheral and central chemoreceptors (4.5%, unperfused, *diagonal hatching*) and stimulation of the peripheral chemoreceptors alone (4.5%, SID = 32 mmol, *cross-hatching*). (Data from Hitzig et al. 1985)

to changes in CO_2/pH per se. It would appear that with the transition from aquatic to aerial respiration, there is an increase in the sensitivity of animals to this complex of stimuli, an increase in the variety of putative receptors possibly involved in eliciting ventilatory responses and an increase in the relative importance of this complex of stimuli to the genesis of resting ventilation.

The variety of CO_2-sensitive chemoreceptors present in the air-breathing species add a degree of complexity to experimental studies of ventilatory responses to CO_2/pH which have not been appreciated by many authors. Because of the locations, discharge characteristics and reflex effects of the different receptor groups, most air-breathing lower vertebrates show different responses to increases in $CO_2/[H^+]$ due to cerebral ischemia, anoxia, metabolic acidosis and environmental hypercarbia (Table 2). In some cases, the differences are only quantitative; turtles respond to both CO_2 inhalation and metabolic acidosis with increases in ventilation of varying magnitude. In other cases, the responses are qualitatively very different. This is exemplified by the numerous species of snakes and lizards which depress ventilation in response to environmental hypercarbia but show a brisk ventilatory response to venous CO_2 loading. These differences appear to reflect differences in the relative strength of the reflexes elicited by the various receptor groups and the net sum of their modulating influences when CO_2/pH are altered via different routes.

In all cases, the input from all of the $CO_2/[H^+]$-sensitive receptors appears to act as a biasing input which summates with other afferent information to modulate respiratory motor output. It does not turn breathing "on and off" per se. Thus, the episodic breathing patterns, which are so common for many of these animals, are not a simple consequence of the oscillation in blood gases and pH which accompany them. This has strong implications for theories of central pattern generation which are outside the scope of this review. It also suggests, however, that although the role of CO_2/pH chemoreceptors in the control of breathing may change quantitatively from fish to mammals, there may be little or no qualitative difference in their role in ventilatory control among the vertebrates.

References

Ackerman RA, White FN (1979) Cyclic carbon dioxide exchange in the turtle *Pseudemys scripta*. Physiol Zool 52: 378–389

Adams WE (1958) The comparative morphology of the carotid body and carotid sinus. Thomas, Springfield

Adams WE (1962) The carotid sinus-carotid body problem in the chelonia (with a note on a Foramen of Panizza in *Dermochelys*). Arch Int Pharmacodyn 139: 28–37

Aota S, Holmgren KD, Gallaugher P, Randall DJ (1990) A possible role for catecholamines in the ventilatory responses associated with internal acidosis or external hypoxia in rainbow trout *Oncorhynchus mykiss*. J Exp Biol 151: 57–70

Ballam GO (1984) Ventilatory response to inspired CO_2 in the lizard, *Tupinambis nigropunctatus*. Comp Biochem Physiol 78A: 757–762

Ballam GO (1985) Breathing response of the tegu lizard to 1–4% CO_2 in the mouth and nose or inspired into the lungs. Respir Physiol 62: 375–386

Ballam GO, Coates EL (1989) Effect of upper airway CO_2 pattern on ventilatory frequency in tegu lizards. Am J Physiol 257: R156–R161

Ballam GO, Donaldson LA (1988) Effect of venous (gut) CO_2 loading on intrapulmonary gas fractions and ventilation in the tegu lizard. J Comp Physiol 158: 591–600

Banzett RB, Burger RE (1977) Response of avian intrapulmonary chemoreceptors to venous CO_2 and ventilatory gas flow. Respir Physiol 29: 63–72

Benchetrit G, Dejours P (1980) Ventilatory CO_2 drive in the tortoise *Testudo horsfieldi*. J Exp Biol 87: 229–236

Benchetrit G, Armand J, Dejours P (1977) Ventilatory chemoreflex drive in the tortoise, *Testudo horsfieldi*. Respir Physiol 31: 183–191

Berger PJ, Tallman RD Jr (1982) Lengthening of inspiration by intrapulmonary chemoreceptor discharge in ducks. J Appl Physiol 53: 1392–1396

Berger AJ, Mitchell RAM, Severinghaus JW (1977) Regulation of respiration. N Engl J Med 297: 194–201

Black EC, Irving L (1937) The effect of carbon dioxide upon the oxygen capacity of the blood of the carp (*Cyprinus carpio* L.). Trans R Soc Can 5: 29–32

Black AMS, McCloskey DI, Torrance RW (1971) The responses of carotid body chemoreceptors in the cat to sudden changes of hypercapnic and hypoxic stimuli. Respir Physiol 13: 36–49

Boutilier RG (1988) Control of arrhythmic breathing in bimodal breathers: Amphibia. Can J Zool 66: 6–19

Boutilier RG (1990) Control and co-ordination of gas exchange in bimodal breathers. In: Boutilier RG (ed) Advances in comparative and environmental physiology, vol 6. Vertebrate gas exchange from environment to cell. Springer, Berlin Heidelberg New York

Boutilier RG, Shelton G (1986) Gas exchange, storage and transport in voluntarily diving *Xenopus laevis*. J Exp Biol 126: 133–155

Boutilier RG, Toews DP (1981) Respiratory, circulatory and acid-base adjustments in a strictly aquatic and predominantly skin-breathing urodele, *Cryptobranchus alleganiensis*. Respir Physiol 46: 177–192

Boutilier RG, Randall DJ, Shelton G, Toews DP (1979) Acid base relationships in the blood of the toad, *Bufo marinus* II. The effects of environmental CO_2. J Exp Biol 82: 331–344

Boutilier RG, McDonald DG, Toews DP (1980) The effects of enforced activity on ventilation, circulation and blood acid-base balance in the aquatic gill-less urodele, *Cryptobranchus alleganiensis;* a comparison with the semi-terrestrial anuran, *Bufo marinus*. J Exp Biol 84: 289–302

Boutilier RG, Iwama GK, Randall DJ (1986) The promotion of catecholamine release in rainbow trout, *Salmo gairdneri*, by acute acidosis: interaction between red cell pH and haemoglobin oxygen carrying-capacity. J Exp Biol 123: 145–157

Branco LGS, Glass ML, Hoffman A (1992) Central chemoreceptor drive to breathing in unanesthetized toads, *Bufo paracnemis*. Respir Physiol 87: 195–204

Burger RE, Osborne JL, Banzett RB (1974) Intrapulmonary chemoreceptors in *Gallus domesticus*: adequate stimulus and functional localization. Respir Physiol 22: 87–97

Burger RE, Barker MR, Nye PCG, Powell FL (1978) Effects of intrapulmonary chemoreceptors in perfused and non-perfused lungs. In: Piiper J (ed) Respiratory function in birds. Adult and embryonic. Springer, Berlin Heidelberg New York

Burggren WW (1979) Bimodal gas exchange during variation in environmental oxygen and carbon dioxide in the air breathing fish *Trichogaster trichopterus*. J Exp Biol 82 197–213

Burggren WW, Doyle M (1986) Ontogeny of gill and lung ventilation in the bullfrog, *Rana catesbeiana*. Respir Physiol 66: 279–291

Burggren WW, Shelton G (1979) Gas exchange and transport during intermittent breathing in chelonian reptiles. J Exp Biol 82: 75–92

Burleson ML, Milsom WK (1990) Propranolol inhibits O_2 sensitive chemoreceptor activity in trout gills. Am J Physiol 27: R1089–R1091

Burleson ML, Smatresk NJ (1990) Effects of sectioning cranial nerves IX and X on cardiovascular and ventilatory reflex responses to hypoxia and NaCN in channel catfish. J Exp Biol 154: 407–420

Burton MD, Johnson DC, Kazemi H (1989) CSF acidosis augments ventilation through cholinergic mechanisms. J Appl Physiol 66: 2565–2572

Butler PJ, Taylor EW (1971) Response of the dogfish (*Scyliorhinus canicula* L) to slowly induced and rapidly induced hypoxia. Comp Biochem Physiol 39A: 307–323

Butler PJ Taylor EW (1975) The effect of progressive hypoxia on respiration in the dogfish (*Scyliorhinus canicula*) at different seasonal temperatures. J Exp Biol 63: 117–130

Butler PJ, Taylor EW, Davison W (1979) The effect of long term, moderate hypoxia on acid-base balance, plasma catecholamines and possible anaerobic end products in the unrestrained dogfish, *Scyliorhinus canicula*. J Comp Physiol 132: 297–303

Bystrzycka EK, Nail BS (1980) CO_2-sensitivity of stretch receptors in the marsupial lung. Respir Physiol 39: 111–119

Cameron JN (1971) Oxygen dissociation characteristics of the blood of the rainbow trout *Salmo gairdneri*. Comp Biochem Physiol 38A: 699–704

Cameron JN (1978) Regulation of blood pH in teleost fish. Respir Physiol 33: 129–144

Cameron JN, Randall DJ (1972) The effect of increased ambient CO_2 on arterial CO_2 tension, CO_2 content and pH in rainbow trout. J Exp Biol 57: 673–680

Chowdhary DS (1951) The carotid labyrinth of *Rana tigrina*. Nature 167: 1074

Coates EL, Ballam GO (1987) Upper airway CO_2 receptors in tegu lizards: localization and ventilatory sensitivity. J Comp Physiol 157: 483–489

Coates EL, Ballam GO (1989) Breathing and upper airway CO_2 in reptiles: role of the nasal and vomeronasal systems. Am J Physiol 25: R91–97

Coates EL, Ballam GO (1990) Olfactory receptor response to CO_2 in bullfrogs. AM J Physiol 27: R1207–R1212

Coates EL, Furilla RA, Ballam GO, Bartlett D Jr (1991) A decrease in nasal CO_2 stimulates breathing in the tegu lizard. Respir Physiol 86: 65–75

Courtice GP (1980) Stimulation of carotid arterial chemoreceptors by hypoxia and hypercapnia in a lizard. Proc 27th Int Cong Physiol Sci Budapest 14: 368

Davies DG (1978) Temperature-induced changes in blood acid-base status in the alligator, *Alligator mississippiensis*.J Appl Physiol 45: 922–926

Davies DG, Kopetzky MT (1976) Effect of body temperature on the ventilatory response to hypercapnia in the awake alligator. Fed Proc 35: 840

Davies DG, Sexton JA (1987) Brain ECF pH and central chemical control of ventilation during anoxia in turtles. Am J Physiol 252: R848–R852

Davies DG, Thomas JL, Smith EN (1982) Effect of body temperature on ventilatory control in the alligator. J Appl Physiol 52: 114–118

Dejours P (1972) Comparison of gas transport by convection among animals. Respir Physiol 14: 96–104

Dejours P (1973) Problems of control of breathing in fishes. In: Bolis L, Schmidt-Nielsen K, Maddrell SHP (eds) Comparative physiology. Elsevier, Amsterdam

De Marneffe-Foulon C (1962) Contribution a l'etude du mechanisme et du controle des movements respiratoires chez *Rana*. Ann Soc Zool Belg 92: 81–132

Dively JL, Mudge JE, Neff WH, Anthony A (1977) Blood PO_2, PCO_2, and pH changes in brook trout (*Salvelinus fontinalis*) exposed to sublethal levels of acidity. Comp Biochem Physiol 57A: 347–351

Dodd GA, Milson WK (1987) Effects of H^+ versus CO_2 on ventilation in the Pekin duck. Respir Physiol 68: 189–201

Douse MA, Mitchell GS (1987) Effects of temperature on CO_2 sensitive intrapulmonary chemoreceptors (IPC) in the tegu lizard (*Tupinambis nigropunctatus*). Fed Proc 46: 344

Douse MA, Mitchell GS (1990) Episodic respiratory related discharge in turtle cranial motoneurons: in vivo and in vitro studies. Brain Res 536: 297–300

Douse MA, Mitchell GS (1992a) Effects of vagotomy on ventilatory responses to CO_2 in alligators. Respir Physiol 87: 63–76

Douse MA, Mitchell GS (1992b) Episodic breathing in alligators: role of sensory feedback. Respir Physiol 87: 77–90
Douse MA, Powell FL, Milsom WK, Mitchell GS (1989) Temperature effects on pulmonary receptor responses to airway pressure and CO_2 in *Alligator mississippiensis*. Respir Physiol 78: 331–34
Eddy FB (1971) Blood gas relationships in the rainbow trout *Salmo gairdneri*. J Exp Biol 55: 695–711
Eddy FB (1973) Oxygen dissociation curves of the blood of the tench *Tinca tinca*. J Exp Biol 58: 281–293
Eddy FB (1976) Acid-base balance in rainbow trout (*Salmo gairdneri*) subjected to acid stresses. J Exp Biol 64: 159–171
Eddy FB, Morgan RIG (1969) Some effects of carbon dioxide on the blood of the rainbow trout *Salmo gairdneri* Richardson. J Fish Biol 1: 361–372
Eldridge FL, Kiley JP, Millhorn DE (1985) Respiratory responses to medullary hydrogen ion changes in cats: different effects of respiratory and metabolic acidosis. J Physiol 358: 285–297
Emilio MG, Shelton G (1974) Gas exchange and its effect on blood gas concentrations in the amphibian, *Xenopus laevis*. J Exp Biol 60: 567–579
Evans BK, Shelton G (1984) Ventilation in *Xenopus laevis* after lung or carotid labyrinth denervation. Proc 1st Int Congr Comp Physiol Biochem Liege 1: A75
Fedde MR, Gatz RN, Slama H, Scheid P (1974a) Intrapulmonary CO_2 receptors in the duck: I. Stimulus specificity. Respir Physiol 22: 99–114
Fedde MR, Gatz RN, Slama H, Scheid P (1974b) Intrapulmonary CO_2 receptors in the duck: II. Comparison with mechanoreceptors. Respir Physiol 22: 115–121
Fedde MR, Kuhlman WD, Scheid P (1977) Intrapulmonary receptors in the tegu lizard: I. Sensitivity to CO_2. Respir Physiol 29: 35–48
Fedde MR, Kiley JP, Powell FL, Scheid P (1982) Intrapulmonary CO_2 receptors and control of breathing in ducks: effects of prolonged circulation time to carotid bodies and brain. Respir Physiol 47: 121–140
Feldman JL, Smith JC, Ellenberger HH, Connelly CA, Liu G, Greer JJ, Lindsay AD, Otto MR (1990) Neurogenesis of respiratory rhythm and pattern: emerging concepts. Am J Physiol 259: R879–R886
Fencl V (1986) Acid-base balance in cerebral fluid. In: Fishman AP, Cherniak NS, Widdicombe JG, Geiger SR (eds) Handbook of physiology, section 3. The respiratory system, vol II. Control of breathing, part 1. Am Physio Soc, Bethesda
Fencl V, Miller TB, Pappenheimer JR (1966) Studies on the respiratory response to disturbances of acid-base balance with deductions concerning the ionic composition of cerebral interstitial fluid. Am J Physiol 210: 459–472
Fidone SJ, Gonzalez C (1986) Intiation and control of chemoreceptor activity in the carotid body. In: Fishman AP, Cherniak NS, Widdicombe JG, Geiger SR (eds) Handbook of physiology, section 3. The respiratory system, vol II. Control of breathing, Part 1. Am Physiol Soc, Bethesda
Fishman AP, Galante RJ, Pack AI (1989) Diving physiology: lungfish. In: Wood SC (ed) Comparative physiology: current concepts, vol 39. Lung biology in health and disease. Dekker, New York
Fitzgerald RS, Lahiri S (1986) Reflex responses to chemoreceptor stimulation. In: Fishman AP, Cherniak NS, Widdicombe JG, Geiger SR (eds) Handbook of physiology, section 3. The respiratory system, vol II. Control of breathing, part 1. Am Physiol Soc, Bethesda
Frankel HM, Spitzer A, Blaine J, Schoener AP (1969) Respiratory response of turtles (*Pseudemys scripts*) to changes in arterial blood gas composition. Comp Biochem Physiol 31: 535–546
Funk GD, Milsom WK (1987) Changes in ventilation and breathing pattern produced by changing body temperature and inspired CO_2 concentration in turtles. Respir Physiol 67: 37–51
Furilla RA, Bartlett D Jr (1987) Intrapulmonary receptors in the greater snake (*Thamnophis sirtalis*). Fed Proc 46: 793

Furilla RA, Bartlett D Jr (1988) Intrapulmonary receptors in the garter snake (*Thamnophis sirtalis*). Respir Physiol 74: 311–322
Furilla RA, Bartlett D Jr (1989) Intrapulmonary CO_2 inhibits inspiration in garter snake. Respir Physiol 78: 207–218
Furilla RA, Coates EL, Bartlett D Jr (1991) The influence of venous CO_2 on ventilation in garter snakes. Respir Physiol 83: 47–60
Gatz RN, Fedde MR, Crawford EC Jr (1975) Lizard lungs: CO_2-sensitive receptors in *Tupinambis nigropunctatus*. Experentia 31: 455–456
Getchell TV, Shepherd GM (1978) Adaptive properties of olfactory receptors analyzed with odour pulses of varying durations. J Physiol 282: 541–560
Glass ML, Johansen K (1976) Control of breathing in *Acrochordus javanicus*, an aquatic snake. Physiol Zool 49: 328–340
Glass ML, Johansen K (1979) Periodic breathing in the crocodile, *Crocodylus niloticus:* consequences for the gas exchange ratio and control of breathing. J Exp Zool 208: 319–336
Glass ML, Wood SC (1983) Gas exchange and control of breathing in reptiles. Physiol Rev 63: 232–260
Glass ML, Burggren WW, Johansen K (1978) Ventilation in an aquatic and a terrestrial chelonian reptile. J Exp Biol 72: 165–179
Glass ML, Ishimatsu A, Johansen K (1986) Responses of aerial ventilation to hypoxia and hypercapania in *Channa argus*, an air-breathing fish. J Comp Physiol B 156: 425–430
Graham JB (1976) Respiratory adaptations of marine air-breathing fishes. In: Hughes GM (ed) Respiration of amphibious vertebrates. Academic Press, New York
Graham JB, Baird TA (1982) The transition to air breathing in fishes I: environmental effects on the facultative air breathing of *Ancistrus chagresi* and *Hypostomus plecostomus* (Loricariidae). J Exp Biol 96: 53–67
Graham JB, Kramer DL, Pineda E (1977) Respiration of the air breathing fish *Piabucina festae*. J Comp Physiol B 122: 295–310
Graham MS, Turner JD, Wood CM (1990) Control of ventilation in the hypercapnic skate *Raja ocellata*: I. Blood and extradural fluid. Respir Physiol 80: 259–277
Gratz RK (1979) Ventilatory response of the diamondback water snake, *Natrix rhombifera* to hypoxia, hypercapnia and increased oxygen demand. J Comp Physiol 129: 105–110
Guimond RW, Hutchison VH (1976) Gas exchange of the giant salamanders of North America. In: Hughes GM (ed) Respiration of amphibious vertebrates. Academic Press, New York, pp 313–337
Hall FG (1931) The respiration of puffer fish. Biol Bull 61: 457–467
Hargis JR (1976) Ventilation and metabolic rate of young rainbow trout (*Salmo gairdneri*) exposed to sublethal environment pH. J Exp Zool 196: 39–44
Hedrick MS, Burleson ML, Jones DR, Milsom WK (1991) An examination of central chemosensitivity in an air-breathing fish (*Amia calva*). J Exp Biol 155: 165–174
Heisler N (1986) Comparative aspects of acid-base regulation. In: Heisler N (ed) Acid-base regulation in animals. Elsevier, Amsterdam
Heisler N (1988) Acid-base regulation. In: Shuttleworth TJ (ed) Physiology of elasmobranch fishes. Springer, Berlin Heidelberg New York
Heisler N, Forcht G, Ultsch GR, Anderson JF (1982) Acid-base regulation in response to environmental hypercapnia in two aquatic salamanders, *Siren lacertina* and *Amphiuma means*. Respir Physiol 49: 141–158
Heisler N, Toews DP, Holeton GF (1988) Regulation of ventilation and acid-base status in the elasmobranch *Scyliorhinus stellaris* during hyperoxia-induced hypercapnia. Respir Physiol 71: 227–246
Heymans C, Neil E (1958) Reflexogenic areas of the cardiovascular system. Churchill, London
Hitzig BM (1982) Temperature-induced change in turtle csf pH and central control of ventilation. Respir Physiol 49: 205–222
Hitzig BM, Jackson DC (1978) Central chemical control of ventilation in the unanesthetized turtle. Am J Physiol 235: R257–R264
Hitzig BM, Nattie EE (1982) Acid-base stress and central chemical control of ventilation in turtles. J Appl Physiol 53: 1365–1370

Hitzig BM, Kneussl MP, Shif V, Brandstetter RD, Kazemi H (1985) Brain amino acid concentrations during diving and acid-base stress in turtles. J Appl Physiol 58: 1751–1754
Hoffman A, Cordeiro de Sousa MB (1982) Cardiovascular reflexes in conscious toads. J Autonom Nervous Syst 5: 345–355
Hornbein TF, Roos A (1963) Specificity of H^+ ion concentration as a carotid chemoreceptor stimulus. J Appl Physiol 18: 580–584
Hughes GM, Shelton G (1962) Respiratory mechanisms and their nervous control in fish. Adv Comp Physiol Biochem 1: 275–364
Hughes GM, Singh BN (1970) Respiration in air-breathing fish, the climbing perch *Anabas testudineus.* II. Respiratory patterns and the control of breathing. J Exp Biol 53: 281–298
Hughes GM, Singh BN (1971) Gas exchange with air and water in an air-breathing catfish, *Saccobranchus* (Heteropneustes) *fossilis.* J Exp Biol 55: 667–682
Infantino R (1990) Ventilatory responses to inspired gas variation in larval bullfrogs. Physiologist 33: A35
Ishii K, Ishii K (1970) Efferent innervation to the chemoreceptor of the carotid labyrinth of the toad. Tohoku J Exp Med 91: 119–129
Ishii K, Ishii K (1973) Fiber composition and derivation of afferent and efferent nerve fibers in the carotid nerve innervating the carotid labyrinth of the toad. Tohoku J Exp Med 109: 323–337
Ishii K, Ishii K, Kusakabe T (1985) Electrophysiological aspects of reflexogenic area in the chelonian, *Geoclemmys reevesii.* Respir Physiol 59: 45–54
Ishii K, Ishii K, Dejours P (1986) Activity of vagal afferent fibers innervating CO_2-sensitive receptors in the tortoise, *Testudo hermanni.* Jpn J Physiol 36: 1015–1026
Jackson DC, Braun BA (1979) Respiratory control in bullfrogs: cutaneous versus pulmonary response to selective CO_2 exposure. J Comp Physiol 129: 339–342
Jackson DC, Palmer SE, Meadow WL (1974) The effects of temperature and carbon dioxide breathing on ventilation and acid-base status of turtles. Respir Physiol 20: 131–146
Jackson DC, Kraus DR, Prange HD (1979) Ventilatory response to inspired CO_2 in the sea turtle: effects of body size and temperature. Respir Physiol 38: 71–81
Janssen RG, Randall DJ (1975) The effects of changes in pH and P_{CO_2} in blood and water on breathing in rainbow trout *Salmo gairdneri.* Respir Physiol 25: 235–245
Jesse MJ, Shub C, Fishman AP (1967) Lung and gill ventilation of the African lungfish. Respir Physiol 3: 267–287
Johansen K (1970) Cardiorespiratory adaptations in the transition from water breathing to air breathing. Fed Proc 29: 1118–1119
Johansen K, Lenfant C (1968) Respiration in the African lungfish, *Protopterus anethiopicus* II. Control of breathing. J Exp Biol 49: 453–468
Johansen K, Lenfant C, Grigg GC (1967) Respiratory control in the lungfish, *Neoceratodus fosteri (krefft).* Comp Biochem Physiol 20: 835–854
Johansen K, Lenfant C, Schmidt-Nielsen K, Petersen J (1968) Gas exchange and control of breathing in the electric eel, *Electrophorus electricus.* Z Vgl Physiol 61: 137–163
Johansen K, Hanson D, Lenfant C (1970) Respiration in a primitive air breather, *Amia calva.* Respir Physiol 9: 162–174
Jones RM (1982) How toads breathe: control of air flow to and from the lungs by the nares in *Bufo marinus.* Respir Physiol 49: 251–265
Jones DR, Chu C (1988) Effect of denervation of carotid labyrinths on breathing in unrestrained *Xenopus laevis.* Respir Physiol 73: 243–256
Jones DR, Milsom WK (1979) Functional characteristics of slowly adapting pulmonary stretch receptors in the turtle (*Chrysemys picta*). J Physiol 291: 37–49
Jones DR, Milsom WK, Butler PJ (1986) Ventilatory response to venous CO_2 loading by gut ventilation in ducks. Can J Zool 63: 1232–1236
Kinkead R, Milsom WK (1994) Chemoreceptors and control of episodic breathing in the bullfrog (*Rana Catesbeiana*) Respir Physiol 95: 81–98
Kinkead R, Perry SF (1991) The effects of catecholamines on ventilation in rainbow trout during hypoxia or hypercapnia. Respir Physiol 84: 77–92
Kinkead R, Fritsche R, Perry SF, Nilsson S (1991) The role of circulating catecholamines in the

ventilatory and hypertensive responses to hypoxia in the Atlantic cod (*Gadus morhua*). Physiol Zool 64: 1087–1109
Kinney JL, White FN (1977) Oxidative cost of ventilation in a turtle, *Pseudemy floridana*. Respir Physiol 31: 327–332
Kobayashi S (1971) Comparative cytological studies 2. Ultrastructure of the synapses on the chief cell. Arch Histol Jpn 33: 397–420
Kuhlmann WD, Fedde MR (1979) Intrapulmonary receptors in the bullfrog: sensitivity to CO_2. J Comp Physiol 132A: 69–75
Kunz AL, Kawashiro T, Scheid P (1976) Study of CO_2 sensitive vagal afferents in the cat lung. Respir Physiol 27: 347–355
Lenfant C, Johansen K (1966) Respiratory function in the elasmobranch *Squalus suckleyi* G.Respir Physiol 1: 13–29
Lenfant C, Johansen K (1968) Respiration in the African lungfish *Protopterus aethiopicus* i. respiratory properties of blood and normal patterns of breathing and gas exchange. J Exp Biol 49: 437–452
Leusen IR (1954) Chemosensitivity of the respiratory center. Influences of changes in the H^+ and total buffer concentrations in the cerebral ventricles on respiration. Am J Physiol 176: 45–51
Lillo RS (1980) Heart rate and blood pressure in bullfrogs during prolonged maintenance in water at low temperature. Comp Biochem Physiol 65A: 251–253
Lloyd R, White WR (1967) Effect of high concentration of carbon dioxide on the ionic composition of rainbow trout. Nature 216: 1341–1342
Loeschcke HH (1982) Central chemosensitivity and the reaction theory. J Physiol 332: 1–24
Lomholt JP, Johansen K (1974) Control of breathing in *Anphipnous cuchia*, an amphibious fish. Respir Physiol 21: 325–340
MacIntyre DG, Toews DP (1976) The mechanics of lung ventilation and the effects of hypercapnia on respiration in *Bufo marinus*. Can J Zool 54: 1364–1374
McArthur MD, Milsom WK (1991) Ventilation and respiratory sensitivity of euthermic Columbian and golden-mantled ground squirrels (*Spermophilus columbianus* and *Spermophilus lateralis*) during the summer and winter. Physiol Zool 64: 921–939
McKenzie DJ, Burleson ML, Randall DJ (1991) The effects of branchial denervation and pseudobranch ablation on cardio-ventilatory control in an air-breathing fish. J Exp Biol 161: 347–345
McLean HA, Mitchell GS, Milsom WK (1989) Effects of prolonged inflation on pulmonary stretch receptor discharge in turtles. Respir Physiol 75: 75–88
Milsom WK (1990) Control and co-ordination of gas exchange in air breathers. In: Boutilier RG (ed) Advances in comparative and environmental physiology, vol 6. Vertebrate gas exchange from environment to cell. Springer, Berlin Heidelberg New York
Milsom WK, Brill RW (1986) Oxygen sensitive afferent information arising from the first gill arch of yellowfin tuna. Respir Physiol 66: 193–203
Milsom WK, Jones DR (1976) Are reptilian pulmonary receptors mechano- or chemosensitive? Nature 261: 327–328
Milsom WK, Jones DR (1977) Carbon dioxide sensitivity of pulmonary receptors in the frog. Experientia 33: 1167–1168
Milsom WK, Jones DR (1979) Pulmonary receptor chemosensitivity and the ventilatory response to inhaled CO_2 in the turtle. Respir Physiol 37: 101–107
Milsom WK, Jones DR (1980) The role of vagal afferent information and hypercapnia in control of the breathing pattern in chelonia. J Exp Biol 87: 53–63
Milsom WK, Jones DR, Gabbott GRJ (1981) On chemoreceptor control of ventilatory responses to CO_2 in unanesthetized ducks. J Appl Physiol 50: 1121–1128
Mitchell GS, Gleeson TT, Bennett AF (1981) Ventilation and acid-base balance during graded activity in lizards. Am J Physiol 240: R29–R37
Mitchell RA (1965) The regulation of respiration in metabolic acidosis and alkalosis. In: Brooks C, Kao F, Lloyd BB (eds) Cerebrospinal fluid and the regulation of respiration. Blackwell, Oxford

Molony V (1974) Classification of vagal afferents firing in phase with breathing in *Gallus domesticus*. Respir Physiol 22: 57–76
Nattie EN (1990) The alphastat hypothesis in respiratory control and acid-base balance. J Appl Physiol 69: 1201–1207
Neubauer JA (1990) Carbonic anhydrase and sensory function in the central nervous system. In: Dodgson SJ, Tashian RG, Gros G, Carter ND (eds) The carbonic anhydrases: cellular physiology and molecular genetics. Plenum Press, New York
Neville CM (1979) Ventilatory response of rainbow trout (*Salmo gairdneri*) to increased H^+ ion concentration in blood and water. Comp Biochem Physiol 63A: 373–376
Nielsen B (1961) On the regulation of the respiration in reptiles. I. The effect of temperature and CO_2 on the respiration of lizards (*Lacerta*). J Exp Biol 38: 301–314
Nielsen B (1962) On the regulation of respiration in reptiles. II. The effect of hypoxia with and without moderate hypercapnia on the respiration and metabolism of lizards. J Exp Biol 39: 107–117
Nolan WF, Frankel HM (1982) Ventilatory responses to CO_2 at different body temperatures in the snake, *Coluber constrictor*. Experientia 38: 943–944
Osborne JL, Mitchell GS, Powell F (1977) Ventilatory responses to CO_2 in the chicken: intrapulmonary and systemic chemoreceptors. Respir Physiol 30: 369–382
Pappenheimer JR, Fencl V, Heisey SR, Held D (1965) Role of cerebral fluids in control of respiration as studied in unanesthetized goats. Am J Physiol 208: 442–250
Perry SF, Kinkead R, Gallaugher P, Randall DJ (1989) Evidence that hypoxemia promotes catecholamine release during hypercapnic acidosis in rainbow trout (*Salmo gairdneri*). Respir Physiol 77: 351–364
Perry SF, Kinkead R, Fritsche R (1992) Are circulating catecholamines involved in the control of breathing by fishes? Rev Fish Biol Fish 2: 65–83
Powell FL, Fedde MR, Gratz RK, Schied P (1978) Ventilatory response to CO_2 in birds. I. Measurements in the unanesthetized duck. Respir Physiol 35: 349–359
Powell FL, Milsom WK, Mitchell GS (1988) Effects of intrapulmonary CO_2 and airway pressure on pulmonary vagal afferent activity in the alligator. Respir Physiol 74: 285–298
Randall DJ (1982) The control of respiration and circulation in fish during exercise and hypoxia. J Exp Biol 100: 275–288
Randall DJ, Jones DR (1973) The effect of deafferentation of the pseudobranch on respiratory response to hypoxia and hyperoxia in the trout (*Salmo gairdneri*.) Respir Physiol 17: 291–301
Randall DJ, Taylor EW (1991) Evidence of a role for catecholamines in the control of breathing in fish. Rev Fish Biol Fish 1: 139–157
Randall WC, Stullken DE, Hiestand WA (1944) Respiration of reptiles as influenced by the composition of the inspired air. Copeia 1944: 136–144
Rogers DC (1967) The structure of the carotid bifurcation in the lizards *Tiliqua occipitalis* and *Trachysaurus rugosus*. J Morphol 122: 115–130
Rovainen CM (1977) Neural control of ventilation in the lamprey. Fed Proc 36: 2386–2389
Sakakibara Y (1978) Localization of CO_2 sensor related to the inhibition of the bullfrog respiration. Jpn J Physiol 28: 721–735
Saunders RL (1962) The irrigation of the gills in fishes. II. Efficiency of oxygen uptake in relation to respiratory flow activity and concentrations of oxygen and carbon dioxide. Can J Zool 40: 817–862
Scheid P, Piiper J (1986) Control of breathing in birds. In: Fishman AP, Cherniak NS, Widdicombe JG, Geiger SR (eds) Handbook of physiology, section 3. The respiratory system, vol II. Control of breathing, part 2. Am Physiol Soc, Bethesda
Scheid P, Gratz RK, Powell FL, Fedde MR (1978) Ventilatory responses to CO_2 in birds. II. Contribution by intrapulmonary CO_2 receptors. Respir Physiol 35: 361–372
Shelton G, Boutilier RG (1982) Apnoea in amphibians and reptiles. J Exp Biol 100: 245–273
Shelton G, Jones DR, Milsom WK (1986) Control of breathing in ectothermic vertebrates. In: Fishman AP, Cherniak NS, Widdicombe JG, Geiger SR (eds) Handbook of physiology, section 3. The respiratory system, vol II. Control of breathing, part 2. Am Physiol Soc, Bethesda, pp 857–909

Silver RB, Jackson DC (1985) Ionic compensation to chronic hypercapnia in turtles: no renal contribution. Physiologist 28: 283

Singh BN, Hughes GM (1973) Cardiac and respiratory responses in the climbing perch *Anabas testudineus*. J Comp Physiol 84: 205–226

Smatresk NJ (1988) Control of the respiratory mode in air-breathing fishes. Can J Zool 66: 144–151

Smatresk NJ (1990a) Respiratory defense reflexes in an air-breathing fish, *Lepisosteus oculatus*. Am Zool 30: 67A

Smatresk NJ (1990b) Chemoreceptor modulation of endogenous respiratory rhythms in vertebrates. An J Physiol 28: R887–R897

Smatresk NJ, Cameron JN (1982a) Respiration and acid-base physiology of the spotted gar, a bimodal breather. II. Responses to temperature change and hypercapnia. J Exp Biol 96: 281–293

Smatresk NJ, Cameron JN (1982b) Respiration and acid-base physiology of the spotted gar, a bimodal breather. III. Response to a transfer from fresh water to 50% sea water, and control of ventilation. J Exp Biol 96: 295–306

Smatresk N, Smits AW (1991) Effects of central and peripheral chemoreceptor stimulation on ventilation in the marine toad, *Bufo marinus*. Respir Physiol 83: 223–238

Smith FM, Jones DR (1982) The effect of changes in blood oxygen carrying capacity on ventilation volume in the rainbow trout (*Salmo gairdneri*). J Exp Biol 97: 325–334

Smyth DH (1939) The central and reflex control of respiration in the frog. J Physiol 95: 305–327

Tallman RD Jr, Grodins FS (1982a) Intrapulmonary CO_2 receptors and ventilatory response to lung CO_2 loading. J Appl Physiol 52: 1272–1277

Tallman RD Jr, Grodins FS (1982b) Intrapulmonary CO_2 receptor discharge at different levels of venous P_{CO_2}. J Appl Physiol 53: 1386–1391

Templeton JR, Dawson WR (1963) Respiration in the lizard *Crotaphytus collaris*. Physiol Zool 36: 104–121

Terni T (1931) Il simpatico cervicale degli amnioti (Ricerche di morfologia comparata). Z Anat Eutwicklungsgesch 96: 289–426

Thomas S (1983) Changes in blood acid-base balance in trout (*Salmo gairdneri* Richardson) following exposure to combined hypoxia and hypercapnia. J Comp Physiol 152B: 53–57

Thomas S, LeRuz H (1982) A continuous study of rapid changes in blood acid-base status of trout during variations of water P_{CO_2}. J Comp Physiol 148B: 123–130

Toews DP (1971) Factors affecting the onset and termination of respiration in the salamander *Amphiuma tridactylum*. Can J Zool 49: 1231–1237

Toews DP, Heisler N (1982) The effects of hypercapnia on intracellular and extracellular acid-base status in the toad *Bufo marinus*. J Exp Biol 97: 79–86

Toews DP, Kirby S (1985) The ventilatory and acid-base physiology of the toad, *Bufo marinus*, during exposure to environmental hyperoxia. Respir Physiol 59: 225–229

Toews D, MacIntyre D (1978) Respiration and circulation in an apodan amphibian. Can J Zool 56: 998–1004

Toews D, Boutilier R, Todd L, Fuller N (1978) Carbonic anhydrase in the amphibia. Comp Biochem Physiol 59A: 211–213

Van Vliet BN, West NH (1986) Cardiovascular responses to electrical stimulation of the recurrent laryngeal nerve in conscious toads (*Bufo marinus*). J Comp Physiol 156B: 363–375

Von Euler C (1986) Brain stem mechanisms for generation and control of breathing pattern. In: Fishman AP, Cherniak NS, Widdicombe JG, Geiger SR (eds) Handbook of physiology, section 3. The respiratory system, vol II. Control of breathing, part 1. Am Physiol Soc, Bethesda

Walker R, Galante RJ, Fishman AP, Pack AI (1990) Effect of GABA on gill and lung ventilation in an in vitro isolated brainstem preparation in the tadpole. Physiologist 33: A35

West NH, Van Vliet BN (1992) Sensory mechanisms regulating the cardiovascular and respiratory systems in amphibians. In: Feder M, Burggren W (eds) Environmental physiology of the amphibians. Univ Chicago Press, Chicago

West NH, Langille BL, Jones DR (1981) Cardiovascular system. In: King AS, McLelland JJ (eds) Form and function in Birds, vol 2. Academic Press, London

West NH, Topor ZL, Van Vilet BN (1987) Hypoxemic threshold for lung ventilation in the toad. Respir Physiol 70: 377–390
West NH, Smits AW, Burggren WW (1989) Factors terminating nonventilatory periods in the turtle, *Chelydra serpentina.* Respir Physiol 77: 337–350
Wood CM, Turner JD, Munger RS, Graham MS (1990) Control of ventilation in the hypercapnic skate *Raja oscellata*: II. Cerebrospinal fluid and intracellular pH in the brain and other tissues. Respir Physiol 80: 279–298
Yamashita S, Evans RE, Hara TJ (1989) Specificity of the gustatory chemoreceptors for CO_2 and H^+ in rainbow trout (*Oncorhynchus mykiss*). Can J Fish Aquat Sci 46: 1730–1734
Yoshii K, Kashiwayanagi M, Kurihara K, Kobatake Y (1979) High sensitivity of the eel palatine receptors to carbon dioxide. Comp Biochem Physiol 66: 327–330

Blood Gas Transport

Chapter 5

Adrenergic Control of Red Cell pH, Organic Phosphate Concentrations and Haemoglobin Function in Teleost Fish

M. Nikinmaa[1] and *R.G. Boutilier*[2]

Contents

1 Introduction

A rapid mechanism for increasing haemoglobin-oxygen affinity in teleost fish is of significant adaptive advantage for two reasons. First, aquatic environments exhibit large temporal and spatial fluctuations in the oxygen tensions. Daily changes of oxygen tension from 10 to 485 torr have been reported in the surface waters of a ditch (Jones 1961) and from 50 to 220 torr in the raceways of a fish

[1] Department of Zoology, University of Helsinki, Box 17, FIN-00014, Finland
[2] Department of Zoology, Cambridge University, Downing Street, Cambridge, CB2, 3EJ, UK

hatchery (Garey and Rahn 1970). There are large oxygen-minimum zones in the oceans, for example in the Pacific ocean off the coast of central and southern Mexico, where the oxygen tension is approximately 1 torr at depths of 100–900 m (see Douglas et al. 1976). Second, the blood of teleost fish is markedly acidified during physical disturbance. The pH of dorsal aortic blood decreases by 0.3–0.5 units during 5–10 min of physical disturbance (Holeton et al. 1983; Jensen et al. 1983; Nikinmaa et al. 1984; Milligan and Wood 1987). Since the sensitivity of most fish haemoglobins to protons is great (for reviews, see e.g. Weber 1982; Brittain 1987; Nikinmaa 1990), this acidification together with the large, rapid fluctuations of oxygen tensions would seriously compromise oxygen binding in gills in the absence of rapid mechanisms which increase blood oxygen affinity/capacity.

The major modulators of haemoglobin-oxygen affinity in teleost fish are protons and the organic phosphates, ATP (adenosine triphosphate) and GTP (guanosine triphosphate; for reviews, see Weber 1982; Weber and Jensen 1988; Weber and Wells 1989). These cofactors bind preferentially to the deoxy conformation of haemoglobin, stabilizing it and thereby decreasing the oxygen affinity of haemoglobin. Whenever present, GTP appears to be more potent than ATP (see e.g. Weber and Jensen 1988; Weber and Wells 1989). Changes in the concentrations of both organic phosphates and protons occur during adaptation to adverse environmental conditions: in teleost fish subjected to hypoxia, the red cell ATP/GTP concentrations decrease and red cell pH increases (Wood and Johansen 1972, 1973), whereby the blood oxygen affinity increases. A mechanism by which the concentrations of these cofactors could be very rapidly regulated within intact erythrocytes was discovered in the early 1980s. Catecholamines increased the red cell pH and decreased the red cell ATP concentration in rainbow trout (*Oncorhynchus mykiss*) within minutes, and caused an increase in haemoglobin-oxygen affinity (Nikinmaa 1982,1983).

In the present chapter we review the organic phosphate metabolism and control of pH in teleost red blood cells, and how these are influenced by catecholamines. Furthermore, we review the regulation of the adrenergic red cell responses, and the physiological situations in which catecholamines play a role in controlling haemoglobin-oxygen affinity.

2 Production of ATP and GTP

Most fish erythrocytes possess adenosine triphosphate (ATP) and guanosine triphosphate (GTP). In some cases as in eel (*Anguilla*), erythrocyte GTP concentrations may exceed those of ATP (Geoghegan and Poluhowich 1974; Weber et al. 1976) whereas in others, as in rainbow trout (e.g. Tetens 1987), GTP levels are much lower than ATP, sometimes below normal detection limits (see review by Weber 1982). ATP in fish erythrocytes is mostly produced as a product of aerobic energy metabolism (Hunter and Hunter 1957; Boutilier and Ferguson 1989; see following sections). Little is known, however, about the metabolic reasons for the high GTP levels in many fish erythrocytes. In addition to its role as an allosteric modulator of Hb O_2-affinity, ATP is also the main energy currency of the cell. When such energy is channelled into the biosynthesis of proteins and other compounds, it occurs by way of GTP. GTP is formed, when guanosine monophosphate (GMP) is first converted to guanosine disphosphate (GDP) in a reaction catalyzed by guanosine monophosphokinase (GMPK), and GDP is subsequently converted to GTP in a reaction catalyzed by nucleoside diphosphokinase (NDPK; see e.g. Weber 1982):

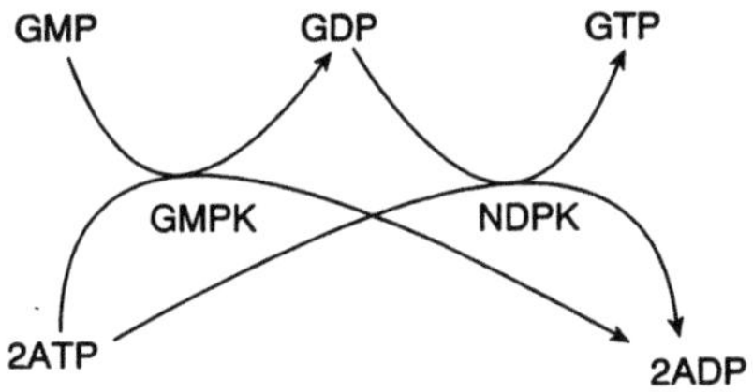

In blood of the eel, *Anguilla rostrata*, activities of GMPK are 100-fold lower than those of NDPK (Parks et al. 1973). Significant activities of NDPK have also been observed in carp blood (Weber 1982). In fish whose haemoglobins are insensitive to organic phosphates, levels of GMPK are either extremely low (dogfish) or undetectable (hagfish). The high concentration of GTP in erythrocytes of eel coincides with a high activity ratio of GMPK to AMPK (adenosine monophosphokinase). In other fish species (e.g. hagfish and dogfish), the GMPK/AMPK ratio and GTP concentration are lower. This suggests that the guanine nucleotide synthetic pathway in eel is more active than in other species. Certainly, a great deal remains to be known about the enzymes and cofactors involved in the production of GTP in fish erythrocytes.

2.1 Aerobic Metabolic Rates of Erythrocytes

Teleost erythrocytes exhibit a high degree of aerobic metabolic activity (Fig. 1). They possess nuclei, mitochondria, endoplasmic reticulum, and various other cell organelles (Schweiger 1962). Aerobic metabolism accounts for more than 90% of the nucleotide triphosphate (NTP) production of oxygenated salmonid erythrocytes in vitro (Ferguson and Boutilier 1988) and in vivo (Boutilier and Ferguson 1989; Ferguson and Boutilier 1989). Similar studies on rainbow trout (Walsh et al. 1990) and sea raven (Sephton et al. 1991) erythrocytes confirm that energy metabolism in these cells is supported almost entirely by aerobic metabolism. This is in sharp constrast to mammalian erythrocytes which produce energy almost entirely via anaerobic glycolysis (e.g. Kim 1985).

Rates of oxygen consumption by erythrocytes of fish are the highest reported for any vertebrate (Boutilier and Ferguson 1989). When taking into account the metabolic rate, haematocrit, and blood volume of a resting fish, one finds that erythrocyte oxygen consumption makes up 2–3% of the metabolic rate of the

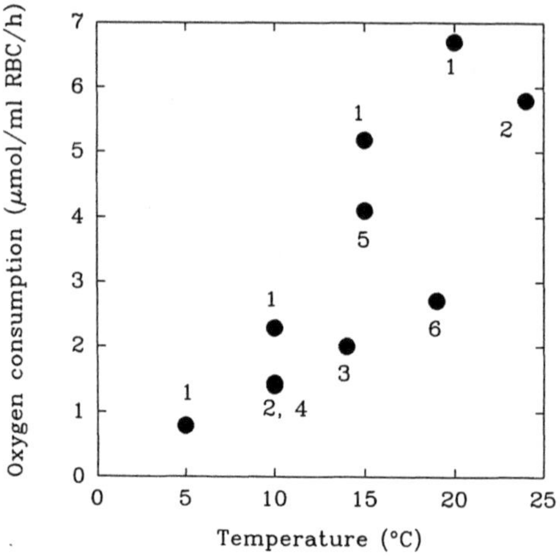

Fig. 1. The oxygen consumption of teleost blood (μmol O_2 ml RBC^{-1} h^{-1}): *1 Oncorhynchus mykiss* (Eddy 1977); *2 Salmo salar* (Ferguson and Boutilier 1988); *3 Salmo trutta* (Bolis et al. 1971); *4 Oncorhynchus mykiss* (Walsh et al. 1990); *5 Hemitripterus americanus* (Sephton et al.1991); *6 Oncorhynchus mykiss.* (Ferguson et al. 1989)

whole animal. Interspecific comparison of aerobic metabolic rates as a function of temperature reveal a Q_{10} value of 2.5, similar to that calculated for rainbow trout by Eddy (1977) and for *Salmo salar* by Ferguson and Boutilier (1988). In addition to temperature, catecholamines stimulate the aerobic metabolic rate of blood cells in vitro (Ferguson and Boutilier 1988; Ferguson et al. 1989). Various authors have reported no direct effect of incubation medium pH on aerobic metabolism of teleost erythrocytes (Ferguson and Boutilier 1988; Ferguson et al. 1989; Sephton et al. 1991), whereas Wood et al. (1990) observed an inhibitory effect caused by respiratory, metabolic and mixed acidoses. Recently, Wang et al. (1994) have demonstrated that a large fraction of the oxygen consumption of trout blood can be attributed to oxygen consumption by leucocytes.

2.2 Metabolic Pathways for Energy Production

Erythrocytes of perch, trout and sea raven have a full complement of glycolytic enzymes (Bachand and Leray 1975; Ferguson and Storey 1991; Sephton et al. 1991) and are thus capable of anaerobic conversion of glucose to lactate. In accordance with their demonstrable oxygen consumption, these cells also possess key enzymes of the tricarboxylic acid (TCA) cycle (Walsh et al. 1990; Ferguson and Storey 1991; Sephton et al. 1991). Enzymes associated with phosphagen and amino acid metabolism, as well as those of the pentose phosphate shunt, have also been detailed for rainbow trout erythrocytes (Walsh et al. 1990; Ferguson and Storey 1991). However, enzymes indicative of glycogenolytic or gluconeogenic potential were found either to be absent, or present at extremely low levels in trout RBCs (Ferguson and Storey 1991). Though stores of intracellular glycogen have been reported for rainbow trout erythrocytes (Walsh et al. 1990), the levels are extremely low and are not thought to contribute significantly to substrate oxidation (Walsh et al. 1990).

2.3 Metabolic Fuels

Metabolic substrate use by erythrocytes of teleost fish has recently become a focus of intense interest. As with other vertebrates, teleost erythrocytes are normally bathed in a plasma that is rich in metabolic fuels, including glucose, lactate, amino acids and fatty acids (Walsh et al. 1990; Ferguson and Storey 1991; Sephton et al. 1991). When freshly sampled whole blood of rainbow trout is incubated in vitro under oxygenated conditions, cellular concentrations of ATP and GTP, as well as oxygen consumption, remain relatively constant over a period of several hours (Ferguson and Boutilier 1988, 1989; Boutilier and Ferguson 1989), indicating that fuel availability is not a limiting factor to cell viability during this time. Experiments testing the long-term viability of rainbow trout erythrocytes incubated in vitro (Houston et al. 1985) show that the nucleotide triphosphate content (NTP/Hb ratio) declines in cells resuspended in glucose containing Eagle's minimal essential medium but remains unchanged over a 24-h incubation if the medium is supplemented with pyruvate. This suggests that the presence of catecholamine hormones or other blood-borne humoral factors is not required for the maintenance of NTP levels and that glucose is not used very effectively by trout red blood cells. In this regard, the CO_2 produced from labelled glucose in rainbow trout erythrocytes accounts for only 10% of the oxygen consumed (Walsh et al. 1990; Wood et al. 1990). Similarly, the difference between oxygen consumptions in native plasma and glucose-containing medium in sea raven erythrocytes (Sephton et al. 1991) may indicate that, when available, other substrates are used preferentially.

Metabolic responses of various fish erythrocytes have been assessed by incubating cells in media containing radiolabelled substrates. In many instances, the substrate first examined has been glucose, because it is the preferred substrate of most, but not all, mammalian erythrocytes (Kim 1985). CO_2 production from ^{14}C-labelled glucose has been measured in erythrocytes of lungfish, nurse shark (Mauro and Isaacks 1989), rainbow trout (Walsh et al. 1990; Wood et al. 1990), sea raven (Sephton et al. 1991) and carp (Tiihonen and Nikinmaa 1991b). In order to compare the relative importance of glucose to other metabolic fuels, several studies have taken the approach of presenting various substrates individually and then directly comparing the rates of $^{14}CO_2$ production between the metabolic fuels. For instance, at normal plasma concentrations of substrate, rates of CO_2 production in rainbow trout erythrocytes are highest for glucose, followed in descending order of importance by lactate, alanine, and oleate (Walsh et al. 1990). The sum total of CO_2 production rates of all four substrates in trout erythrocytes is, however, approximately ten-fold lower than the measured whole blood oxygen consumption, indicating that either other substrates are important or a significant proportion of radioactive substrates is not oxidized to $^{14}CO_2$. In sea raven erythrocytes, the rates of oxidation of exogenous glucose, determined from $^{14}CO_2$ production, are about 1000-fold lower than rates calculated from either oxygen consumption or direct measurement of the decrease in extracellular glucose concentration. Sephton et al. (1991) suggest that this shortfall may result from exogenous ^{14}C-labelled glucose being highly diluted before entering the TCA cycle. Alternatively, the low rates of glucose oxidation, calculated from the rate of $^{14}CO_2$ liberation, may be the consequence of low glucose permeability of the cell membrane (Ingermann et al. 1985; Tiihonen and Nikinmaa 1991a) and sluggish glycolysis (Bachand and Leray 1975). During the relatively short (120 min) incubation time, very little of the radioactive glucose may have reached the citric acid cycle, whereby the apparent rate of glucose oxidation is much lower than that measured from the disappearance of unlabelled glucose which is present at the different stages of the metabolic path throughout the incubation period.

Exogenous glucose is able to fuel the aerobic metabolism of sea raven erythrocytes, as rates of oxygen uptake and glucose disappearance are closely matched (Sephton et al. 1991). This demonstrates, for the first time, that glucose can be the dominant metabolic fuel in a fish erythrocyte. Glucose is also an important fuel of rainbow trout erythrocytes, given that it is preferred over a number of other substrates (Walsh et al. 1990). It would, however, be useful to test whether pyruvate is preferred over glucose as a substrate (cf. above and Houston et al. 1985). In carp erythrocytes (Tiihonen and Nikinmaa 1991b), the monocarboxylates, L-lactate and pyruvate, are quantitatively the most important metabolic fuels, followed by glutamine and D-glucose.

The relative rates of glucose and lactate oxidation in rainbow trout erythrocytes in vivo vary considerably as a result of exercise (Wood et al. 1990). Although plasma glucose levels of trout rise following severe exercise, rates of glucose oxidation are depressed (Wood et al. 1990), partly due to the known inhibitory effect of elevated lactate levels (Walsh et al. 1990). Lactate oxidation rates, on the other hand, are markedly stimulated at 0–2 h post-exercise, closely tracking the increases in plasma lactate concentration during this time (Wood et al.1990). Post-exercise rates of lactate oxidation are, however, much higher than can be predicted solely on the basis of substrate availability (Walsh et al. 1990). Thus, in addition to the stimulatory effects caused by the rise in plasma lactate (i.e. by substrate availability), oxidation of lactate in vivo is enhanced further by separate effects of increased P_{CO_2} and catecholamine concentrations, the latter only being effective under acidotic conditions (Wood et al. 1990).

2.4 Membrane Permeability of Substrates

There are now data on the glucose permeability of red cell membrane in the eel (*Anguilla*), rainbow trout, and paddyfield eel (*Monopterus albus*; Tse and Young 1990), carp (Tiihonen and Nikinmaa 1991a), and *Embiotoca lateralis* (Ingermann et al. 1985). Apart from the eel, glucose transport is exceedingly slow in all of these species, with half-times of equilibration of several hours. In *Embiotoca lateralis* and eel, a facilitated pathway is present (i.e. glucose uptake shows saturation kinetics), while in the others no carrier appears to exist. It is notable, however, that despite the low permeability of most fish erythrocytes to glucose, the high plasma glucose concentration generates an adequate flux of glucose to support a significant utilization of glucose as a metabolic fuel.

In tuna (*Katsuwonus pelamis*; Moon et al. 1987) erythrocytes, lactate uptake evidently occurs by a simple diffusion process with no sign of a carrier molecule at a concentration range of 0–30 mM. In carp erythrocytes, lactate transport is largely carrier-mediated (Tiihonen and Nikinmaa 1993), i.e. it shows saturation kinetics and is inhibited by para-chloromercuriphenylsulphonic acid (pCMPS). In both simple and carrier-mediated diffusion, however, as plasma lactate concentrations increase following acute exercise, the elevated transmembrane gradient supports greater uptake of lactate. Following severe exercise in rainbow trout, for example, a diffusion-mediated increase in erythrocyte lactate concentrations (Ferguson and Storey 1991) presumably facilitates concomitant increases in lactate oxidation (Wood et al. 1990).

3 Control of Red Cell pH in the Absence of Significant Secondarily Active Transport of Acid-Base Relevant Ions

In the absence of significant secondarily active transport of protons, bicarbonate and chloride across the red cell membrane, all of these ions are distributed according to the membrane potential at steady state, whereby the distribution ratio for protons is the inverse of the chloride distribution ratio (i.e. $H_{out}/H_{in} = Cl_{in}/Cl_{out}$), and the intracellular pH can be calculated from the extracellular pH and chloride distribution ratio using the formula

$$pH_{in} = pH_{out} + \log [Cl^-]_{in} - \log [Cl^-]_{out}. \qquad (1)$$

The distribution ratio for exchangable anions (chloride and bicarbonate) is given from the relation (for derivation of the formula, see Hladky and Rink 1977)

$$A_{in}/A_{out} = 2\{[\text{cation}] + z[\text{imp-ion}]\}/\{2\cdot[\text{cations}] + (z+1)\cdot[\text{imp-ion}]\}, \qquad (2)$$

in which z is the charge of impermeable polyions (imp-ion). Based on the above equations, the proton distribution ratio (i.e. the relationship between extracellular and intracellular pH) depends on the amount of cations, organic phosphates and haemoglobin within the cell, and the net charge of organic phosphates and haemoglobin.

Movement of acid and base equivalents across the red cell membrane after acid-base disturbances is achieved via the Jacobs–Stewart cycle (Jacobs and Stewart 1942; Hladky and Rink 1977). In the case of extracellular acid loads, protons react with bicarbonate ions to form carbonic acid, which is dehydrated extracellularly, in an uncatalyzed reaction, to carbon dioxide and water. CO_2 then diffuses down its concentration gradient into the red cell and is hydrated to bicarbonate and protons in a reaction catalyzed by carbonic anhydrase. The protons are taken up largely by intracellular buffers, whereas the bicarbonate ions are exchanged for chloride via the anion exchanger, completing the cycle.

The net result of the Jacobs-Stewart cycle is thus the transport of protons from the extracellular to the intracellular compartment. The rate-limiting step of this passive acid equilibration is the extracellular, uncatalyzed dehydration reaction of carbonic acid to carbon dioxide, with apparent half-times of approx. 1 min at 15 °C and 3–4 min at 5 °C (Heming 1984).

Since both haemoglobin (German and Wyman 1937; Jensen 1989) and organic phosphates (Dawson et al. 1987) have ionizable groups within the physiological pH range, changes in the extracellular pH, which are transferred to the intracellular compartment via the Jacobs-Stewart cycle, will affect their net charge. A decrease in extracellular pH decreases intracellular pH and, consequently, the negative charge of haemoglobin and organic phosphates, whereby the anion ratio increases, as can be seen from Eq. (2). Because the anion ratio increases, the pH gradient across the red cell membrane decreases and the cells swell (Funder and Wieth 1966; Dalmark 1975; Albers 1986). Opposite changes occur if the extracellular pH increases.

Oxygenation-deoxygenation reactions of haemoglobin are also associated with changes in red cell pH. Oxyhaemoglobin is more acidic than deoxyhaemoglobin within the physiological pH range (Jensen 1989). As a consequence, protons are taken up upon deoxygenation, whereby the charge on the haemoglobin molecule decreases, and the anion ratio, red cell volume and intracellular pH increase. Deoxygenation-induced increases in intracellular pH at constant extracellular pH (Haldane effect) are pronounced in many teleost fish: in carp (*Cyprinus carpio*) deoxygenated erythrocytes have a higher pH of ca. 0.25 units than their oxygenated counterparts at an extracellular pH range 7.4 to 8.0 (Albers et al. 1983), and in tench (*Tinca tinca*) the difference (at extracellular pH 7.9) is even greater, 0.35 pH units (Jensen 1986). This is one of the factors contributing to the increase in erythrocyte pH in hypoxic fish.

In addition to the charge of haemoglobin and organic phosphates, their intracellular concentrations influence the pH gradient across the red cell membrane. A decrease in the amount of intracellular organic phosphates causes an increase in the cell volume and in the red cell pH (Duhm 1972). Such a decrease in the total amount of organic phosphates occurs in hypoxic fish (Tetens 1987), thus contributing to the observed increase in intracellular pH in hypoxia (Wood and Johansen 1973). Similarly, adrenergic stimulation decreases cellular ATP concentrations (see Sect. 5).

4 Stimulation of the Sodium/Proton Exchanger by Catecholamines and Its Effects on Red Cell pH

Stimulation of rainbow trout red cells with adrenaline in physiological medium causes an immediate, dose-dependent increase in red cell volume, and a reduction of the pH gradient across the red cell membrane (Nikinmaa 1982). Adrenergic effects of varying magnitude have subsequently been observed at least in the red cells of Salmoniformes (Ferguson and Boutilier 1988; Salama and Nikinmaa 1989; Cossins and Kilbey 1991), in the red cells of the Perciformes, *Morone saxatilis* and *Stizostedion lucioperca* (Nikinmaa and Huestis 1984; Salama and Nikinmaa 1989) and in the red cells of the cyprinids, carp, *Cyprinus carpio* (Fuchs and Albers 1988; Salama and Nikinmaa 1988) and tench, *Tinca tinca* (Cossins and Kilbey 1991). Generally, the responses are large in the salmonids, and much more modest in the other species studied. In the cyprinids, the responses are only observed under hypoxic conditions or at lowered pH values (Salama and Nikinmaa 1988). Eel erythrocytes appear to be devoid of adrenergic responses under physiological conditions (see e.g. Hyde and Perry 1990).

4.1 Properties of the Adrenergically Activated Sodium/Proton Exchanger

The red cell swelling and the increase in intracellular pH are the consequence of the β-adrenergic activation of the sodium/proton exchanger (Baroin et al. 1984; Nikinmaa and Huestis 1984; Cossins and Richardson 1985; Borgese et al. 1986). The sodium gradient ($Na_{out}/Na_{in} = 4.5$ in unstimulated erythrocytes; Nikinmaa et al. 1990b), which is actively maintained by the sodium pump, can be used to transport protons away from electrochemical equilibrium. As expected for a β-adrenergic system, cyclic AMP is the second messenger for the activation. The sodium/proton exchanger of rainbow trout red cells can be activated either by forskolin, which activates adenylate cyclase without interacting with the adrenergic receptor, or by adding 8-bromo cyclic AMP to the extracellular medium (Mahé et al. 1985). Similar treatments also cause cell swelling in the carp (Salama and Nikinmaa 1990).

Noradrenaline is a much more potent activator of the adrenergic proton extrusion than adrenaline: the concentration required for half-maximal response in rainbow trout erythrocytes is ca. 3×10^{-8} M for noradrenaline and 3–5×10^{-7} M for adrenaline (Tetens et al. 1988; Cossins and Kilbey 1989). In carp erythrocytes, a significant adrenergic elevation of the cAMP concentration is observed at 10^{-8} M noradrenaline, but only at 10^{-6} M adrenaline (Salama and Nikinmaa 1990).

The maximal activity of the sodium/proton exchanger in the membrane of oxygenated rainbow trout erythrocytes at 20 °C, as judged by the initial sodium influx after adrenergic stimulation (Nikinmaa et al. 1990b) or by the initial influx of $^{22}Na^+$ (Borgese et al. 1987), is ca. 30 mmol/kg cell dry weight/min. Although this value is 10–100 times greater than that for osmotically activated sodium/proton exchange in other red cell types, it is much smaller than the chloride exchange fluxes via the anion exchanger, which in rainbow trout erythrocytes at 15 °C can be as great as 5 mol/kg cell dry weight/min (calculated from the data of Romano and Passow 1984), i.e. a value ca. 200 times greater than that for the adrenergic sodium influx. Thus, it is clear that the adrenergic sodium/proton exchanger could not generate measurable proton disequilibrium across the red cell membrane if the rate-limiting step of passive proton equilibration were the anion exchange.

The maximal activity of the sodium/proton exchanger is influenced by both pH and oxygen tension. In the salmonids, maximal adrenergic sodium fluxes (Borgese et al. 1987) and accumulation of sodium (Salama and Nikinmaa 1989) are observed at around extracellular pH 7.3, whereas in the carp and pikeperch, there is a continuous increase in the activity with decreasing pH at least down to pH 7.1 (Salama and Nikinmaa 1989). With regard to oxygen, a decrease in the oxygen tension increases the activity of the exchanger in rainbow trout erythrocytes—the maximal flux of sodium in a nitrogen atmosphere is approximately three times greater than that in an oxygen atmosphere (Motais et al. 1987). Similarly, at pH 7.5, the maximal adrenergic volume increase in carp erythrocytes in 30 min was three to four times greater under hypoxic than in normoxic conditions (Salama and Nikinmaa 1990). The effect of oxygen on the sodium/proton exchanger is probably mediated via its effects on haemoglobin conformation. The maximal activity of the sodium/proton exchanger is also reduced in carbon monoxide-treated erythrocytes (in which haemoglobin assumes a stable structure similar to that of oxyhaemoglobin) which are incubated under a nitrogen atmosphere (Motais et al. 1987). Furthermore, the adrenergic proton extrusion from the red cells is inhibited if the cells are treated with nitrite which converts haemoglobin to methaemoglobin (Nikinmaa and Jensen 1992).

Temperature has a pronounced effect on the activity of the adrenergic sodium/proton exchanger in rainbow trout red cells. Although the exchanger is activated by adrenergic drugs at a temperature range 4–20 °C, the maximal proton efflux increases ten-fold between these temperatures (Cossins and Kilbey 1990).

4.2 The Mechanism of Adrenergic pH Changes

The factors affecting the pH gradient across the red cell membrane after adrenergic stimulation are the intra- and extracellular hydration/dehydration reaction sequence between carbon dioxide, bicarbonate and protons, the activity of the sodium/proton exchanger and its inactivation after the initial maximal activity (Garcia-Romeu et al. 1988), and the anion exchange. The intracellular hydration/dehydration reactions between carbon dioxide, bicarbonate and protons, and the anion exchange are much faster than the extracellular reactions between carbon dioxide, bicarbonate and protons, and the adrenergic sodium/proton exchange. Thus, the adrenergic pH changes can be modelled to depend mainly on the activity of the sodium/proton exchange and the rate of the extracellular reaction sequence from bicarbonate and protons to carbon dioxide. Results of such modelling, at a haematocrit value of 50, are given in Fig. 2. In the model, the system was allowed to reach a steady state before adrenergic stimulation, whereby the rates of carbon dioxide hydration and carbonic acid dehydration were equal. Intra- and extracellular non-bicarbonate buffer values were

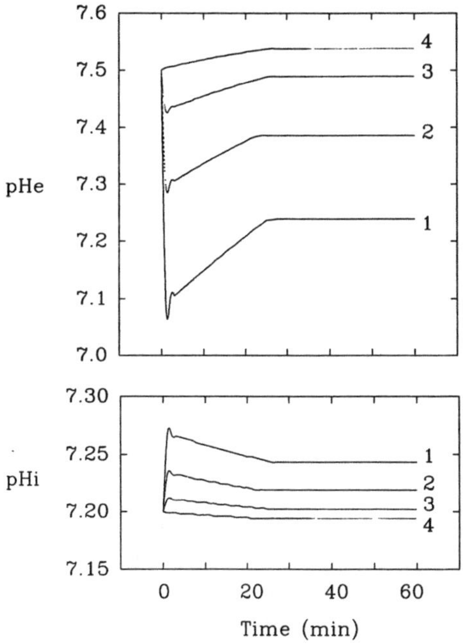

Fig. 2. Modelling of intracellular (pH_i) and extracellular pH (pH_e) values as a function of time after adrenergic stimulation of teleost erythrocytes. The adrenergic pH changes were modelled taking into account (i) chloride/bicarbonate exchange, (ii) adrenergic proton extrusion via the sodium/proton exchanger and the self-inhibition of the extrusion as a function of time, (iii) extracellular hydration/dehydration reaction sequence between bicarbonate and carbon dioxide at different levels of extracellular catalysis and (iv) intra- and extracellular non-bicarbonate buffering (buffer values, 30 and 5 mmol kg^{-1} pH^{-1}, respectively). For details of the model and applied values, see text. The cells were at the steady state before adrenergic proton extrusion ensued at time 0. *1* Adrenergic proton extrusion 30 mmol kg dry $cells^{-1}$ min^{-1}; *2* adrenergic proton extrusion halved from trace 1; *3* adrenergic proton extrusion as in trace 1, extracellular catalysis increased five-fold; *4* adrenergic proton extrusion as in trace 1, extracellular catalysis increased ten-fold

taken as 30 and 5 $mmol \cdot l^{-1} \cdot pH^{-1}$, respectively. Initial intra- and extracellular pHs were 7.2 and 7.5, intra- and extracellular $[HCO_3^-]$4 and 8 mM, the carbon dioxide tension 5 mmHg, and internal and external $[Cl^-]$ 60 and 120 mM, respectively. The rate constant for chloride/bicarbonate exchange was taken to be $2 s^{-1}$ (Romano and Passow 1984). It was further supposed that the carbon dioxide tension did not change during the incubation time and that the intracellular bicarbonate concentration varied as a consequence of changes in intracellular pH, generating a net efflux of bicarbonate. Also, it was supposed that the extracellular bicarbonate concentration could be maintained throughout the incubation. In fact, the extracellular bicarbonate concentration decreases during adrenergic stimulation, as shown by the decrease in total CO_2 content of the incubation in vitro (Motais et al. 1989). Such a decrease would slow net formation of carbon dioxide from bicarbonate and protons by the extracellular reaction, thereby making the adrenergic pH changes larger than those given by the model.

After the initial steady state, the cells were adrenergically stimulated, the activity of the sodium/proton exchange being 30 mmol (kg dry cells)$^{-1}$ min^{-1} (Nikinmaa et al. 1990b). The adrenergic proton extrusion generates a disequilibrium for the extracellular dehydration/hydration reaction sequence. The rate of CO_2 formation, which equals the rate of proton buffering by external bicarbonate ions, is given by the equation (see Roughton 1964)

$$J_{H}^{+} = (k_1 \times [H^+][HCO_3^-]/K_A - k_2 \times [CO_2]) \times (1 + 1 \times A), \qquad (3)$$

in which k_1 = rate constant for carbonic acid dehydration (20 s^{-1}) and k_2 = rate constant for carbon dioxide hydration (0.04 s^{-1}), taken from Edsall (1969), K_A = the acid constant for carbonic acid, 3×10^{-4}, I = the catalysis effect by extracellular anions (or carbonic anhydrase), and A is the concentration of extracellular anions with a catalytic effect (in the present model, the value given for I was initially 2, i.e. the same as for dihydrogen phosphate ion, and the concentration of the anion was taken as 3 mM). Adrenergic stimulation caused an initial increase of 0.07 units in the intracellular pH (trace 1 in Fig. 2), and an initial decrease of 0.45 units in the extracellular pH. After the initial activation, the sodium/proton exchanger was self- inactivated (Garcia-Romeu et al. 1988) in the model, so that the activity was reduced to 55% in 25 min (whereafter, in the results presented in the figure, it was supposed that the activity remained constant). As a consequence of the decrease in the activity of the sodium/proton exchange, the rate of the extracellular reaction sequence from bicarbonate and protons to carbon dioxide became faster than the rate of the sodium/proton exchange, and the extracellular pH started to increase. The time course of the observed pH changes in the model is similar to that observed in red cell suspensions incubated in carbon dioxide/bicarbonate-buffered media (Motais et al. 1989).

The pH changes observed after adrenergic stimulation were reduced when the activity of the sodium/proton exchanger at the initial stages of stimulation was reduced. Trace 2 in Fig. 2 indicates the effect of reducing the activity by 50%, to 15 mmol·kg dry weight^{-1}·min^{-1}. Also, the pH effects in both the intra- and the extracellular compartments were reduced (trace 3 in Fig. 2) and finally abolished (trace 4 in Fig. 2) by increasing the effectiveness of extracellular catalysis, five-fold in trace 3 and ten-fold in trace 4. When the sodium/proton exchange was inhibited in the model at 200 s, a rapid increase in the extracellular pH was observed (Fig. 3). All these changes are similar to those observed in red cell suspensions (see Motais et al. 1989; Nikinmaa et al. 1990b).

The similarity of the observed pH changes given by the model and measured in experimental work after the different treatments indicates that the major factors affecting the behaviour of the system after adrenergic stimulation are the activity of the sodium/proton exchange and the rate of the reaction sequence from bicarbonate and protons to carbon dioxide, i.e. the speed by which protons are buffered extracellularly.

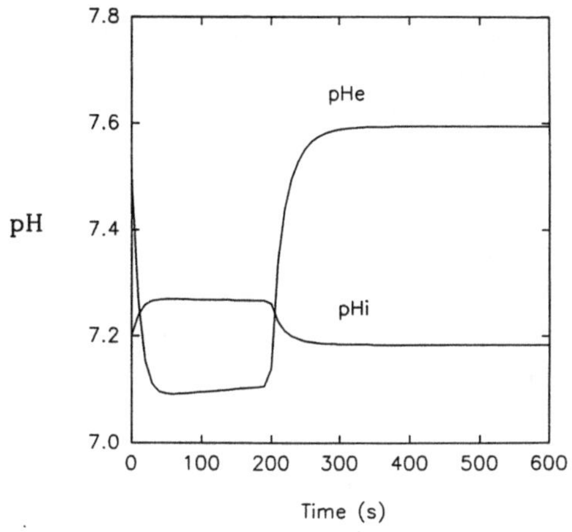

Fig. 3. Modelling of intracellular (pH_i) and extracellular (pH_e) pH values as a function of time after adrenergic stimulation of teleost red cells followed by full inhibition of the adrenergic proton extrusion 200 s after stimulation. Adrenergic proton extrusion, 30 mmol (kg dry cells)$^{-1}$ min^{-1}. For details of the model, see text and legend to Fig. 2

An important factor, not taken into account in the model, is the increase in the oxygenation of the haemoglobin molecule as the consequence of the adrenergic increase in the intracellular pH. As outlined in Section 3, protons are liberated from haemoglobin upon oxygenation, which tends to decrease the intracellular pH. The observed net change in intracellular pH after adrenergic stimulation is thus the result of these two opposing influences. Since the negative charge of haemoglobin increases upon oxygenation, the distribution ratio for permeable anions at a given intracellular cation and impermeable anion concentration decreases. This phenomenon favours a net efflux of permeable anions (bicarbonate and chloride) from the cell via conductive pathways, and would speed up the "recovery" of external pH after the initial marked decrease.

At a given activity of the sodium/proton exchange and for given rate constants for the extracellular hydration/dehydration reactions, the apparent net proton efflux after adrenergic stimulation depends on the bicarbonate concentration of the medium at any given instant. Other things staying constant, the increase in intracellular pH will generate a disequilibrium for the anion exchanger, whereby a net efflux of bicarbonate occurs. This will tend to increase the external bicarbonate concentration. In contrast, the formation of carbon dioxide in the extracellular dehydration reaction will reduce the extracellular bicarbonate concentration. This reduction will, however, by generating a further disequilibrium for the anion exchanger, speed up the efflux of bicarbonate and counteract the effect of dehydration. As a consequence of these events, the flux of radiolabelled carbon from externally added ^{14}C-labelled bicarbonate to carbon dioxide is reduced in the initial stages of adrenergic stimulation, because bicarbonate influx, and consecutive rapid conversion of bicarbonate to carbon dioxide intracellularly, is slowed down (Perry et al. 1991b; Wood and Perry 1991).

Inside the fish, in both the gills and the tissues, the situation is more complex. During the 1–3S (Hughes and Koyama 1975) passage of erythrocytes through the gill secondary lamellae, there is a pronounced partial pressure gradient of carbon dioxide from the animal to the environment (where the carbon dioxide tension is close to zero). Consequently, the rapid dehydration reaction within the red cell proceeds from bicarbonate to carbon dioxide and generates a disequilibrium for the anion exchanger, leading to a net influx of bicarbonate. During the passage of blood through the gills, the carbon dioxide tension decreases by 15% (Soivio et al. 1981) and the total CO_2 content by 25% (Nikinmaa and Jensen 1986) in resting rainbow trout. If the red cells are simultaneously adrenergically stimulated, the intracellular dehydration is to some degree opposed by the activation of the sodium/proton exchange and the consecutive increase in intracellular pH, which would favour production of bicarbonate in the intracellular reaction. However, it is likely that the net direction of the reaction is from bicarbonate to carbon dioxide, because the diffusion of carbon dioxide out of the cell and to the environment is very rapid. As a consequence, a net bicarbonate influx into the red cells will occur in the gill capillaries even if the cells are adrenergically stimulated. Thus, the net proton efflux via the sodium/proton exchanger will occur unopposed and the increase in red cell pH will be great. In the tissues, the conditions favour the intracellular hydration of carbon dioxide to bicarbonate and protons, and, consequently, net efflux of bicarbonate from the cells. This reduces the adrenergic effects on the intracellular pH. In accordance with these speculations, slowing down the net rate of intracellular carbon dioxide hydration by acetazolamide markedly increased the intra- and extracellular pH changes of adrenergically stimulated rainbow trout erythrocytes (Nikinmaa et al. 1990b). Furthermore, the pH gradient across the red cell membrane decreased much more after physical disturbance in postbranchial than in prebranchial blood of rainbow trout (Milligan and Wood 1986).

4.3 Interactions Between Catecholamines and Other Hormones

Cortisol treatment increases the total pool of β-adrenergic receptors in rainbow trout erythrocytes (Reid and Perry 1991). The increase in the number under normoxic conditions is mainly of the low-affinity type, possibly representing internalized receptors. However, under hypoxic conditions, in which the number of high-affinity β-adrenergic binding sites increases (Marttila and Nikinmaa 1988; Reid and Perry 1991) cortisol markedly increases the apparent number of high-affinity binding sites (Reid and Perry 1991). Also, cortisol increases the accumulation of cyclic AMP after adrenergic stimulation under hypoxic conditions (Reid and Perry 1991). These results suggest that animals subjected to chronic stresses, e.g. crowding, in which cortisol levels are elevated, may become "preadapted" to acute stresses in which the adrenergic responses are required. At present, these are the only data showing interactions between catecholamines and other hormones at the red cell level.

5 Effects of Catecholamines on Red Cell Organic Phosphate Concentrations

Many membrane ion exchange mechanisms that participate in volume and pH regulation in teleost erythrocytes (see Sect. 4) depend on an inwardly directed electrochemical Na^+ gradient which is maintained by the active transport of Na^+ and K^+ via the energy-consuming Na^+/K^+ pump. The Na^+/K^+ pump accounts for approximately 20% of the total oxygen consumption of rainbow trout erythrocytes under resting conditions (Tufts and Boutilier 1991). This percentage is roughly similar to the energy allocated to the Na^+/K^+ pump in rabbit reticulocytes (Rapoport 1985).

Adrenergic activation of Na^+/H^+ exchange (see Sect. 4) causes an increase in erythrocyte oxygen consumption, the magnitude of which is directly proportional to the degree of activation of the sodium/proton exchanger, as indicated by the relationship between cellular oxygen consumption and the adrenergic increase in cellular Na^+ concentration (Fig. 4). The metabolic response appears to be one of increased ATP turnover via oxidative phosphorylation. In salmonids, NTP levels decrease within the first 10 min of β-adrenergic stimulation (Nikinmaa 1983), and are thereafter maintained at a relatively constant level (Ferguson and Boutilier 1988), as long as the cells are well oxygenated (Ferguson and Boutilier 1989; Ferguson et al. 1989). Under anoxic conditions, β-adrenergic stimulation leads to a rapid and continual fall in NTP levels (Ferguson and Boutilier 1989; Tufts and Boutilier 1991). Notably, erythrocyte NTP levels fall rapidly under anoxic conditions (79% decrease in 2 h) even in the absence of β-stimulation (Tufts and Boutilier 1991).

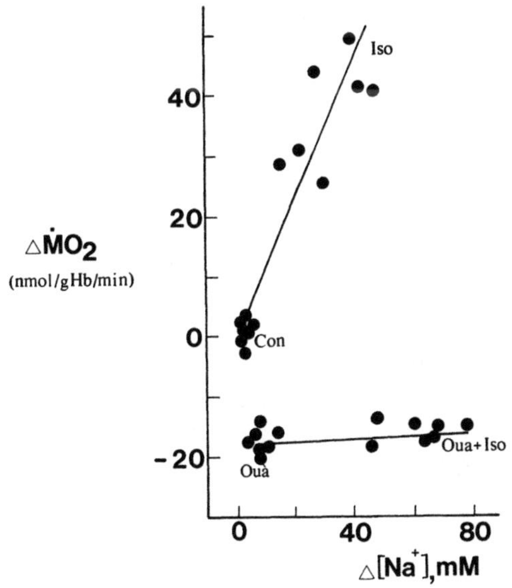

Fig. 4. Relationship between the change in erythrocyte Na^+ concentration ($\Delta[Na^+]$) and the corresponding change in erythrocyte oxygen consumption ($\Delta\dot{M}_{O2}$) in cells incubated in the presence of saline vehicle (*Con*) or isoproteronol (*Iso*) alone, or in combination with the Na^+, K^+-ATPase inhibitor ouabain (*Oua*; *Oua* + *Iso*). (R.G. Boutilier and B.L. Tufts, unpubl. data)

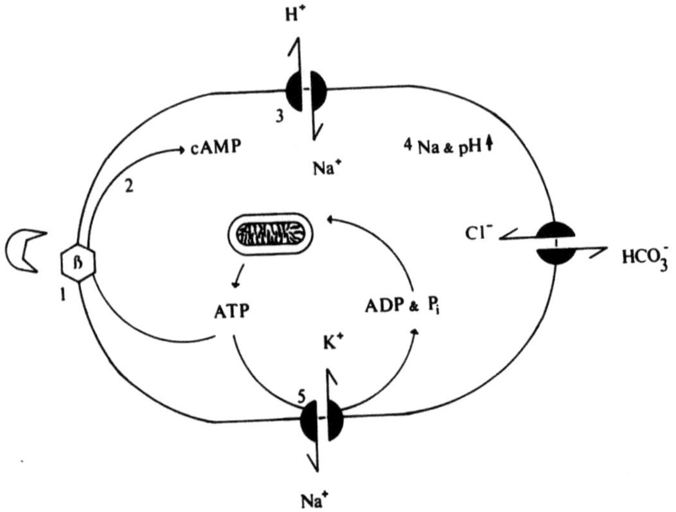

Fig. 5. A model of the metabolic and membrane events during adrenergic stimulation of a salmonid erythrocyte. First, β-adrenoceptor stimulation (*1*), leads to a rise in intracellular concentration of cyclic AMP (*2*), which triggers activation of the Na^+/H^+ exchanger such that protons are extruded from the cell in exchange for Na^+ (*3*). The net effect is a cell alkalinization and an increase in intracellular Na^+ concentration (*4*). The rise in intracellular $[Na^+]$ stimulates Na^+/K^+ ATPase and leads to increased consumption of energy (*5*). (Ferguson and Boutilier 1989)

The increase in red cell oxygen consumption and decrease in red cell NTP levels appear to be mainly due to the following sequence of events (Fig. 5; Ferguson and Boutilier 1989). The activation of sodium/proton exchange by catecholamines results in an increase in cell sodium. The increased cell sodium concentration causes an increase in the activity of the Na^+/K^+ pump. The increased energy consumption by the Na^+/K^+ pump results in the somewhat reduced ATP levels, and in the pronounced increase in cellular oxygen consumption. When rainbow trout erythrocytes are incubated in the presence of adrenaline, the marked increases of both active and passive K^+ influx (Bourne and Cossins 1982) indicate that the Na^+/K^+-ATPase activity is elevated. The increase in oxygen consumption following β-adrenergic stimulation is inhibited entirely by ouabain (Fig. 4; Tufts and Boutilier 1991), demonstrating that the increased energy requirement in stimulated cells is entirely due to increased Na^+, K^+-ATPase activity. The energy requirements of the sodium/potassium pump can rise to more than 40% of the cellular energy budget in adrenergically stimulated rainbow trout erythrocytes (Tufts and Boutilier 1991) or to more than 50% in adrenergically stimulated *Salmo salar* erythrocytes (Ferguson and Boutilier 1988).

6 In Vivo Regulation of Haemoglobin Function by Catecholamines

The major thrust of work on adrenergic control of haemoglobin-oxygen affinity has concentrated on physically disturbed and hypoxic rainbow trout. However, it is apparent that catecholamines are also liberated to the bloodstream and increase haemoglobin-oxygen affinity of rainbow trout in other situations in which blood oxygen content is rapidly reduced, e.g. during hypercapnic acidosis (Vermette and Perry 1988). It is further likely that if two disturbances which affect oxygen transport or oxygen demand occur simultaneously, the release of catecholamines is greater, and the adrenergic effects larger, than if the disturbances occur alone. For example, mild hypoxia during acid exposure caused a pronounced liberation of catecholamines in the bloodstream, and a much larger change in the red cell status than similar acid or hypoxia exposure alone (Nikinmaa et al. 1990a). Also, the measured catecholamine concentrations in swimming, hypoxic rainbow trout are greater than in resting, hypoxic animals (Fievet et al. 1987; Tetens and Christensen 1987; Ristori and Laurent 1989).

6.1 Factors Affecting Catecholamine Release

Catecholamines are released into the circulation of fish upon decreases in arterial oxygen tension such as during environmental hypoxia (Fievet et al. 1987; Tetens and Christensen 1987; Ristori and Laurent 1989), or when blood is acidified as a result of physical disturbance and exhaustive exercise (Primmett et al. 1986). The elevation in plasma catecholamine concentration of individual fish after chasing correlates with the decrease in the extracellular pH (Tang and Boutilier 1988). Catecholamine release can also be induced by infusing acid into the circulation (Boutilier et al. 1986). In the absence of other responses, both the reduced arterial oxygen tension and the plasma acidosis decrease arterial oxygen content. The catecholamine release is not caused by acidosis as such: when normoxic and hyperoxic (water oxygen tension, 650 mmHg) fish were made acidotic by hypercapnia (carbon dioxide tension of water, 7 mmHg), the plasma and red cell pH decreased more in the hyperoxic animals than in the normoxic ones, whereas the plasma catecholamine concentration only increased in the normoxic animals in which the acidosis caused a reduction in the arterial oxygen content (Perry et al. 1989). Further, the blood oxygen tension as such is not the controlling factor behind catecholamine release: when the arterial oxygen content of normoxic (arterial oxygen tension, 120 mmHg) and hyperoxic (arterial oxygen tension, 300–400 mmHg) animals was reduced by repetitive blood removal, a similar, marked increase in plasma catecholamine concentration was observed in both groups of animals, despite the very different arterial oxygen tensions. Thus, the controlling factor behind catecholamine release appears to be the blood oxygen content. It appears also that the effect of reduced oxygen

Table 1. Plasma adrenaline and noradrenaline concentrations of rainbow trout at rest and after different types of disturbance

Resting values (nM)		Type of disturbance	Disturbed values (nM)		
Adrenaline	Noradrenaline		Adrenaline	Noradrenaline	Reference[a]
1.4	10.2	Swimming (1 bl/s[b], 30 min)	0.3	2.5	1
		Swimming (2 bl/s, to exhaustion)	14.48	22.8	
		Repeated burst swimming	212	85	
1.2	1.1	Swimming (2 bl/s, 30 min)	1.3	1.1	2
0.9	0.7	Burst swimming to exhaustion	37	270	3
1.5	1.1	Tail grabbing (3 min)	7.9	4.5	2
11.4	14.2	5-min chasing (summer)	186	202	4
5.2	16.9	5-min chasing (winter)	248	58.5	4
2.2	2.7	6-min chasing	29.7	36.4	5
3.8	2.9	20-min hypoxia (blood oxygen tension, 70–110 mmHg	5.1	5.6	6
3.8	5.9	(blood oxygen tension, 30–47 mmHg)	14.7	20.5	6
0.6	1.2	120-min hypoxia (water oxygen tension, 40 mmHg)	4.51	14.5	7
5.0	11.6	5-min hypoxia (water oxygen tension, 40 mmHg)	25.1	13.7	8

[a] References: 1, Butler et al. (1986); 2, Ristori and Laurent (1985); 3, Primmett et al. (1986); 4, Milligan et al. (1989); 5, Milligan and Wood (1987); 6, Ristori and Laurent (1989); 7, Tetens and Christensen (1987); 8, Fievet et al. (1987).
[b] bl/s, Body lengths per second.

content has a local stimulatory effect on catecholamine release at the level of chromaffin tissue (Perry et al. 1991a). Therefore, the factor causing catecholamine release may be a reduction in the oxygenation of the chromaffin tissue.

There is wide variation as to which catecholamine, noradrenaline or adrenaline is the predominant one in resting situations, and which is released to the circulation more during stresses (see Table 1). The reasons behind the differences in the patterns of release are unknown at present. However, because of the much higher apparent affinity of red cell β-adrenergic receptors to noradrenaline than to adrenaline (see Sect. 4.1), noradrenaline elicits most of the red cell responses observed in vivo (Tetens et al. 1988; Nikinmaa 1992).

6.2 Regulation of Haemoglobin-Oxygen Affinity by Catecholamines in Burst Exercise

Exhaustive exercise in teleost fish normally leads to a large extracellular lactacidosis and a marked rise in circulating catecholamines (Ristori and Laurent 1985; Primmett et al. 1986; Milligan and Wood 1987; Tang and Boutilier 1988). Circulating catecholamines activate the sodium/proton exchange (see Sect. 4) with the consequence that red cell pH can either be maintained nearly constant, as in coho salmon (*Oncorhynchus kisutsch*) and rainbow trout (Van den Thillart et al. 1993; Boutilier et al. 1986; Primmett et al. 1986; Milligan and Wood 1987), or it decreases less in non-treated than in β-blocked animals, as in striped bass (*Morone saxatilis*; Nikinmaa et al. 1984), in the face of an exercise- or metabolic acid-induced extracellular acidosis. Prior to these studies, it was implicitly assumed that stress-induced extracellular acidoses would decrease red cell pH and, consequently, decrease Hb-O_2 affinity and blood O_2-carrying capacity due to the Bohr and Root effects of fish haemoglobin. However, because the pH of the haemoglobin environment can be maintained, blood O_2-carrying capacity and Hb-O_2 affinity are unaffected by the plasma acidosis (Boutilier et al. 1986; Primmett et al. 1986; Milligan and Wood 1987), and the arterial oxygen content can be maintained. Thus, burst-exercised animals can achieve similar levels of aerobic performance after the burst as before (Randall et al. 1987).

The adrenergic red cell response is pronounced in salmonids, whereas in other species of teleost fish, e.g. tench (*Tinca tinca*; Jensen 1987) and starry flounder (*Platichtys stellatus*; Milligan and Wood 1986), normoxic disturbances do not elicit adrenergic red cell responses. Adrenergic effects on erythrocyte pH are sometimes absent even from salmonids which are exercised to exhaustion (e.g. Nikinmaa and Jensen 1986; Ferguson et al. 1989; Tufts et al. 1991) even though catecholamine levels are markedly elevated. Nikinmaa and Jensen (1986) suggest that β-adrenergic effects may dominate in summer and diminish in winter due to down-regulation of receptor numbers and/or shifts in agonist- receptor sensitivity. Cossins and Kilbey (1989) demonstrated that the adrenergic proton extrusion from rainbow trout erythrocytes is reduced during winter months.

In addition to activating sodium/proton exchange across the red cell membrane, catecholamines cause an increase in red cell number per unit volume (e.g. Nikinmaa 1982). This can be the result of either the release of erythrocytes from storage organs or haemoconcentration. In the rainbow trout, splenic release of erythrocytes can be induced by adrenaline infusion (Kita and Itazawa 1989). During anaerobic 'burst' exercise, many teleost fish species recruit erythrocytes from splenic reservoirs and this results in significant increases in blood haemoglobin concentration and haematocrit values (Yamamoto et al. 1980; Kita and Itazawa 1989; Yamamoto and Itazawa 1989; Wells and Weber 1990; Pearson and Stevens 1991). The resulting increase in blood O_2-carrying capacity works to offset the deleterious effects of an erythrocyte acidosis and a subsequent decrease in haemoglobin oxygenation, which occur if, for some reason, the sodium/proton exchange is poorly activated by catecholamines (e.g. Tufts et al. 1991).

6.3 Regulation of Haemoglobin-Oxygen Affinity by Catecholamines in Acute Hypoxia

The adrenergic response is very well suited for responding to acute hypoxia for three reasons. First, catecholamines are released into the circulation in response to a reduced oxygen content as discussed in Section 6.1. Second, the β-adrenergic receptor density of the red cell membrane increases in acutely hypoxia-exposed animals both in vitro and in vivo (Marttila and Nikinmaa 1988; Reid and Perry 1991). This increase probably represents recycling of internalized receptors back to the cell membrane (Reid and Perry 1991). As a consequence of the increased receptor density in acute hypoxia, more cyclic AMP is accumulated, and lower concentrations of catecholamines are required to elicit an adrenergic response in hypoxic than in normoxic carp erythrocytes (Salama and Nikinmaa 1990). Third, the sodium/proton exchanger activity is greater at low than at high oxygen tensions (Motais et al. 1987; Salama and Nikinmaa 1988). All of these factors maximize adrenergic effects on red cells in acute hypoxia.

A rapid (within minutes) increase in haemoglobin oxygen affinity, coinciding with the swelling of erythrocytes in acutely hypoxia-exposed animals was observed by Soivio and Nikinmaa (1981). After discovering the effects of catecholamines on erythrocyte pH, Nikinmaa (1982, 1983) suggested that the rapid oxygen affinity increase in hypoxia would be caused by catecholamines. This was, indeed, shown to be the case: Tetens and Christensen (1987) showed that a very rapid increase in haemoglobin-oxygen affinity of hypoxia-exposed rainbow trout coincided with the elevation of plasma noradrenaline concentration, an increase in intracellular pH and a rise in cellular sodium concentration. The increases in haemoglobin-oxygen affinity, red cell pH and sodium concentration were all inhibited by treating the animals with propranolol, a β-antagonist, before hypoxia exposure. Similar results on rainbow trout were obtained by Fievet et al. (1988), Claireaux et al. (1988) and Thomas et al. (1988). In carp, the red cells of which are unresponsive to catecholamine stimulation at high pH values at atmospheric oxygen tension, a propranolol-inhibitable increase in red cell pH, sodium concentration and cellular water content was observed when the animals were subjected to hypoxia (Nikinmaa et al. 1987). The threshold value in arterial oxygen tension for the propranolol-inhibitable increases in cell volume and sodium concentration was 10 mmHg. In these acute hypoxia exposures, the rapid increase in haemoglobin-oxygen affinity was due solely to an adrenergic increase in intracellular pH, as indicated by the observations that NTP/Hb molar ratios did not decrease (Nikinmaa et al. 1987; Tetens and Christensen 1987).

In longer term hypoxia which the animals are able to tolerate, plasma catecholamine concentrations start to decrease after the peak values are reached (Thomas et al. 1991). Since the activity of the sodium/proton exchanger is related to the catecholamine dose (Tetens et al. 1988), and since the sodium/proton exchange activity even at a given dose of catecholamines is reduced at continuous exposure to β-agonists (Motais et al. 1990), it is apparent that the long-term increases in haemoglobin-oxygen affinity in hypoxic animals (Wood and Johansen 1972, 1973; Greaney and Powers 1978; Soivio et al. 1980; Tetens and Lykkeboe 1981) are not caused by the adrenergic mechanisms. The increase in haemoglobin-oxygen affinity coincides with the reduction of cellular NTP concentration; whenever both GTP and ATP are present, it is always the GTP concentration which shows the more dramatic changes (e.g. Lykkeboe and Weber 1978; Jensen and Weber 1985).

7 Summary

In teleost fish, haemoglobin-oxygen affinity can be rapidly increased via adrenergic activation of the sodium/proton exchange across the red cell membrane. The system is uniquely suited for controlling oxygen affinity in situations in which environmental oxygen availability is reduced, diffusion of the oxygen through the gill epithelium is hindered, blood oxygen transport capacity is reduced or oxygen demand by the animal is increased. Catecholamines are released into the circulation whenever the blood oxygen content is reduced, acute reduction in the oxygen tension of the red cell environment increases the apparent number of β-adrenergic receptors on the cell membrane and the activity of the exchanger as such is higher at low than at high oxygen tensions. The activity of the exchanger appears to be controlled by the conformation of haemoglobin. If haemoglobin is predominantly in R-like conformations, the exchanger is inhibited. For this reason, the intrinsic properties of haemoglobin affect the adrenergic response, and the response at atmospheric oxygen tension is restricted to species with low intrinsic haemoglobin-oxygen affinity. Although adrenergic activation of the sodium/proton exchange also causes a decrease in the cellular ATP concentration, the increase in haemoglobin-oxygen affinity is mainly due to the proton extrusion via the sodium/proton exchanger.

References

Albers C (1986) Donnan equilibrium in fish erythrocytes. In: Heisler N, Meyer M (eds) Topics in respiratory and comparative physiology; Funktionsanalyse biologischer Systeme 16: 127–131. Gustav Fischer, Stuttgart

Albers C, Goetz K-H, Hughes GM (1983) Effect of acclimation temperature on intraerythrocytic acid-base balance and nucleoside triphosphates in the carp, *Cyprinus carpio*. Respir Physiol 54: 145–159

Bachand L, Leray C (1975) Erythrocyte metabolism in the yellow perch (*Perca flavescens* Mitchill). I. Glycolytic enzymes. Comp Biochem Physiol 50 B: 567–570

Baroin A, Garcia-Romeu F, Lamarre T, Motais R (1984) A transient sodium-hydrogen exchange system induced by catecholamines in erythrocytes of rainbow trout, *Salmo gairdneri*. J Physiol (Lond) 356: 21–31

Bolis L, Luly P, Baroncelli V (1971) D(+)-glucose permeability in brown trout *Salmo trutta* L. erythrocytes. J Fish Biol 3: 273–275

Borgese F, Garcia-Romeu F, Motais R (1986) Catecholamine-induced transport systems in trout erythrocyte. Na^+/H^+ countertransport or NaCl cotransport? J Gen Physiol 87: 551–566

Borgese F, Garcia-Romeu F, Motais R (1987) Ion movements and volume changes induced by catecholamines in erythrocytes of rainbow trout: effect of pH. J Physiol (Lond) 382: 145–157

Bourne PK, Cossins AR (1982) On the instability of K^+ influx in erythrocytes of the rainbow trout *S. gairdneri*, and the role of catecholamine hormones in maintaining in vivo influx. J Exp Biol 101: 93–104

Boutilier RG, Ferguson RA (1989) Nucleated red cell function: metabolism and pH regulation. Can J Zool 67: 2986–2993

Boutilier RG, Iwama GK, Randall DJ (1986) The promotion of catecholamine release in rainbow trout, *Salmo gairdneri*, by acute acidosis: interactions between red cell pH and haemoglobin oxygen-carrying capacity. J Exp Biol 128: 145–157

Brittain T (1987) The Root effect. Comp Biochem Physiol 86B: 473–481

Butler PJ, Metcalfe JD, Ginley SA (1986) Plasma catecholamines in the lesser spotted dogfish and in rainbow trout at rest and during different levels of exercise. J Exp Biol 123: 409–421

Claireaux G, Thomas S, Fievet B, Motais R (1988) Adaptive respiratory responses of trout to hypoxia. II. Blood oxygen carrying properties during hypoxia. Respir Physiol 74: 91–98

Cossins AR, Kilbey RV (1989) The seasonal modulation of Na^+/H^+ exchanger activity in trout erythrocytes. J Exp Biol 144: 463–478

Cossins AR, Kilbey RV (1990) The temperature dependence of the adrenergic Na^+/H^+ exchanger of trout erythrocytes. J Exp Biol 148: 303–312

Cossins AR, Kilbey RV (1991) Adrenergic responses and the Root effect in erythrocytes of freshwater fish. J Fish Biol 38: 421–429

Cossins AR, Richardson PA (1985) Adrenalin-induced Na^+/H^+ exchange in trout erythrocytes and its effects upon oxygen- carrying capacity. J Exp Biol 118: 229–246

Dalmark M (1975) Chloride and water distribution in human red cells. J Physiol (Lond) 250: 65–84

Dawson RMC, Elliott DC, Elliott WH, Jones KM (1987) Data for biochemical research, 3rd edn. Clarendon Press, Oxford, 580 pp

Douglas EL, Friedl WA, Pickwell GV (1976) Fishes in oxygen-minimum zones: blood oxygenation characteristics. Science 191: 957–959

Duhm J (1972) The effect of 2,3-DPG and other organic phosphates on the Donnan equilibrium and the oxygen affinity of human blood. In: Rorth M, Astrup P (eds) Oxygen affinity of hemoglobin and red cell acid-base status. Alfred Benzon Symp IV, Munksgaard, Copenhagen, pp 583–594

Eddy FB (1977) Oxygen uptake by rainbow trout blood, *Salmo gairdneri*. J Fish Biol 10: 87–90

Edsall JT (1969) Carbon dioxide, carbonic acid, and bicarbonate ion: physical properties and kinetics of interconversion. In: Forster RE, Edsall JT, Otis AB, Roughton FJW (eds) Co_2: chemical, biochemical and physiological aspects. NASA, Washington, DC, pp 15–27

Ferguson RA, Boutilier RG (1988) Metabolic energy production during adrenergic pH regulation in red cells of the Atlantic salmon, *Salmo salar*. Respir Physiol 74: 65–76
Ferguson RA, Boutilier RG (1989) Metabolic-membrane coupling in red blood cells of trout: the effects of anoxia and adrenergic stimulation. J Exp Biol 143: 149–164
Ferguson RA, Storey KB (1991) Glycolytic and associated enzymes of rainbow trout (*Oncorhynchus mykiss*) red cells: in vitro and in vivo studies. J Exp Biol 155:469–485
Ferguson RA, Tufts BL, Boutilier RG (1989) Energy metabolism in trout red cells: consequences of adrenergic stimulation in vivo and in vitro. J Exp Biol 143: 133–147
Fievet B, Motais R, Thomas S (1987) Role of adrenergic-dependent H^+ release from red cells in acidosis induced by hypoxia in trout. Am J Physiol 252: R269–R275
Fievet B, Claireaux G, Thomas S, Motais R (1988) Adaptive respiratory responses of trout to acute hypoxia. III. Ion movements and pH changes in the red blood cell. Respir Physiol 74: 99–114
Fuchs DA, Albers C (1988) Effect of adrenaline and blood gas conditions on red cell volume and intraerythrocytic electrolytes in the carp, *Cyprinus carpio*. J Exp Biol 137: 457–477
Funder J, Wieth JO (1966) Chloride and hydrogen ion distribution between human red cells and plasma. Acta Physiol Scand 68: 234–245
Garcia-Romeu F, Motais R, Borgese F (1988) Desensitization by external Na of the cyclic AMP-dependent Na^+/H^+ antiporter in trout red blood cells. J Gen Physiol 91: 529–548
Garey GF, Rahn H (1970) Gas tensions in tissues of trout and carp exposed to diurnal changes in oxygen tension of water. J Exp Biol 52: 575–582
Geoghegan WD, Poluhowich JJ (1974) The major erythrocytic organic phosphates of the American eel, *Anguilla rostrata*. Comp Biochem Physiol 49B: 281–290
German B, Wyman J (1937) The titration curves of oxygenated and reduced hemoglobin. J Biol Chem 117: 533–550
Greaney GS, Powers DA (1978) Allosteric modifiers of fish hemoglobins: in vitro and in vivo studies of the effect of ambient oxygen and pH on erythrocyte ATP concentrations. J Exp Zool 203: 339–350
Heming TA (1984) The role of fish erythrocytes in transport and excretion of carbon dioxide. PhD Thesis, Univ British Columbia, Vancouver, 177 pp
Hladky SB, Rink TJ (1977) pH equilibrium across the red cell membrane. In: Ellory JC, Lew VL (eds) Membrane transport in red cells. Academic Press, London, pp 115–135
Holeton GF, Neumann P, Heisler N (1983) Branchial ion exchange and acid-base regulation after strenuous exercise in rainbow trout (*Salmo gairdneri*). Respir Physiol 51: 303–318
Houston AH, McCullough CAM, Keen J, Maddalena C, Edwards J (1985) Rainbow trout red cells in vitro. Comp Biochem Physiol 81A: 555–565
Hughes GM, Koyama T (1975) Gas exchange of single red blood cells within secondary lamellae of fish gills. J Physiol 246: P82–P83
Hunter AS, Hunter FR (1957) A comparative study of erythrocyte metabolism. J Cell Comp Physiol 49: 479–502
Hyde DA, Perry SF (1990) Absence of red cell pH and oxygen content regulation in American eel (*Anguilla rostrata*) during hypercapnic acidosis in vivo and in vitro. J Comp Physiol 159 B: 687–693
Ingermann RL, Bissonnette JM, Hall RE (1985) Sugar uptake by red blood cells. In: Gilles R (ed) Circulation, respiration, and metabolism. Current comparative approaches. Springer, Berlin Heidelberg New York, pp 290–300
Jacobs MH, Stewart DR (1942) The role of carbonic anhydrase in certain ionic exchanges involving the erythrocyte. J Gen Physiol 25: 539–552
Jensen FB (1986) Pronounced influence of Hb-O_2 saturation on red cell pH in tench blood in vivo and in vitro. J Exp Zool 238: 119–124
Jensen FB (1987) Influences of exercise stress and adrenaline upon intra- and extracellular acid-base status, electrolyte composition and respiratory properties of blood in tench (*Tinca tinca*) at different seasons. J Comp Physiol 157 B: 51–60
Jensen FB (1989) Hydrogen ion equilibria in fish haemoglobins. J Exp Biol 143: 225–234
Jensen FB, Weber RE (1985) Kinetics of the acclimational responses of tench to combined hypoxia and hypercapnia. I. Respiratory responses. J Comp Physiol 156B: 197–203

Jensen FB, Nikinmaa M, Weber RE (1983) Effects of exercise stress on acid-base balance and respiratory function in blood of the teleost *Tinca tinca*. Respir Physiol 51: 291–301

Jones JD (1961) Aspects of respiration in *Planorbis corneus* L. and *Lymnaea stagnalis* L. (Gastropoda: Pulmonata). Comp Biochem Physiol 4: 1–29

Kim HD (1985) ATP metabolism in mammalian red blood cells. In: Gilles R (ed) Circulation, respiration and metabolism: current comparative approaches. Springer, Berlin Heidelberg New York, pp 312–321

Kita J, Itazawa Y (1989) Release of erythrocytes from the spleen during exercise and splenic constriction by adrenaline infusion in the rainbow trout. Jpn J Ichthyol 36: 48–52

Lykkeboe G, Weber RE (1978) Changes in the respiratory properties of the blood in the carp, *Cyprinus carpio*, induced by diurnal variation in ambient oxygen tension. J Comp Physiol 128: 117–125

Mahé Y, Garcia-Romeu F, Motais R (1985) Inhibition by amiloride of both adenylate cyclase activity and the Na^+/H^+ antiporter in fish erythrocytes. Eur J Pharmacol 116: 199–206

Marttila ONT, Nikinmaa M (1988) Binding of β-adrenergic antagonists ^{3}H-DNA and ^{3}H-CGP 12177 to intact rainbow trout (*Salmo gairdneri*) and carp (*Cyprinus carpio*) red blood cells. Gen Comp Endocrinol 70: 429–435

Mauro NA, Isaacks RE (1989) Relative oxidation of glutamate and glucose by vertebrate erythrocytes. Comp Biochem Physiol 94A: 95–97

Milligan CL, Wood CM (1986) Intracellular and extracellular acid-base status and H^+ exchange with the environment after exhaustive exercise in the rainbow trout. J Exp Biol 123: 93–121

Milligan CL, Wood CM (1987) Regulation of blood oxygen transport and red cell pHi after exhaustive activity in rainbow trout (*Salmo gairdneri*) and starry flounder (*Platichthys stellatus*). J Exp Biol 133: 263–282

Milligan CL, Graham MS, Farrell AP (1989) The response of trout red cells to adrenaline during seasonal acclimation and changes in temperature. J Fish Biol 35: 229–236

Moon TW, Brill RW, Hochachka PW, Weber J-M (1987) L-(+)-lactate translocation into the red blood cells of the skipjack tuna (*Katsuwonus pelamis*). Can J Zool 65: 2570–2573

Motais R, Garcia-Romeu F, Borgese F (1987) The control of Na^+/H^+ exchange by molecular oxygen in trout erythrocytes. A possible role of hemoglobin as a transducer. J Gen Physiol 90: 197–207

Motais R, Fievet B, Carcia-Romeu F, Thomas S (1989) Na^+/H^+ exchange and pH regulation in red blood cells: role of uncatalyzed H_2CO_3 dehydration. Am J Physiol 256: C728–C735

Motais R, Scheuring U, Borgese F, Garcia-Romeu F (1990) Characteristics of β-adrenergic-activated Na-proton transport in red blood cells. Prog Cell Res 1: 179–193

Nikinmaa M (1982) Effects of adrenaline on red cell volume and concentration gradient of protons across the red cell membrane in the rainbow trout, *Salmo gairdneri*. Mol Physiol 2: 287–297

Nikinmaa M (1983) Adrenergic regulation of haemoglobin oxygen affinity in rainbow trout red cells. J Comp Physiol 152B: 67–72

Nikinmaa M (1990) Vertebrate red blood cells. Springer, Berlin Heidelberg New York, 262 pp

Nikinmaa M (1992) Membrane transport and the control of haemoglobin-oxygen affinity in nucleated erythrocytes. Physiol Rev 72: 301–321

Nikinmaa M, Huestis WH (1984) Adrenergic swelling in nucleated erythrocytes: cellular mechanisms in a bird, domestic goose, and two teleosts, striped bass and rainbow trout. J Exp Biol 113: 215–224

Nikinmaa M, Jensen FB (1986) Blood oxygen transport and acid-base status of stressed trout (*Salmo gairdnerii*): pre- and postbranchial values in winter fish. Comp Biochem Physiol 84A: 391–396

Nikinmaa M, Jensen FB (1992) Inhibition of the adrenergic sodium/proton exchange activity in rainbow trout red cells by nitrite-induced methaemoglobinaemia. 162: 424–429

Nikinmaa M, Cech JJ Jr, McEnroe M (1984) Blood oxygen transport in stressed striped bass (*Morone saxatilis*): role of beta-adrenergic responses. J Comp Physiol 154B: 365–369

Nikinmaa M, Cech JJ Jr, Ryhänen E-L, Salama A (1987) Red cell function of carp (*Cyprinus carpio*) in acute hypoxia. Exp Biol 47: 53–58

Nikinmaa M, Salama A, Tuurala H (1990a) Respiratory effects of environmental acidification in perch (*Perca fluviatilis*) and rainbow trout (*Salmo gairdneri*). In: Kauppi P, Anttila P, Kenttämies K (eds) Acidification in Finland. Springer, Berlin Heidelberg New York, pp 929–940

Nikinmaa M, Tiihonen K, Paajaste M (1990b) Adrenergic control of red cell pH in salmonid fish: roles of the sodium/proton exchange, Jacobs-Stewart cycle and membrane potential. J Exp Biol 154: 257–271

Parks RE Jr, Brown PR, Cheng Y-C, Agarwal KC, Kong CM, Agarwal RP, Parks CC (1973) Purine metabolism in primitive erythrocytes. Comp Biochem Physiol 45B: 355–364

Pearson MP, Stevens ED (1991) Size and hematological impact of the splenic erythrocyte reservoir in rainbow trout, *Oncorhynchus mykiss*. Fish Physiol Biochem 9: 39–50

Perry SF, Kinkead R, Gallaugher P, Randall DJ (1989) Evidence that hypoxemia promotes catecholamine release during hypercapnic acidosis in rainbow trout (*Salmo gairdneri*). Respir Physiol 77: 351–364

Perry SF, Fritsche R, Kinkead R, Nilsson S (1991a) Control of catecholamine release in vivo and in situ in the Atlantic cod (*Gadus morhua*) during hypoxia. J Exp Biol 155: 549–566

Perry SF, Wood CM, Thomas S, Walsh PJ (1991b) Adrenergic inhibition of carbon dioxide excretion by trout red blood cells in vitro is mediated by activation of Na^+/H^+ exchange. J Exp Biol 157: 367–380

Primmett DRN, Randall DJ, Mazeaud M, Boutilier RG (1986) The role of catecholamines in erythrocyte pH regulation and oxygen transport in rainbow trout (*Salmo gairdneri*) during exercise. J Exp Biol 122: 139–148

Randall DJ, Mense D, Boutilier RG (1987) The effects of burst swimming on aerobic swimming in chinook salmon (*Oncorhynchus tshawytscha*). Mar Behav Physiol 13: 77–88

Rapoport SM (1985) Mechanisms of the maturation of the reticulocyte. In: Gilles R (ed) Circulation, respiration and metabolism: current comparative approaches. Springer, Berlin Heidelberg New York, pp 333–342

Reid SD, Perry SF (1991) The effects and physiological consequences of raised levels of cortisol on rainbow trout (*Oncorhynchus mykiss*) erythrocyte ß-adrenoreceptors. J Exp Biol 158: 217–240

Ristori MT, Laurent P (1985) Plasma catecholamines and glucose during moderate exercise in the trout: comparisons with bursts of violent activity. Exp Biol 44: 247–253

Ristori MT, Laurent P (1989) Plasma catecholamines in rainbow trout (*Salmo gairdneri*) during hypoxia. Exp Biol 48: 285–290

Romano L, Passow H (1984) Characterization of anion transport system in trout red blood cell. Am J Physiol 246: C330–C338

Roughton FJW (1964) Transport of oxygen and carbon dioxide. In: Fenn WO, Rahn H (eds) Handbook of physiology, respiration, vol I. Am Physiol Soc, Washington DC, pp 767–825

Salama A, Nikinmaa M (1988) The adrenergic responses of carp (*Cyprinus carpio*) red cells: effects of Po_2 and pH. J Exp Biol 136: 405–416

Salama A, Nikinmaa M (1989) Species differences in the adrenergic responses of fish red cells: studies on whitefish, pikeperch, trout and carp. Fish Physiol Biochem 6: 167–173

Salama A, Nikinmaa M (1990) Effect of oxygen tension on catecholamine-induced formation of cAMP and on swelling of carp red blood cells. Am J Physiol 259: 723–C726

Schweiger HG (1962) Pathways of metabolism in nucleate and anucleate erythrocytes. Int Rev Cytol 13: 135–201

Sephton DH, MacPhee WL, Driedzic WR (1991) Metabolic enzyme activities, oxygen consumption and glucose utilization in sea raven (*Hemitripterus americanus*) erythrocytes. J Exp Biol 159: 407–418

Soivio A, Nikinmaa M (1981) The swelling of erythrocytes in relation to the oxygen affinity of the blood of the rainbow trout, *Salmo gairdneri* Richardson. In: Pickering AD (ed) Stress and fish. Academic Press, London, pp 103–119

Soivio A, Nikinmaa M, Westman K (1980) The blood oxygen binding properties of hypoxic *Salmo gairdneri*. J Comp Physiol 136B: 83–87

Soivio A, Nikinmaa M, Nyholm K, Westman K (1981) The role of gills in the responses of *Salmo gairdneri* during moderate hypoxia. Comp Biochem Physiol 70A: 133–139

Tang Y, Boutilier RG (1988) Correlation between catecholamine release and degree of acidotic stress in trout. Am J Physiol 255: R395–R399

Tetens V (1987) Regulation of blood O_2 affinity during acute hypoxic exposure of rainbow trout, *Salmo gairdneri*: organismal and cellular processes. PhD Thesis, Aarhus Univ, Denmark, 12 pp

Tetens V, Christensen NJ (1987) ß-adrenergic control of blood oxygen affinity in acutely hypoxia exposed rainbow trout. J Comp Physiol 157B: 667–675

Tetens V, Lykkeboe G (1981) Blood respiratory properties of rainbow trout *Salmo gairdneri*: responses to hypoxia acclimation and anoxic incubation of blood in vitro. J Comp Physiol 145B: 117–125

Tetens V, Lykkeboe G, Christensen NJ (1988) Potency of adrenaline and noradrenaline for β-nadrenergic proton extrusion from red cells of rainbow trout, *Salmo gairdneri*. J Exp Biol 134: 267–280

Thomas S, Fievet B, Claireaux G, Motais R (1988) Adaptive respiratory responses of trout to acute hypoxia. I. Effects of water ionic composition on blood acid-base status response and gill morphology. Respir Physiol 74: 77–90

Thomas S, Kinkead R, Walsh PJ, Wood CM, Perry SF (1991) Desensitization of adrenaline-induced red blood cell H^+ extrusion in vitro after chronic exposure of rainbow trout to moderate environmental hypoxia. J Exp Biol 156: 233–248

Tiihonen K, Nikinmaa M (1991a) D-glucose permeability in river lamprey (*Lampetra fluviatilis*) and carp (*Cyprinus carpio*) erythrocytes. Comp Biochem Physiol 100 A: 581–584

Tiihonen K, Nikinmaa M (1991b) Substrate utilization by carp (*Cyprinus carpio*) erythrocytes. J Exp Biol 161: 509–514

Tiihonen K, Nikinmaa M (1993) Membrane permeability and utilization of L-lactate and pyruvate in carp red blood cells. J Exp Biol 178: 161–172

Tse C-M, Young JD (1990) Glucose transport in fish erythrocytes: variable cytochalasin-B-sensitive hexose transport activity in the common eel (*Anguilla japonica*) and transport deficiency in the paddyfield eel (*Monopterus albus*) and rainbow trout (*Salmo gairdneri*). J Exp Biol 148: 367–383

Tufts BL, Boutilier RG (1991) Interactions between ion exchange and metabolism in erythrocytes of the rainbow trout *Oncorhynchus mykiss*. J Exp Biol 156: 139–151

Tufts BL, Tang Y, Tufts K, Boutilier RG (1991) Exhaustive exercise in 'wild' Atlantic salmon (*Salmo salar*): acid-base regulation and blood gas transport. Can J Fish Aquat Sci 48: 868–874

Van Den Thillart G, Randall D, Hoa-Ren L (1983) Co_2 and H^+ excretion by swimming coho salmon, *Oncorhynchus kisutch*. J Exp Biol 107: 169–180

Vermette MG, Perry SF (1988) Adrenergic involvement in blood oxygen transport and acid-base balance during hypercapnic acidosis in the rainbow trout, *Salmo gairdneri*. J Comp Physiol B 158: 107–115

Walsh PJ, Wood CM, Thomas S, Perry SF (1990) Characterization of red blood cell metabolism in rainbow trout. J Exp Biol 154: 475–489

Wang T, Nielsen OB, Lykkeboe G (1994) The relative contributions of red and white blood cells to whole-blood energy turnover in trout. J Exp Biol 190: 43–54

Weber RE (1982) Intraspecific adaptation of hemoglobin function in fish to oxygen availability. In: Addink ADF, Spronk N (eds) Exogenous and endogenous influences on metabolic and neural control. Pergamon, Oxford, pp 87–102

Weber RE, Jensen FB (1988) Functional adaptations in hemoglobins from ectothermic vertebrates. Annu Rev Physiol 50: 161–179

Weber RE, Wells RMG (1989) Hemoglobin structure and function. In: Wood SC (ed) Comparative pulmonary physiology. Current concepts. Dekker, New York, pp 279–310

Weber RE, Lykkeboe G, Johansen K (1976) Physiological properties of eel haemoglobin: hypoxic acclimation, phosphate effects and multiplicity. J Exp Biol 64: 75–88

Wells RMG, Weber RE (1990) The spleen in hypoxic and exercised rainbow trout. J Exp Biol 150: 461–466

Wood CM, Perry SF (1991) A new in vitro assay for carbon dioxide excretion by trout red blood cells: effects of catecholamines. J Exp Biol 157: 349–366

Wood CM, Walsh PJ, Thomas S, Perry SF (1990) Control of red blood cell metabolism in rainbow trout after exhaustive exercise. J Exp Biol 154: 491–507
Wood SC, Johansen K (1972) Adaptation to hypoxia by increased HbO_2 affinity and decreased red cell ATP concentration. Nature New Biol 237: 278–279
Wood SC, Johansen K (1973) Organic phosphate metabolism in nucleated red cells: influence of hypoxia on eel HbO_2 affinity. Neth J Sea Res 7: 328–338
Yamamoto K, Itazawa Y (1989) Erythrocyte supply from the spleen of exercised carp. Comp Biochem Physiol 92A: 139–144
Yamamoto K, Itazawa Y, Kobayashi H (1980) Supply of erythrocytes into the circulating blood from the spleen of exercised fish. Comp Biochem Physiol 65A: 5–11

Central Vascular Circulation

Chapter 6

Mechanisms of Intracardiac Shunting in Reptiles: Pressure vs Washout Shunting

J.W. Hicks[1] and *G.M. Malvin*[2]

Contents

1 Introduction

In adult mammals and birds, the right and left sides of the heart are normally separate. Intracardiac shunting, which is usually detrimental, results only from congenital abnormalities of the heart and great vessels. In contrast, intracardiac shunting normally occurs in noncrocodilian reptiles. The complex and unique anatomy of these animals can allow both the systemic venous blood to bypass the lungs as well as the pulmonary venous blood to bypass the systemic circulation. It is not uncommon for these bypasses to comprise most of the cardiac output. The success of this vertebrate class indicates that intracardiac shunting is not detrimental to these animals, and may, in fact, provide certain survival advantages (Burggren 1985).

The purpose of the chapter is to review briefly the cardiac anatomy of noncrocodilian reptiles and to discus two principal hypotheses that describe the mechanism of intracardiac shunting. Previous experimental evidence supporting these hypotheses along with results from current studies will be discussed.

[1] Department of Ecology and Evolutionary Biology, University of California, Irvine CA 92717, USA. Present address: School of Biological Sciences, University of California, Irvine CA, USA
[2] Oxygen Transport Program, Lovelace Medical Foundation, Albuquerque, NM, USA

Finally, a possible intracardiac mechanism for regulation of shunting will be considered.

2 Reptilian Cardiovascular Anatomy

The reptilian cardiovascular system has intrigued biologists for over 100 years (see Griel 1903 for review of 19th century work, Goodrich 1916, 1919; O'Donoghue 1918; Rau 1924; Mathur 1946; White 1959, 1968; Webb et al 1974; Van Mierop and Kutsche 1985). The reptilian heart, unlike the mammalian and avian heart, shows considerable variation in its morphology (see references above). Despite this anatomical variability, a basic reptilian cardiac plan emerges.

The reptilian heart consists of two atrial chambers, separated by a complete septum and a single ventricle that is subdivided into three anatomically interconnected chambers or cava (Fig. 1). A distinctive feature of the ventricular anatomy is the presence of a septum-like structure referred to as the muscular ridge or *Muskelleiste* (Brücke 1852). The muscular ridge originates from the ventral ventricular wall, running from apex to base, and divides the ventricle into a smaller cavum pulmonale (CP) and larger cavum dorsale (Van Mierop and Kutsche 1981, 1985). The dorsolateral border of the muscular ridge is free, allowing potential communication between the CP and cavum dorsale. Opposite the free border of the muscular ridge is another ridge-like structure called the *Bulbuslamelle* by German authors (Van Mierop and Kutsche 1981). In fixed hearts, the free edge of the muscular ridge abuts perfectly with the Bulbuslamelle and completely separates the CP from the cavum dorsale (Van Mierop and Kutsche 1985). The development of the muscular ridge is variable among different species, being the least developed in turtles (Van Mierop and Kutsche 1981) and the most developed in varanid lizards. An example of this variation is demonstrated by comparing the muscular ridge in the ventricle of the turtle

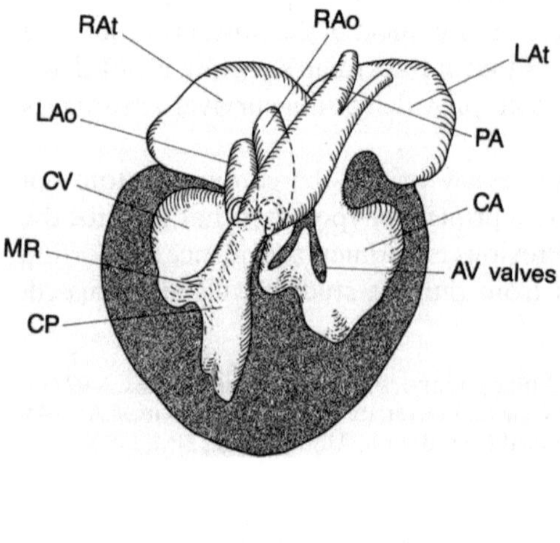

Fig. 1. Representation of a reptilian heart. *RAt* Right atrium; *LAt* left atrium; *RAo* right aortic arch; *LAo* left aortic arch; *PA* pulmonary artery; *AV* atrioventricular; *CV* cavum venosum; *CA* cavum arteriosum; *CP* cavum pulmonale. (Hicks 1993)

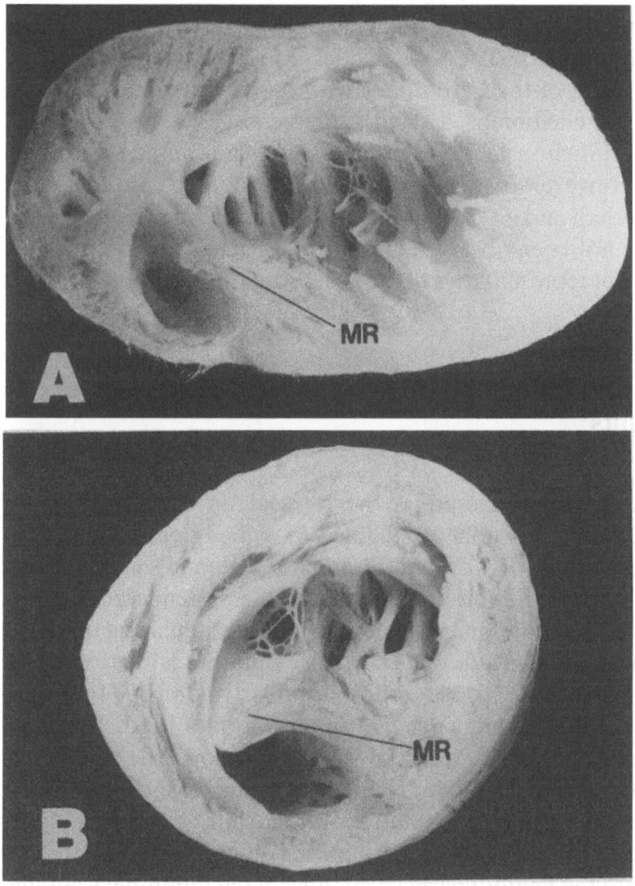

Fig. 2A, B. Comparison of muscular ridge development in the reptilian heart. **A** *Pseudemys scripta;* **B** *Varanus niloticus. MR* Muscular ridge. (Photograph courtesy of A. Ishimatsu)

Pseudemys scripta and the varanid lizard, *Varanus niloticus* (Fig. 2). Associated with the muscular ridge is a small nodule of hyaline cartilage situated between the bases of the pulmonary artery and left aortic arch. This anatomical structure, the *Bojanusscher Knorpel* (Benninghoff 1933) has received little attention. White (1959) suggested the cartilage might aid in giving rigidity to the muscular ridge during systole.

A second incomplete vertical septum, which originates from the dorsal aspects of the muscular ridge to the dorsal wall of the cavum dorsale, further subdivides the cavum dorsale into the cavum arteriosum (CA) and the cavum venosum (CV). However, the presence of this vertical septum is not a consistent finding among investigators (Van Mierop and Kutsche 1985).

In all reptiles, three great vessels arise from the ventricle: the pulmonary artery (PA), the right aortic arch (RAo) and left aortic arch (LAo). The pulmonary artery emerges to the left of the two aortic arches and originates from the CP. The RAo and LAo arise from the CV. The RAo divides into subclavian and carotid arteries, and a third branch of the RAo unites with the LAo mediocaudally to form the dorsal aorta. In chelonians, a third systemic arch has been reported emerging from the CV (Shelton and Burggren 1976; Johansen and Burggren 1980; Burggren 1985). However, in most studies of the chelonian cardiovascular system (Mathur 1964; Khali and Zaki 1966; Van Mierop and Kutsche 1985), this third systemic arch has not been observed. This third systemic arch may depict a rare condition that is not representative of the general anatomical design (White et al. 1989).

3 Intracardiac Shunts

These anatomical features of the reptilian heart result in the potential for intracardiac shunting. Cardiac shunts are defined as, either right-to-left shunt (R-L) or left-to-right (L-R). A R-L shunt represents systemic venous blood bypassing the pulmonary circulation and reentering the systemic circulation. A L-R shunt is defined as pulmonary venous blood recirculating into the pulmonary circulation. The magnitude of intracardiac shunting can be expressed as the ratio of pulmonary to systemic blood flow (Qp/Qs) (Rudolf 1974). A ratio less than 1 indicates a net R-L shunt, while a ratio greater than 1 indicates a net L-R shunt. A R-L shunt decreases O_2 transport by the addition of deoxygenated blood to the arterial circulation and can result in systemic blood flows being greater than pulmonary blood flows. In contrast, a L-R shunt has minor effects on O_2 transport but may influence CO_2 elimination (White 1985). In addition, a L-R shunt will increase the hemodynamic demands on the pulmonary circulation.

3.1 Effects of Lung Ventilation on Intracardiac Shunting

A characteristic feature of chelonians and squamates is an intermittent ventilatory pattern, where brief periods of ventilation are interspersed among breath holds of variable duration (Shelton et al. 1986). Millen et al. (1964) used an indicator dilution technique and reported the first correlation of shunt direction with ventilatory state in reptiles. In the turtles, *Pseudemys scripta* and *Pseudemys floridana,* a net L-R shunt was apparent when the animals were breathing ambient air. In contrast, during inhalation of N_2 or prolonged diving, a relatively large R-L shunt developed and the L-R shunt was diminished.

In reptiles, quantitative assessment of the flow ratios between the pulmonary and systemic circulation during ventilation and apneas is lacking. This may be due to (1) the anatomical arrangement of the great vessels and (2) the required number of flow probes to simultaneously determine systemic and pulmonary

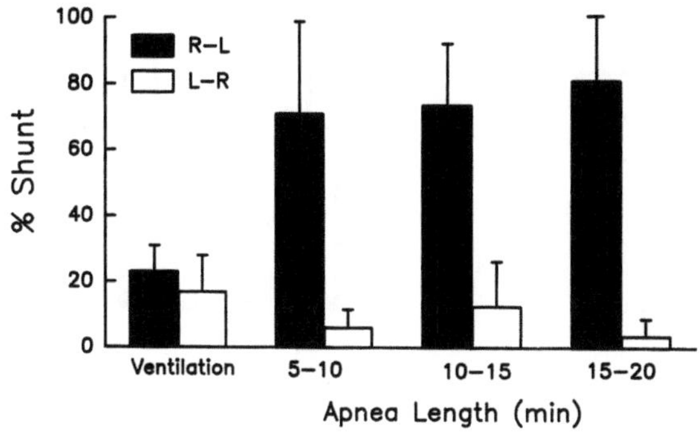

Fig. 3. Change in R-L and L-R shunts determined from central vascular blood oxygen measurements in the freely diving turtle, *Pseudemys scripta* at 15 °C. (Data from White et al. 1989)

blood flows within single animals. In the turtle, *Pseudemys scripta,* blood flow measurements from selected central vascular sites have suggested that a net R-L shunt develops during apnea and a net L-R shunt occurs during ventilation (White and Ross 1966; Shelton and Burggren 1976). Shelton and Burggren (1979) measured the flows between the outflow vessels and determined that total systemic flow was estimated by multiplying the flow in the right branch of the right aortic arch by a factor of 2.75. In addition, the total pulmonary blood flow was estimated by doubling the flow through the left pulmonary artery. Using these factors and the blood flows reported, the Qp/Qs is approximately 0.75 during apnea and increases to 1.5 during ventilation.

In the turtle, analysis of central vascular blood gases revealed that both the right aortic arch (RAo) and left aortic arch (LAo) PO_2 were reduced relative to the pulmonary venous levels (Burggren and Shelton 1979). These changes in blood gases indicated the development of a R-L shunt. A recent study confirmed the intensification of R-L shunting during apnea in the turtle *Pseudemys scripta* (White et al 1989). During ventilation, analysis of central vascular O_2 levels indicated bidirectional shunting, with both R-L and L-R shunts being approximately equal (Fig. 3). As apnea progressed, the R-L shunt increased (Fig. 3). An increase in L-R shunting during ventilation and its subsequent reduction during apnea could not be confirmed with oxygen measurements (Fig. 3). Recent analyses have suggested that measurements of central vascular blood oxygen levels alone may not accurately reveal the absolute level of pulmonary bypass in reptiles (Ishimatsu et al. 1988; Hicks 1993).

4 Mechanism of Intracardiac Shunting

The mechanism of intracardiac shunting in reptiles is controversial (Burggren 1985; Heisler and Glass 1985; Hicks and Wood 1988). There are two principal hypotheses addressing the mechanism(s) of intracardiac shunts, termed *pressure shunting* and *washout shunting* (Heisler and Glass 1985). In terms of both R-L and

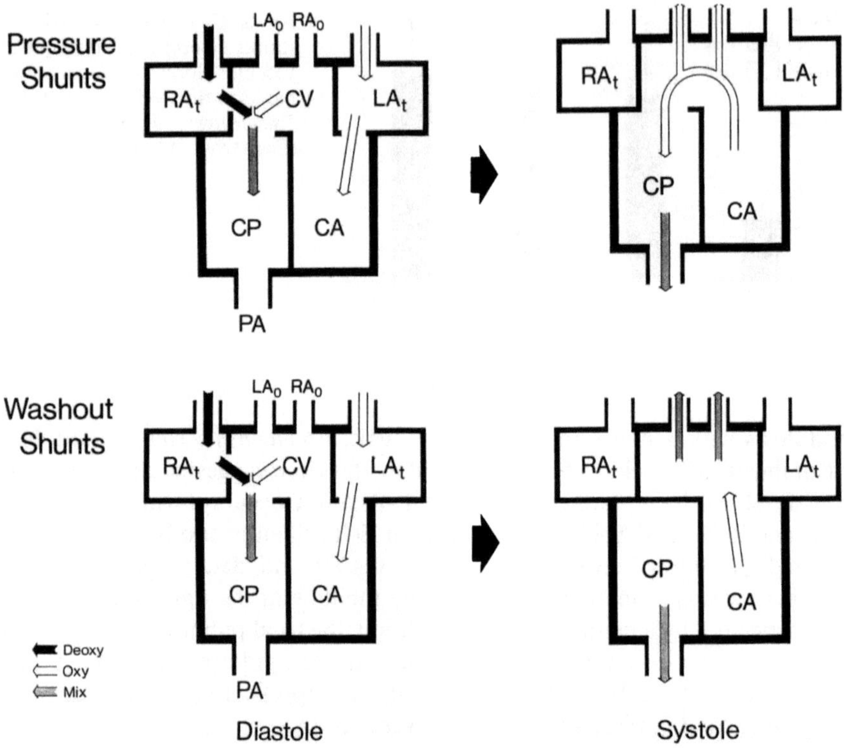

Fig. 4. Basic features of the pressure and washout hypotheses for R-L intracardiac shunting

L-R shunting, the critical difference between these two hypotheses is the ability of the muscular ridge to separate the CP from the CV and CA during systole (Figs. 4 and 5). Both mechanisms have appealing features in terms of regulation. Pressure shunting is regulated by alterations in pulmonary and systemic vascular resistances. Washout shunting is regulated by factors that modify the end-systolic and end-diastolic volumes of the ventricular cava.

4.1 Previous Evidence for Pressure Shunting

Experimental support for the pressure shunting hypothesis has consisted of measurments of ventricular and central vascular pressures. Steggerda and Essex (1957) measured central vascular pressures in the turtle, *Chelydra serpentina.* These measurements revealed no difference in the systolic pressures between the CP and cavum dorsale. In addition, the systolic pressures in the aortas and pulmonary arteries were similar, with only diastolic pressure in the pulmonary artery being significantly lower. It was concluded that the pressure profiles in the turtle heart resembled those found in humans with ventricular septal defects. Johansen (1959) studied the central vascular pressure profiles in the snake,

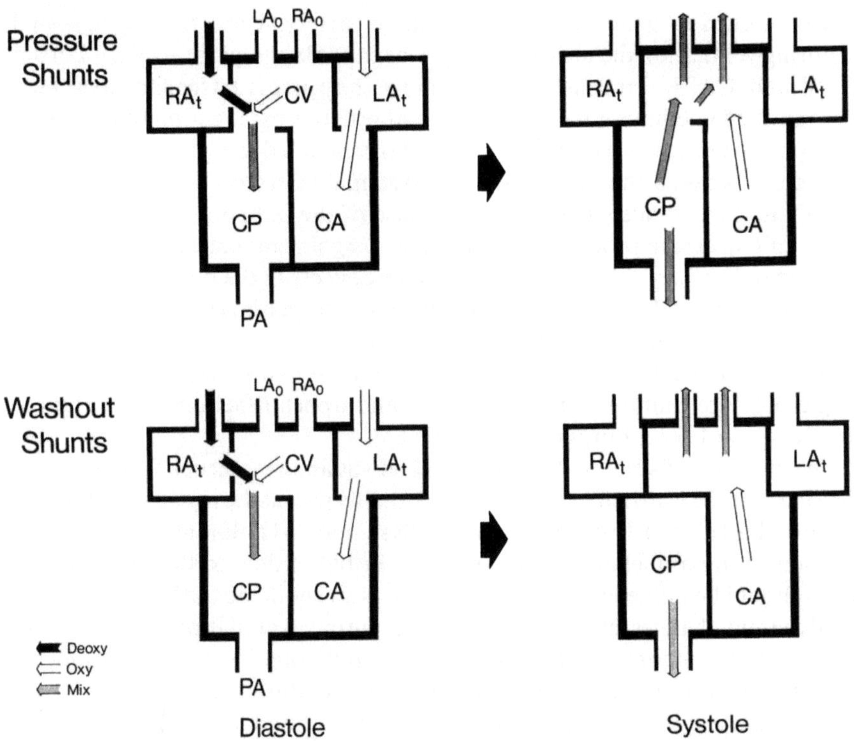

Fig. 5. Basic features of the pressure and washout hypotheses for L-R intracardiac shunting

Tripodonotus natrix. The results were similar to those reported in the turtle, *Chelydra serpentina.* The systolic pressures in the systemic and pulmonary arteries were equal, while the diastolic pressure in the pulmonary artery was lower. It was also noted that the pressure rise in the pulmonary artery occurred 40 to 50 ms before the pressure rise in the aorta. This suggested that during ventricular systole, the ejection of blood must start through the pulmonary trunk, due to its lower resistance and that a L-R shunt was favored (Johansen 1959). As in the study on the turtle, the pressure profiles of the snake were said to be similar to a ventricular septal defect in mammals (Johansen 1959). The comparison to a ventricular septal defect was perhaps unfortunate. However, the initial studies on the reptilian heart advanced the concept that the muscular ridge was incapable of separating the pulmonary and systemic circulations and that the ventricular cava were functionally patent during all phases of the cardiac cycle.

White and Ross (1966), who investigated circulatory adaptations in the freely diving turtle, *Pseudemys scripta,* reported that during ventilation a systolic pressure gradient existed between pulmonary and systemic arteries of 2–6 mmHg. In contrast, during diving, the systolic pressures in the pulmonary and systemic arteries were equal, although the diastolic pressure of the pulmonary artery continued to be lower than the systemic artery. These results suggested that during ventilation the muscular ridge was capable of separating the CP from the CV and CA. In contrast, during apnea the three ventricular cava were interconnected and functionally patent. An alternative explanation to the above findings has been advanced. Burggren (1977a) concluded that the systolic pressure gradient between the pulmonary and systemic artery resulted from a small band of muscular tissue surrounding the base of the pulmonary artery (*bulbus cordis*). This structure produces a transient but significant increase in resistance across the pulmonary valve. The hemodynamic effects of the bulbus cordis have been demonstrated both in vitro and in vivo (Burggren 1977a). Subsequent studies involving simultaneous pressure measurements in all three cava or outflow tracts indicated that the hearts of chelonians and squamates function as a single pressure pump during ventilation and apnea (Shelton and Burggren 1976; Burggren 1977b; Burggren et al. 1977).

The majority of studies in chelonians and squamates (except *Varanus*) have suggested that the ventricular cava are functionally patent during all phases of the cardiac cycle (Table 1). In such a circuit the systemic and pulmonary circulations are arranged in parallel and the distribution of blood flow to the two vascular circuits should be influenced by the relative resistances in the outflow tracts. The hemodynamic changes associated with intermittent ventilation in turtles are consistent with this concept. During ventilation, pulmonary vascular resistance is 25% lower than the systemic resistance. In contrast, during apnea, the systemic arterial resistance increases only 10–15%, while pulmonary arterial resistance can increase 150% (Shelton and Burggren 1976). The asymmetrical changes in vascular resistance during apnea result in the systemic circuit becoming the "*favored*" route for ventricular ejection (Burggren 1985). In contrast, during ventilation, the relationship between pulmonary and systemic vascular resistances becomes reversed, and thus the "*favored*" route for ventricular ejection is the pulmonary circulation (Burggren 1985).

The alterations in shunt patterns during ventilation and apnea may be mediated by the reciprocal interplay of cholinergic pulmonary vasoconstriction during apnea and adrenergic pulmonary vasodilation during ventilation (Lillywhite and Donald 1989; Comeau 1992). Vagal stimulation and the injection of acetylcholine mimic the hemodynamic changes observed in the pulmonary circulation during apnea (Burggren 1977a; Milsom et al. 1977; White 1978; Comeau 1992). In contrast, injection of epinephrine results in an increase in pulmonary blood flow and reduction in pulmonary vascular resistance (Comeau 1992). Finally, the increase in the pulmonary blood flow during ventilation is diminished by adrenergic blockade (Lillywhite and Donald 1989; Comeau and Hicks 1994).

Support for the pressure shunting hypothesis has not been conclusive. Past analyses have concentrated on general descriptions of pressure profiles within the ventricle and great vessels that have led to conclusions about direction and magnitude of flow during the ventilatory cycle. Stronger support for the pressure hypothesis requires quantitative assessments of the timing, magnitude and

Table 1. Central vascular blood pressures in reptilies (mmHg)

Species	Measurement site				References
	Pulmonary systolic	Pulmonary diastolic	Systemic systolic	Systemic diastolic	
Turtles					
Pseudemys scripta	24	19	25	10	Shelton and Burggren (1976)
Pseudemys scripta	19	7	22	19	White and Ross (1966)
Chelydra serpentina	30	12	32	23	Steggerda and Essex (1957)
Testudo gracea	35	12	34	23	Shelton and Burggren (1976)
Snakes					
Thamnophis radix	40	20	55	37	Burggren (1977b)
Thamnophis elegans	35	12	40	27	Burggren (1977b)
Thamnophis sirtalis	39	23	51	34	Burggren (1977b)
Lizards					
Varanus niloticus	27	16	96	82	Millard and Johansen (1973)
Varanus niloticus	28	16	100	76	Ishimatsu et al. (1988)
Varanus exanthematicus	42	22	83	58	Burggren and Johansen (1982)

duration of both flow and pressure development within the ventricle and great vessels during a variety of physiological states.

4.2 Previous Evidence for Washout Shunting

The washout model for intracardiac shunting was elucidated by Heisler et al. (1983). Based on the distribution of microspheres and on post-mortem anatomical observations, it was suggested that during systole tight adpression of the muscular ridge against the dorsal ventricular wall prevented translocation of systemic venous blood from CP into the RAo or LAo (see Fig. 4). Conversely, blood from the CV and CA was prevented from flowing into the CP during systole. Consequently, the R-L shunt was dependent on the volume of blood in the CV at end-diastole (Vcv_{ED}) and the volume of blood ejected from the CA during systole (Vca_S). The ratio, Vcv_{ED}/Vca_S, determines the O_2 level of systemic arterial blood. In addition, blood remaining in the CV at end-systole (Vcv_{ES}) would be *washed* into the CP during the following diastole, accounting for the L-R shunt (Fig. 5). This would increase the O_2 level of pulmonary arterial blood. The capacity of the muscular ridge to separate the CP from the CV and CA is not altered by ventilation or apnea (Heisler and Glass 1985). Variability in intracardiac shunting between individuals or during different physiological states could be accounted for by the volumes of the CV during the cardiac cycle (Heisler et al. 1983).

The washout model was developed from investigations primarily on varanid lizards. These lizards are considered unique among noncrocodilian reptiles. Burggren and Johansen (1982) studied cardiac pressures in the varanid lizard, *Varanus exanthematicus*. In contrast to the chelonians and other squamates, there was marked pressure separation between the pulmonary and systemic circulation. Systemic pressures in varanid lizards tend to be higher than in chelonians and other squamates (Table 1). Consequently, the varanid heart is considered a dual pressure pump, suggesting that the CP is functionally isolated from the CV and CA during systole. Burggren (1985) suggested that the varanids exhibit cardiovascular adaptations that are on a physiological continuum between the single pressure pumps of chelonians and squamates and the hearts of birds and mammals. The well-developed muscular ridge of varanid lizards is consistent with this notion. In addition to the cardiovascular differences observed, varanid lizards exhibit higher metabolic scopes as compared to other reptiles (Bennett 1972; Wood et al. 1977). Therefore, the washout hypothesis has not been generally accepted for the chelonians or other squamates.

4.2.1 Recent Evidence for Washout Shunting

The mechanisms of intracardiac shunting can be directly tested by visualizing or tracing blood flow patterns during various physiological conditions. A variety of techniques can provide such information, including angiocardiography, ultrasound, and indicator dilution techniques.

Angiography. Angiocardiography was used previously to deduce blood flow patterns through the central circulation of lower vertebrates (Foxon and Walls 1947; Foxon 1951; Johansen 1963). A recent study in the tegu lizard, *Tupinambis teguxin,* used angiocardiography to determine blood flow patterns during ventilation and apnea (Johansen et al. 1987). Injection of contrast medium into the vena cava, sinus venosus, right atrium, and CP demonstrated that the pulmonary circulation was completely isolated from the systemic circulation during ventilation. Injection of contrast medium directly into the CA showed no systolic admixture from this chamber into the CP. However, during diastole, a "small" L-R shunt was observable. During apnea, contrast medium injected into either the vena cava or sinus venosus appeared in the pulmonary outflow tracts and the LAo. However, contrast medium was not observable in the RAo. Injection of contrast medium directly into the CP demonstrated flow into the pulmonary arteries and LAo during apnea. In this study, the CP injection required that the catheter be advanced through the right atrium, around the muscular ridge and into the CP. The potential effects of such a procedure on the ability of the muscular ridge to effectively separate the pulmonary and systemic circulations were not considered.

In the turtle, *Pseudemys scripta,* L-R shunting may result from washout of the CV during the early phases of diastole. Fig. 6 represents an angiocardiographic sequence in which radiopaque material (Renografin-76) was injected into the left atrium. The animal was anesthetized, supine, mechanically ventilated, and a small (3 × 3cm) opening made in the plastron for placement of infusion catheters. Blood flow was video taped, digitized, and enhanced utilizing commer-

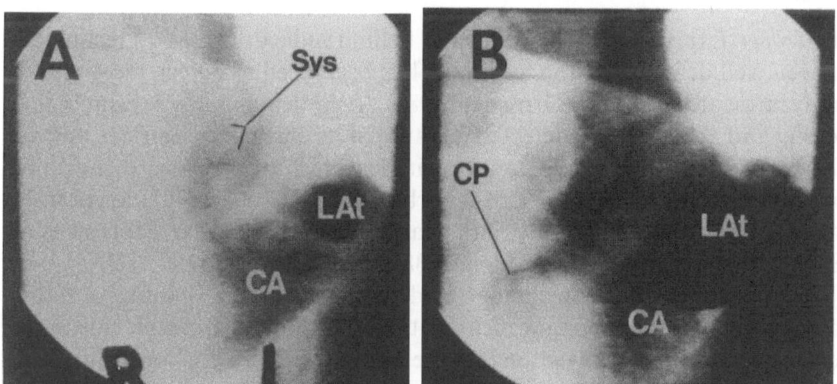

Fig. 6A,B. Ventral view of contrast medium (Renografin-76) injected into the left atrium of *Pseudemys scripta* **A** Mid-systole; note contrast medium in left atrium (*LAt*), cavum arteriosum (*CA*), and the systemic arteries (*Sys*). **B** Entry diastole; note the admixture in the cavum pulmonale (*CP*), most likely representing flow from cavum venosum. (unpubl. observ.)

cial software (JAVA, Jandel Inc.). During mid-systole (Fig. 6A), flow of contrast medium from the CA is clearly discernible in the systemic vessels. As the next diastole begins (Fig. 6B), contrast medium is seen to enter into the CP. These results are consistent with the washout hypothesis.

A problem with previous studies and the sequence shown in Fig. 6 is that the flow patterns were not quantified. The washout hypothesis predicts that intracardiac shunting results from alterations in ventricular volumes. These volumes are theoretically obtainable from angiocardiographic studies (Greene et al. 1967). Studies quantifying these ventricular volumes during various physiological conditions are required.

Ultrasound. Echocardiographic methods can be used for the detection and localization of intracardiac shunts (Grossman 1986). There are no published reports of echocardiographic studies in lower vertebrates. This technique can provide detailed images of intraventricular structures during the cardiac cycle. In addition, blood flow patterns can be deduced by the injection of an agitated saline solution into the cardiac chambers. These solutions contain microbubbles that function as a contrast agent (Grossman 1986). We have recently attempted to image the ventricular cava during the cardiac cycle in *Pseudemys scripta* using ultrasound technology. As in the angiocardiographic studies, the animal studied was supine, anesthetized, and mechanically ventilated. During diastole (Fig. 7A), the muscular ridge is visible and communication between the CP, CV, and CA is apparent. As systole begins, the muscular ridge appears to be pressed against the dorsal wall of the ventricle, separating the CP from CV and CA (Fig. 7B). These images suggest that in the turtle ventricle, the muscular ridge separates the ventricle into two distinct chambers; a prerequisite for the washout hypothesis. However, whether the entire free edge of the muscular ridge is pressed against the ventricular wall or is in just partial contact cannot be definitely determined from these images.

Blood-Tracer Experiments. The principles behind indicator dilution techniques have been well defined and the technique has been used to detect, localize, and quantify intracardiac shunts (Grossman 1986). An indicator is any substance that can easily and accurately be detected, because of its chemical, electrical, optical, or thermal properties. Several types of indicators have been used including indocyanine green (Swan et al. 1953), ascorbic acid (Levy et al. 1967), krypton-85 (Long et al. 1960), hydrogen gas (Clark et al. 1960), nitrous oxide (Morrow et al. 1958), and ether (Benenson and Hitzig 1938).

The application of this technique to detect intracardiac shunting is rather simple. Assuming no pulmonary recirculation, detection of the indicator in the systemic arches following injection into the right atrium, will demonstrate a R-L shunt. Detection of indicator in the pulmonary artery, following injection into the left atrium, will demonstrate a L-R shunt. In reptiles, the mechanism of R-L shunting is tested by injection of indicator into the CP. Detection of the indicator in the systemic arches after injection into the CP supports the pressure shunting hypothesis. In contrast, absence of the indicator in the systemic arch, following

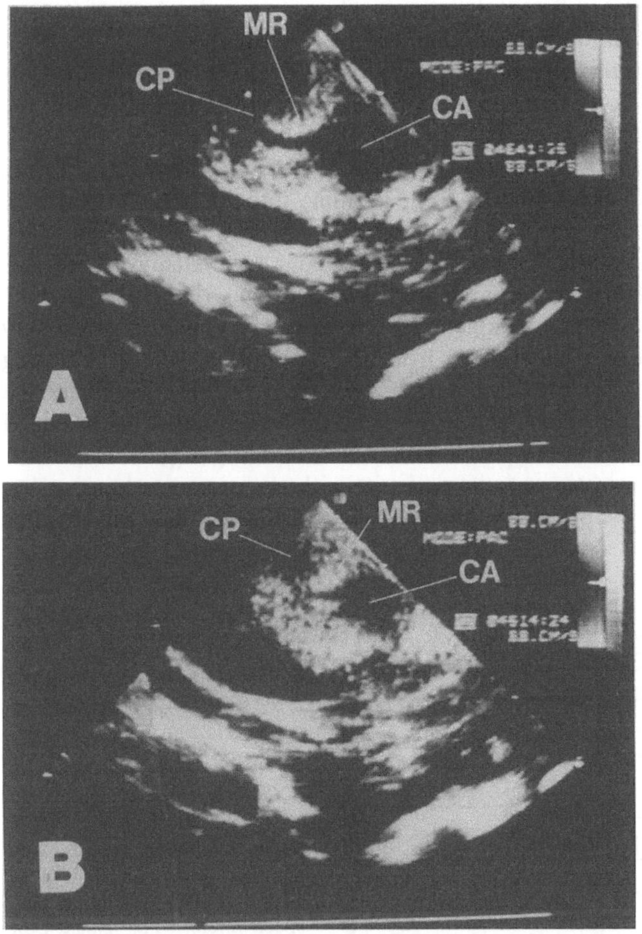

Fig. 7A,B. Echocardiographic view of the ventricle in the turtle, *Pseudemys scripta.* Ultrasonic flow probe (5 MHz; *top* of each figure) was placed on the ventral surface of the heart, near the apex. **A** In diastole, the cavum pulmonale (*CP*) and cavum arteriosum (*CA*) appear to be in communication. Note the muscular ridge (*MR*). **B** In systole, the *MR* appears to press against the dorsal wall of the ventricle and the *CP* and *CA* appear as two distinct chambers

injection into CP, supports the washout hypothesis. The mechanism of L-R shunting cannot be easily determined by applying this technique.

Recent studies (Hicks and Malvin 1992) have used a mass spectrometer to monitor central vascular blood gases and to detect an inert tracer gas following its injection into the cardiac chambers of the turtle, *Pseudemys scripta.* These studies employed a quadrupole mass spectrometer equipped with a multiple blood gas probe inlet system that allowed sequential gas measurements from three different blood gas probes. The advantage of such a system is that both central vascular oxygen and the tracer gas, in this case He, can be measured simultaneously at three different locations. Thus, two indicators of central vascular shunting are simultaneously detected.

These experiments pharmacologically manipulated the level of R-L intracardiac shunting by infusion of epinephrine (EPI) or acetylcholine (ACh) into the venous circulation. Injection of ACh resulted in an intensification of the R-L shunt, as indicated by reductions in the systemic arch PO_2 from control values (Fig. 8.). During this period, injection of He-saline into the right atrium was always detected in the systemic arches (Hicks and Malvin 1992). In contrast, injection of He-saline into the CP was never detected in the systemic arches (Fig. 8). Based on these He tracer patterns, Hicks and Malvin (1992) concluded that the muscular ridge functioned as a complete septum between CP and CV, effectively separating the pulmonary circulation from the systemic circulations during systole. Interestingly, following administration of EPI, He-saline injected into the right atrium was never detected in the systemic outflow tracts. This provided evidence that adrenergic stimulation eliminates R-L shunting (Fig. 8). The results following ACh administration are inconsistent with the pressure shunting hypothesis and strongly support the washout mechanism for R-L intracardiac shunting.

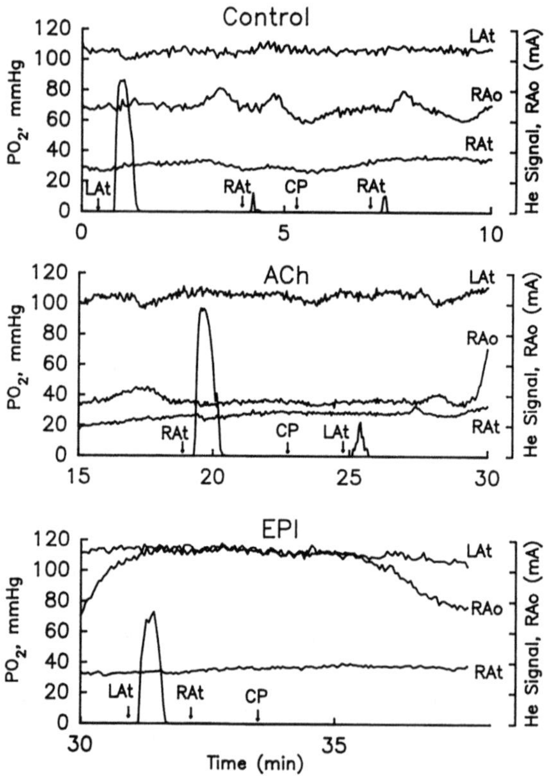

Fig. 8. Continuous monitoring of blood gases and He in RAt, LAt, and RAo utilizing a mass spectrometer. *Top panel* Control conditions. The He-saline injected into RAt and LAt was detected in Rao. The He-saline injected into CP was not detected in RAo. *Middle panel* Central vascular PO_2 following injection of ACh (0.1 mg/kg). The He-saline injected into the RAt and LAt was detected in the RAo but was not detected following injection into CP. *Bottom panel* Central vascular PO_2 following injection of EPI (1 μg/kg). The He-saline injected into RAt and CP was not detected in RAo. The He-saline injected into LAt was detected in the RAo. (Hicks and Malvin 1992)

An alternative to the mass spectrometry system described above is the use of hydrogen electrodes. The potentiometric use of a platinum electrode to detect hydrogen and localize intracardiac shunts has been previously reported (Clark et al. 1960). Basically, a small platinum wire (30 g) can be nonocclusively implanted into central vascular sites and blood flow patterns are deduced as described above. The advantage of this technique is that the hydrogen electrodes are small, inexpensive, and exhibit a very fast response time (<100 ms; Clark et al. 1960). In addition, the small size and flexibility of the platinum wire allows for sufficient instrumentation of the animal for the measurement of additional physiological parameters, i.e. blood flow and pressure.

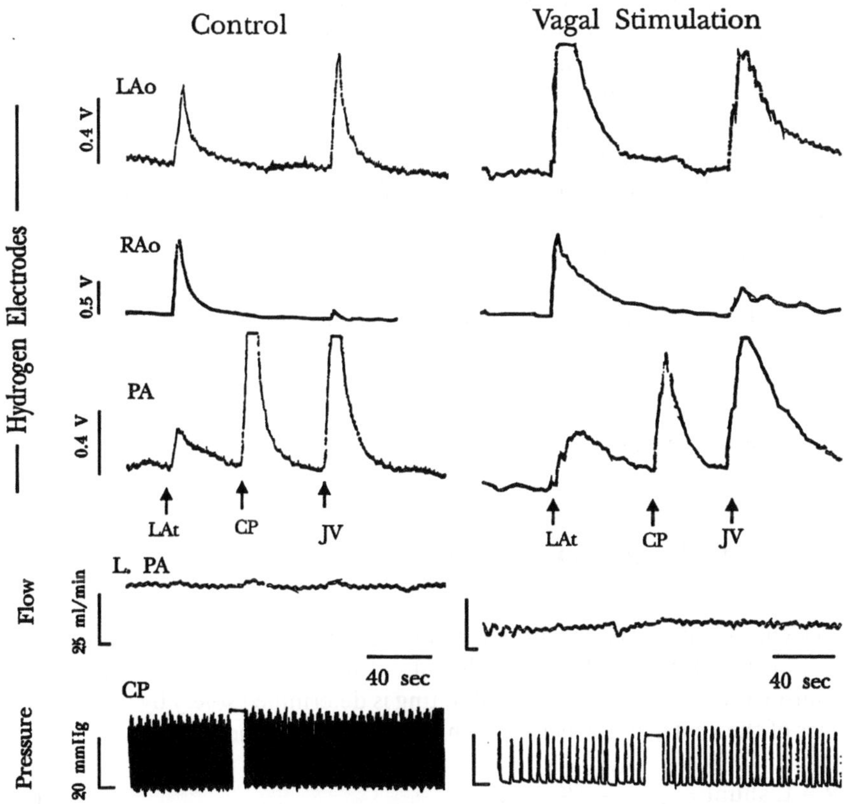

Fig. 9. Simultaneous detection of H_2-saline in the three great vessels of the turtle *Pseudemys scripta*. H_2 electrodes are implanted into left aortic arch (*LAo*), right aortic arch (*RAo*), and common pulmonary artery (*PA*). Blood flow is measured by a transient-time ultrasonic flow probe applied to the left pulmonary artery (*LPA*). Ventricular pressure is continuously recorded in the cavum pulmonale (*CP*). *Arrows* indicate sequential injection of an indicator (H_2-saline) into left atrium (*LAt*), cavum pulmonale (*CP*), and jugular vein (*JV*) during control and during vagal stimulation. Note interruption of pressure trace represents injection of indicator into the CP. (Hicks and Comeau 1994)

Recent studies (Hicks and Comeau 1994) have investigated the effects of vagal stimulation on central vascular blood gases and blood flow patterns in the red-eared turtles, *Pseudemys scripta*. These experiments used a very similar preparation to that described by Hicks and Malvin (1992), the primary difference being the use of vagal nerve stimulation to modulate intracardiac shunt levels. During control, the H_2 injected into the left atrium or jugular vein was detected in RAo, LAo, and PA (Fig. 9). H_2 infused into CP was never detected in LAo or RAo during control (Fig. 9). This detection pattern indicates bidirectional shunting during control. During stimulation of the right vagal efferent nerve, a hemodynamic response, similar to that observed during diving, occurred. There was an approximately 70% reduction in both heart rate and pulmonary blood flow and pulmonary vascular resistance increased approximately 107%. This was accompanied by an increase in the R-L intracardiac shunt, indicated by a reduction in the PO_2 of both RAo and LAo. During these conditions, H_2 infused into the jugular vein and left atrium was detected in both RAo and LAo (Fig. 9). However, H_2 infused into the CP was detected in the systemic arches in only two of seven animals. It is of interest to note that under these conditions, H_2 injected into left atrium was simultaneously detected in the pulmonary artery, demonstrating the presence of bidirectional intracardiac shunting during vagal stimulation (Fig. 8). It was concluded that vagally induced intracardiac R-L shunting occurs primarily by the washout mechanism.

These indicator dilution experiments provide a strong test of these two hypotheses for the mechanism of R-L intracardiac shunting. Further studies utilizing a combination of these techniques are required to more fully evaluate the mechanism of L-R intracardiac shunting. Although these experimental techniques are rather invasive, they provide a powerful tool to study the mechanism of intracardiac shunting in the reptilian heart.

5 Regulation of Washout Shunting

The evidence cited above supports the concept that washout of the CV is the primary mechanism for R-L intracardiac shunting in reptiles. According to this mechanism, the level of intracardiac shunting is determined solely by the relative volumes of the ventricular cava and can be described by the ratio;

$$\mathrm{R-L\ shunt} = \frac{\mathrm{Vcv_{ED}}}{(\mathrm{Vcv_{ED}} + \mathrm{Vca_S})}, \tag{1}$$

where Vcv_{ED} = end-diastolic volume of CV, Vcv_{ES} = end-systolic volume of CV, and Vca_S = stroke volume of CA. This equation assumes a steady state in which $Qs = Vcv_{ED} + Vca_S$ and $Qp = (Qs - Vcv_{ED}) + Vcv_{ES}$. Any factor that affects these ventricular volumes will determine the level of R-L shunting. For example, during apnea, an increase in vagal tone promotes a reduction in heart rate, reduction in pulmonary blood flow, and a large increase in pulmonary vascular resistance.

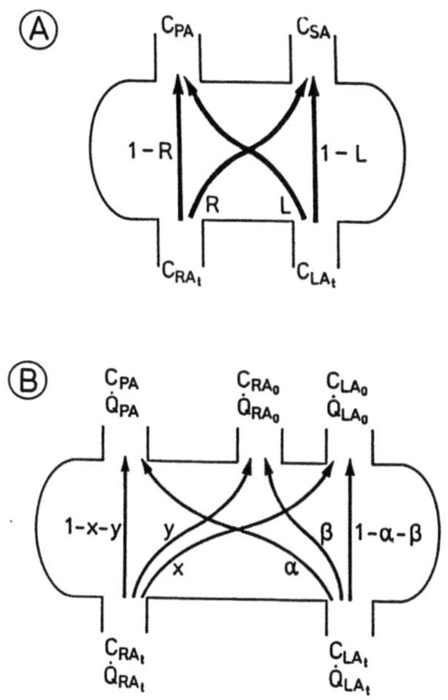

Fig. 10. A The two-vessel model on which the standard shunt equations are based. *PA* Pulmonary artery; *SA* systemic artery; *RAt* right atrium; *LAt* left atrium. For all vessels $C = O_2$ concentration. Intracardiac shunts are designated as *R* (right-to-left) and *L*(left-to-right). **B** The three-vessel model that describes intracardiac shunting in the presence of dual aortic arches. *RAo* Right aortic arch; *LAo* left aortic arch. The fractions of blood distributed to the various outflow tracts are designated as *x*, *y*, and α. See text for description. (Adapted from Ishimatsu et al. 1988)

These hemodynamic changes will reduce the volume of blood ejected from the CP and promote a rise in the end-diastolic pressure of the CP. An increase in diastolic "backpressure" will reduce the effective filling of the CP during subsequent diastoles resulting in an increase in the Vcv_{ED}. The reduction in pulmonary venous return combined with vagal reductions in myocardial contractility reduces the Vca_S. These combined events result in an increase in the ratio, $Vcv_{ED}/(Vcv_{ED} + Vca_S)$, and a desaturation of arterial blood.

The above equations are based on a standard two-compartment model for intracardiac shunting (Fig. 10A). However, recent studies on noncrocodilian reptiles (Ishimatsu et al. 1988) have suggested that flow patterns within the ventricle should be represented by a three-vessel model (Fig. 10B). In this model, there are three outflow vessels (RAo, LAo, and PA) and six potential flow patterns (Fig. 10B). Analysis of this model indicates that the Vcv_{ED} can be differentially distributed between the two aortic arches. These flows are designated as x and y (Fig. 10B) and represent the fraction of Vcv_{ED} entering the LAo and RAo, respectively. The fraction of Vcv_{ES} recirculating into the PA (α) does not differ from that predicted by the standard two-vessel model. The three-vessel model illustrates that the oxygen levels within a single aortic arch are insufficient to analyze shunting in reptiles.

There is little information on the distribution of the Vcv_{ED} into the aortic arches (Table 2). In the turtle, *Chelydra serpentina*, no difference was found in the O_2 saturation levels of RAo or LAo (Steggerda and Essex 1957). In contrast, Khalil and Zaki (1964) found consistently higher O_2 levels in the RAo than in the LAo in the blood of the turtle, *Testudo leithii*. In the monitor lizard, *Varanus*

Table 2. Oxygen content measurements from central vascular sites in reptiles

Species	Oxygen content (vol%) Central vascular site					Reference
	RAo	LAo	LAt	RAt	PA	
Turtles						
Chelydra serpentina	3.6	...	6.1	3.1	3.2	Steggerda and Essex (1957)
Testudo leithii	7.9	5.7	8.9	3.05	3.6	Khalil and Zaki (1964)
Pseudemys scripta	6.9	6.5	7.2	4.9	...	Hicks (unpubl.)
Snakes						
Coluber constrictor	7.2	7.2	6.9	...	...	White (1959)
Lytorhynchus (Zamenis) diadema	6.7	6.9	8.1	4.1	4.0	Khalil and Zaki (1964)
Lizards						
Iguana iguana	3.7	3.7	3.8	...	...	White (1959)
Agama stellio	...	...	...	2.1	4.6	Khalil and Zaki (1964)
Varanus gresius	6.3	...	7.45	4.1	5.3	Khalil and Zaki (1964)
Varanus exanthematicus	9.5	8.8	10.9	4.1	5.7	Burggren and Johansen (1982)
Varanus niloticus	6.1	5.05	6.1	3.1	3.2	Ishimatsu et al. (1988)

exanthematicus, Burggren and Johansen (1982) reported higher PO_2 and O_2 content levels in the RAo than the LAo. Recently, Ishimatsu et al. (1988) studied the O_2 distribution in the aortic arches of the Nile monitor lizard, *Varanus niloticus,* as a function of body temperature. This study indicated that the RAo PO_2 was higher than the PO_2 in the LAo at both 25 and 35 °C, with the difference increasing at the lower temperature.

Clearly, further studies are needed to evaluate the changes in both the hemodynamic parameters promoting intracardiac shunting and the resulting distribution of ventricular blood flow into the major outflow vessels. Describing these events during a variety of physiological conditions will provide insights into the functional significance of this complex cardiovascular arrangement.

Acknowledgments. The authors express their appreciation to the Departments of Radiology and Cardiology, School of Medicine, Creighton University. The expert technical assistance of Mr. Joe Rysavy and Ms. Cheri Lawrence is greatly appreciated. Funding was provided by the NSF grant DCB-9004570 to J.W. Hicks and NIH Grant, HL-38942 to G.M. Malvin.

References

Benenson W, Hitzig WM (1938) Diagnosis of venous arterial shunt by ether circulation time method. Proc Soc Exp Biol 38: 256–258

Bennett AF (1972) The effect of activity on oxygen consumption, oxygen debt, and heart rate in the lizards *Varanus gouldii* and *Souromalus hispidus.* J Comp Physiol 79: 259–280

Benninghoff VA (1933) Herz. In: Bolk L, Göppert E, Kallius E, Lubosch W (eds) Handbuch der vergleichenden Anatomie der Wirbeltiere. Urban and Schwarzenberg, Berlin, pp 467–556, reprinted 1967 by A Asher and Co, Amsterdam

Brücke E (1852) Beiträge zur vergleichenden Anatomie und Physiologie des Gefäßsystems. Denkschr Akad Wien 3: 335–367

Burggren WW (1975) A quantitative analysis of ventilation tachycardia and its control in two chelonians, *Pseudemys scripta* and *Testudo graeca.* J Exp Biol 63: 367–380

Burggren WW (1977a) The pulmonary circulation of the chelonian reptile: morphology, pharmacology and hemodynamics. J Comp Physiol B 116: 303–324

Burggren WW (1977b) Circulation during intermittent lung ventilation in the garter snake *Thamnophis.* Can J Zool 55: 1720–1725

Burggren WW (1985) Hemodynamics and regulation of central cardiovascular shunts in reptiles. In: Johansen K, Burggren WW (eds) Cardiovascular shunts: phylogenetic, ontogenetic and clinical aspects. Munksgaard, Copenhagen, pp 121–142

Burggren WW, Johansen K (1982) Ventricular hemodynamics in the monitor lizard, *Varanus exanthematicus:* pulmonary and systemic pressure separation. J Exp Biol 96: 343–354

Burggren WW, Shelton B (1979) Gas exchange and transport during intermittent breathing in chelonian reptiles. J Exp Biol 82: 75–92

Burggren WW, Glass ML, Johansen K (1977) Pulmonary ventilation: perfusion relationships in terrestrial and aquatic chelonian reptiles. Can J Zool 55: 2024–2034

Clark LC, Bargeron LM, Lyons C, Bradley MN, McArthur KT (1960) Detection of right-to-left shunts with an arterial potentiometric electrode. Circulation 22: 949–955

Comeau S (1992) Vagal regulation of central vascular blood flow in the turtle, *Pseudemys scripta.* MS Thesis, Creighton University, Omaha, NE

Comeau SG, Hicks JW (1994) Regulation of central vascular blood flow in the turtle. Am J Physiol 267: R569–R578

Foxon GEH (1951) A radiographic study of the passage of the blood through the heart in the frog and toad. Proc Zool Soc Lond 121: 529–538
Foxon GEH, Walls EW (1947) The radiographic demonstration of the mode of action of the heart of the frog. J Anat Lond 81: 111–117
Goodrich ES (1916) On the classification of the Reptilia. Proc R Soc B 89: 261–276
Goodrich ES (1919) Note on the reptilian heart. J Anat Lond 53: 298–304
Greene DG, Carlisle R, Grant C, Bunnell IL (1967) Estimation of left ventricular volume by one-plane cineangiography. Circulation 35: 61–69
Griel A (1903) Beiträge zur vergleichenden Anatomie und Entwicklungsgeschichte des Herzens und des Truncus arteriosis der Wirbelthiere. Morphol Jahrb 31: 123–310
Grossman W (1986) Shunt detection and measurement. In: Grossman W (ed) Cardiac catheterization and angiography. Lea and Febiger, Philadelphia, pp 155–169
Heisler N, Glass ML (1985) Mechanisms and regulation of central vascular shunts in reptiles. In: Johansen K, Burggren WW (eds) Cardiovascular shunts: phylogenetic, ontogenetic and chinical aspects. Munksgaard, Copenhagen, pp 334–353
Heisler N, Neumann P, Maloiy GMO (1983) The mechanism of intracardiac shunting in the lizard *Varanus exanthematicus.* J Exp Biol 105: 15–31
Hicks JW (1993) Regulation of intracardiac shunting in reptiles: anatomic vs effective shunting. In: Bicudo JE, Glass ML (eds) The vertebrate gas transport cascade: adaptations to environment and mode of life. CRC Press, Boca Raton, pp 252–265
Hicks JW, Comeau SG (1994) Vagal regulation of intracardiac shunting in the turtle, *Pseudemys scripta.* J Exp Biol 186: 109–126
Hicks JW, Malvin GM (1992) Mechanism of intracardiac shunting in *Pseudemys scripta.* Am J Physiol 262: R986–R992
Hicks JW, Wood SC (1988) Oxygen homeostasis in lower vertebrates: The impact of external and internal hypozia. In: Wood SC (ed) Lung Biology in Health and Disease Comparative Pulmonary Physiology: Current Concepts. Dekker, New York, pp 311–341
Ishimatsu A, Hicks JW, Heisler N (1988) Analysis of intracardiac shunting in the lizard, *Varanus niloticus:* a new model based on blood oxygen levels and microsphere distribution. Respir Physiol 71: 83–100
Johansen K (1959) Circulation in the three-chambered snake heart. Circ Res 7: 828–832
Johansen K (1963) Cardiovascular dynamics in the amphibian, *Amphiuma tridactylum.* Acta Physiol Scand 60: 1–82
Johansen K, Burggren WW (1980) Cardiovascular function in lower vertebrates. In: Bourne G (ed) Hearts and heart-like organs, vol I. Academic Press, New York, pp 61–117
Johansen K, Abe A, Andresen JH (1987) Intracardiac shunting revealed by angiocardiography in the lizard *Tupinambis teguixin.* J Exp Biol 130: 1–12
Khalil F, Zaki K (1964) Distribution of blood in the ventricle and aortic arches in Reptilia. Z Vgl Physiol 48: 663–689
Levy AM, Monroe RG, Hugenholtz PG, Nada AS (1967) Clinical use of ascorbic acid as an indicator of right-to-left shunt. Br Heart J 29: 22–29
Lillywhite HB, Donald JA (1989) Pulmonary blood flow regulation in an aquatic snake. Science 245: 293–295
Long RTL, Braunwald E, Morrow AG (1960) Intracardiac injection of radioactive krypton: clinical applications of new methods for characterization of circulatory shunts. Circulation 21: 1126–1133
Mathur PN (1946) The anatomy of the reptilian heart. Part II. Serpentes, Testudinata and Loricata. Proc Indian Acad Sci 20: 1–29
Millard RW, Johansen K (1973) Ventricular outflow dynamics in the lizard, *Varanus niloticus:* responses to hypoxia, hypercarbia and diving. J Exp Biol 60: 871–880
Millen JE, Murdaugh HV, Bauer CB, Robin D (1964) Circulatory adaptation to diving in the freshwater turtle. Science 145: 591–593
Milsom WK, Langille BL, Jones DR (1977) Vagal control of pulmonary vascular resistance in the turtle, *Chrysemys scripta.* Can J Zool 55: 359–367
Morrow AG, Sanders RJ, Braunwald E (1958) The nitrous oxide test: an improved method for the detection of left-to right shunts. Circulation 17: 284–291

O'Donoghue CH (1918) The heart of the leathery turtle, *Dermochelys (Sphargis) coriacea.* With a note on the septum ventriculorum in the Reptilia. J Anat Lond 52: 823–890
Rau AS (1924) Observations of the anatomy of the heart of *Tiliqua scincoides* and *Eunectes murinus.* J Anat Lond 59: 60–71
Rudolf AM (1974) Congenital diseases of the heart: clinical-physiologic considerations in diagnosis and management. Year Book Medical Publishers Inc, Chicago
Shelton G, Burggren WW (1976) Cardiovascular dynamics of the Chelonia during apnea and lung ventilation. J Exp Biol 64: 323–343
Shelton G, Jones DR, Milsom WK (1986) Control of breathing in ectothermic vertebrates. In: Fishman E (ed) Handbook of physiology, section 3: the respiratory system, vol II. Control of breathing, part 2. Am Physiol Soc, Bethesda, pp 857–909
Steggerda FR, Essex HE (1957) Circulation and blood pressure in the great vessels and heart of the turtle *Chelydra serpentina.* Am J Physiol 190: 320–326
Swan HJC, Zapata-Diaz J, Wood EH (1953) Dye dilution curves in cyanotic congenital heart disease. Circulation 8: 70–76
Van Mierop LHS, Kutsche LM (1981) Comparative anatomy of the ventricular septum. In: Wenink ACG (ed) The ventricular septum of the heart, Nijhoff, The Hague, pp 35–46
Van Mierop LHS, Kutsche LM (1985) Some aspects of comparative anatomy of the heart. In: Johansen K, Burggren WW (eds) Cardiovascular shunts: phylogenetic, ontogenetic and clinical aspects. Munksgaard, Copenhagen, pp 38–56
Webb GJ, Heatwolfe H, de Bavay J (1974) Comparative cardiac anatomy of the Reptilia. II. A critique of the literature on the Squamata and Rhynchocephalia. J Morphoe 142: 1–20
White FN (1959) Circulation in the reptilian heart (Squamata). Anat Rec 135: 129–134
White FN (1968) Functional anatomy of the heart of reptiles. Am Zool 8: 211–219
White FN (1978) Circulation: a comparison of reptiles, mammals and birds. In: Piiper J (ed) Respiratory function in birds, adult and embryonic. Springer, Berlin Heidelberg New York, pp 51–60
White FN (1985) Role of intracardiac shunts in pulmonary gas exchange in chelonian reptiles. In: Johansen K, Burggren WW (eds) Cardiovascular shunts: phylogenetic, ontogenetic and clinical aspects. Munksgaard, Copenhagen, pp 296–309
White FN, Ross G (1966) Circulatory changes during experimental diving in the turtle. Am J Physiol 211: 15–18
White FN, Hicks JW, Ishimatsu A (1989) Respiratory states and intracardiac shunts in turtles. Am J Physiol 256: R240–R247
Wood SC, Hicks JW (1985) Oxygen homeostasis in vertebrates with cardiovascular shunts. In: Johansen K, Burggren WW (eds) Cardiovascular shunts: phylogenetic, ontogenetic and clinical aspects. Munksgaard, Copenhagen, pp 354–362
Wood SC, Johansen K, Gatz RN (1977) Pulmonary blood flow, ventilation/perfusion ratio, and oxygen transport in a varanid lizard. Am J Physiol 233: R89–R93

Chapter 7

Central Cardiovascular Dynamics in Reptiles

S. Nilsson

Contents

1 Introduction

The ability of the heart to keep hypoxic ("venous") blood that returns to the heart via the large systemic veins separate from the oxygenated blood from gills and/or lungs increases the efficiency of gas transport for the organism. A phylogenetic trend towards a higher degree of blood separation within the heart, and a more sophisticated control of respiratory and systemic blood flow distribution, is readily discernible among the vertebrates. The classical view, based primarily on deductions from anatomical studies, regarded the situation found in birds and mammals as the "final" stage, with a heart completely separated into four compartments with no possibility of intracardiac mixing of oxygenated and hypoxic blood. While preserving the separation of the flow of the two bloodstreams, the arrangement does not allow reduction of pulmonary flow during periods of apnoea. The avian and mammalian heart can be regarded as a set of two pumps arranged in series, and any change in blood flow in either the pulmonary or systemic circuit must obviously be replicated in the other circuit (Johansen and Burggren 1980; Burggren 1985, 1987; Johansen 1985).

Department of Zoophysiology, University of Göteborg, Medicinaregt. 18, S-413 90 Göteborg, Sweden

Advances in Comparative and Environmental Physiology, Vol. 21

In animals with intermittent breathing, such as most reptiles, the use of cardiac energy to maintain pulmonary perfusion during prolonged periods of breath-holding is clearly wasteful, although the energy expenditure may be small. Furthermore, plasma filtration in the reptilian lung has been shown to be large compared to the situation in mammals, and a reduced pulmonary perfusion will diminish, or even reverse, fluid leakage (Burggren 1982, 1985).

In amphibians and noncrocodilian reptiles, blood entering the anatomically single ventricle from the right and left atria can be pumped into both the pulmonary arterial outflow tract and the systemic aortic arches. The direction of the intraventricular shunt is determined by the relative vascular resistances between the two circuits, and allows a partial pulmonary bypass during periods of apnoea (for a detailed evaluation of intracardiac shunts, see Hicks and Malvin, this Vol.). During periods of ventilation, a rapid and substantial increase in pulmonary perfusion is essential to provide adequate pulmonary gas exchange (White 1976; Burggren and Shelton 1979).

Thus, an "ideal" heart for an intermittently breathing animal would (1) provide a high degree of separation between oxygenated and hypoxic bloodstreams, while (2) preserving the option of a right-to-left shunt (pulmonary bypass) during periods of apnoea. In addition, (3) a lower pulmonary arterial blood pressure would reduce the plasma filtration in the reptilian lung, and (4) adjustment of cardiac work to the breathing pattern is advantageous. As we shall see, three different stages of this "ideal" heart can be recognized in the reptilian groups.

2 Chelonians and Nonvaranid Squamates

All reptiles have a heart with a complete atrial septum, while the organization of the ventricle shows some differences. In non crocodilian reptiles the different parts of the ventricle are anatomically continuous, and blood pressures are identical in all parts of the chelonian and non varanid squamate ventricular cavities during all phases of the cardiac cycle (Shelton and Burggren 1976; Burggren 1977).

The presence of ridges and septa within the ventricle produces three distinguishable compartments (Fig. 1). A more or less distinct vertical septum divides the dorsal part of the ventricle into the *cavum arteriosum* and the *cavum venosum*. The *cavum arteriosum* receives oxygenated blood from the left atrium, and the *cavum venosum* collects hypoxic "venous" blood from the right atrium (and also to some degree oxygenated overflow from the *cavum arteriosum*). The passage between these two chambers, the interventricular canal (White 1959), is open throughout the cardiac cycle (Webb et al. 1971).

A third chamber, the *cavum pulmonale*, is incompletely yet relatively effectively separated from the *cavum venosum* by a transverse miscular ridge (horizontal septum) often described by the German term "Muskelleiste". Opposite this

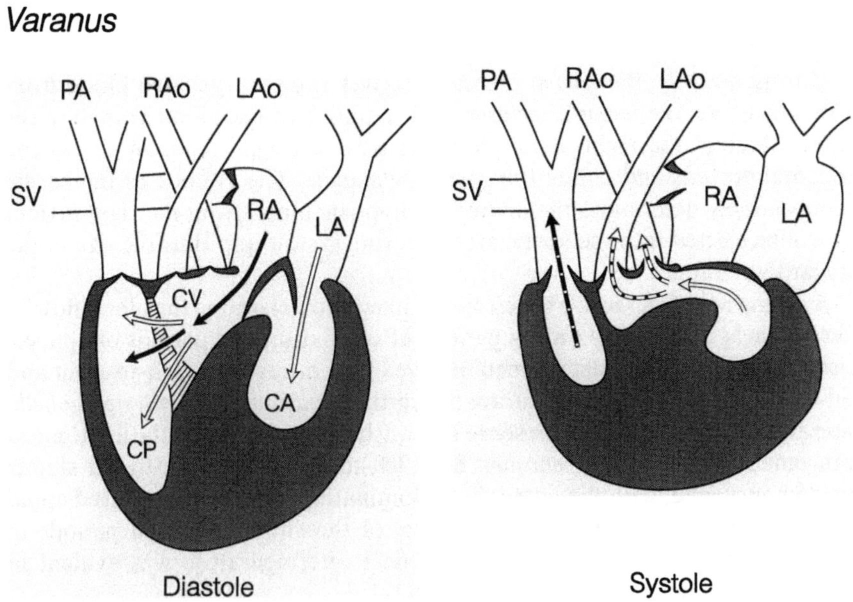

Fig. 1. Diagrammatic representation of the ventricular compartments and major arteries of the turtle, *Chrysemys picta belli*, showing the flow of hypoxic ("venous"; *filled arrows*), oxygenated ("arterial"; *open arrows*) and mixed (*broken arrows*) blood during diastole (*left*) and systole (*right*). During diastole, hypoxic blood from the right atrium (*RA*) enters the *cavum venosum* (*CV*) and *cavum pulmonale* (*CP*). Oxygenated blood from the left atrium (*LA*) enters the *cavum arteriosum* (*CA*) and may, due to incomplete occlusion of the interventricular canal, also flow into the *cavum venosum*. Oxygenated blood remaining in the *cavum venosum* from the previous systole will also flow to the *cavum pulmonale*. During systole, venous blood from the sinus venosus (*SV*) fills the right artrium, and blood from the pulmonary veins enters the left atrium (*LA*). Mixed blood in the *cavum pulmonale* is ejected into the pulmonary arteries (*PA*), and from the *cavum venosum* into the left (*LAo*) and right (*RAo*) aortic arch. The *cavum venosum* is refilled from the *cavum arteriosum*, and mixed blood may also flow over the incompletely occluded muscular ridge into the *cavum pulmonale*. (Redrawn from Heisler and Glass 1985)

septum is another ridge, the "Bulbuslamelle", which meets the horizontal septum and in effect separates the *cavum pulmonale* from the rest of the ventricle (Webb et al. 1971; van Mierop and Kutsche 1985). The *cavum pulmonale* receives the mixed (mainly hypoxic) blood from the *cavum venosum* and feeds the pulmonary outflow tract during systole (Fig. 1; Webb et al. 1971; van Mierop and Kutsche 1985).

During systole, the ventricle ejects blood into the two aortic arches and the pulmonary outflow tract. Both aortic arches issue from the *cavum venosum*, with the right arch closest to the *cavum arteriosum*, which is the recipient of oxygenated blood returning from the lungs via the left atrium (Fig. 1). The right arch (including the brachycephalic artery) feeds the head and anterior part of the body, while the gut and posterior body receives blood from the left arch (Shelton and Burggren 1976). Contrary to the situation in birds and mammals, with their pulmonary and systemic circuits coupled in series, the corresponding circuits in chelonians and squamates are thus arranged in parallel (Fig. 1; Johansen and Burggren 1980).

During diastole, the *cavum venosum* receives some oxygenated blood from the left atrium via the *cavum arteriosum* and, mainly, hypoxic blood from the right atrium. Most of the right atrial blood passes the *cavum venosum* across the horizontal septum and enters the *cavum pulmonale*. The volume of the *cavum venosum* largely determines the amount of hypoxic blood from the right atrium that will be ejected into the aortic arches during systole and thus the size of the intracardiac shunts.

By using injected, radiolabelled microspheres to determine the blood flow to different parts of the body during periods of ventilation and periods of apnoea, Heisler and Glass (1985) determined the size of the intracardiac left-to-right and right-to left shunts that occur during the cardiac cycle in *Chrysemys picta belli*. These authors concluded the presence of a washout shunt, due to the anatomical arrangement of the *cavum venosum*. Both left-to-right and right-to-left shunts were demonstrated, with the latter shunt dominating. The technique used could not demonstrate any difference in the size of the shunts between periods of ventilation and apnoea, although some effect of temperature was evident in *Chrysemys* (see Hicks and Malvin, this Vol.).

2.1 Pulmonary Blood Flow in Chelonians

Heart rate and pulmonary blood flow increase substantially during respiration in both the turtle, *Trachemys* (= *Pseudemys*, = *Chrysemys*) *scripta*, and the tortoise, *Testudo graeca*, and the intraventricular right-to-left shunt (partial pulmonary bypass) seen during apnoea (diving) is reversed to a significant left-to- right shunt (White and Ross 1966; Burggren 1975; Burggren and Shelton 1979; White 1985). Anoxia (nitrogen breathing) also produces major redistribution of blood flows in the turtle, in this case away from the viscera into the central nervous system (Davies 1989).

Apart from the more obvious increase in cardiac output (20- to 30-fold; White and Ross 1966; Shelton and Burggren 1976) at the onset of normal ventilation, a major basis for the changes in pulmonary blood flow is the adjustment of pulmonary vascular resistance. Clearly, the ability to shunt blood between the ventricular compartments with little or no resistance makes blood distribution directly dependent on changes in the relative vascular resistance between the pulmonary and systemic circuits (White 1968, 1970; Johansen et al. 1970; Shelton and Burggren 1976; Burggren 1985).

In addition, a reduction in pulmonary arterial blood pressure occurs due to the actions of a vascular sphincter in the pulmonary outflow tract (extrinsic pulmonary artery). A cholinergic excitatory vagal innervation of the reptilian pulmonary vasculature has been long recognized (Luckhardt and Carlson 1921), and a vascular smooth muscle sphincter in the proximal portion of the pulmonary outflow tract has been demonstrated in chelonians (*Trachemys scripta*,

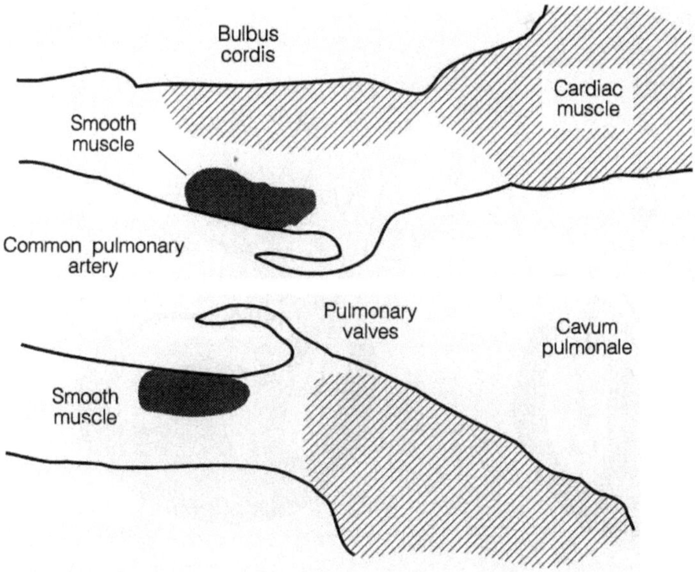

Fig. 2. Sagittal section through the proximal part of the pulmonary artery of a chelonian reptile. Note the smooth muscle sphincter immediately distal to the pulmonary arterial valves. (Redrawn from Burggren 1985)

Testudo graeca (Fig. 2; Burggren 1977). This sphincter provides a high and variable resistance in the pulmonary arterial vasculature, and is controlled by autonomic nerve fibres that are cholinergic and run in the vagus nerve. A similar pulmonary arterial sphincter occurs also in other reptiles (White 1968; Berger 1974, cited in Milsom et al. 1977; Burggren 1977; Milsom et al. 1977; Smith and MacIntyre 1979), and in amphibians (Campbell 1971; Smith and Campbell 1976; de Saint-Aubain and Wingstrand 1979; de Saint- Aubain 1982). Nervous control of the intrinsic vasculature of the lungs appears to be relatively insignificant in the reptiles studied, and an adrenergic vasomotor innervation may even be lacking (Burggren 1977; Milson et al. 1977).

3 Varanids

The horizontal septum is well developed in varanids, and there is a discrete vertical septum. Mixing of left and right atrial blood during the diastolic filling of the ventricular compartments is reduced or abolished by the large atrioventricular valves which obstruct the interventricular canal. The volume of the *cavum venosum* is much reduced compared to the chelonians and non varanid squamates. Mixing of oxygenated blood from the *cavum arteriosum* and hypoxic blood remaining in the *cavum venosum* from the previous diastole is therefore

Chrysemys

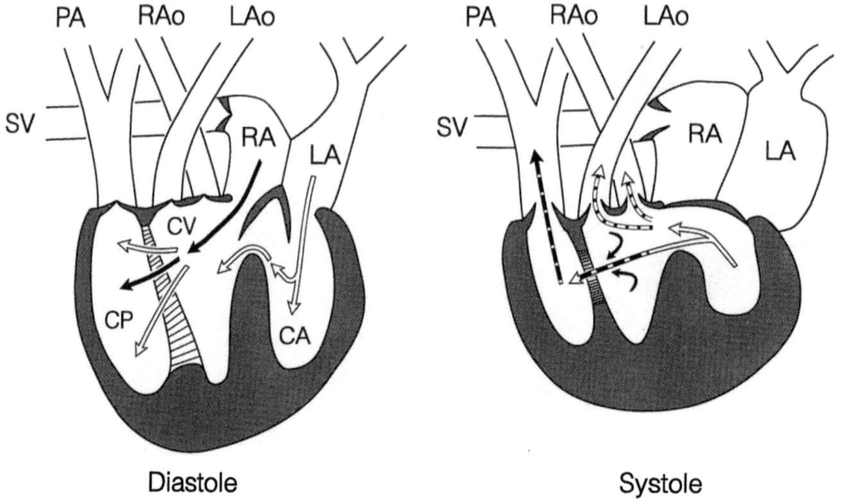

Fig. 3. Diagrammatic representation of the ventricular compartments and major arteries of the goanna (monitor lizard), *Varanus*, showing the flow of hypoxic ("venous"; *filled arrows*), oxygenated ("arterial"; *open arrows*) and mixed (*broken arrows*) blood during diastole (*left*) and systole (*right*). During diastole, the situation resembles that in *Chrysemys* except that the interventricular canal is likely to be occluded by the large atrioventricular values, preventing blood flow from the *cavum arteriosum* into the *cavum venosum*. During systole, a complete separation of the systemic (*cavum venosum* and *cavum arteriosum*) and pulmonary (*cavum pulmonale*) ventricular compartments occurs, allowing different intracardiac systolic peak blood pressures in the pulmonary and systemic circuits. Mixing of oxygenated and hypoxic blood in the *cavum venosum*, and therefore in the *cavum pulmonale*, results from the events during diastole. Due to the small volume of the *cavum venosum*, this shunt is smaller than in the chelonians and nonvaranid squamates. *Abbreviations* as in Fig. 1. (Redrawn from Heisler et al. 1983)

modest, which means that the washout shunt is smaller in the varanids (Fig. 3; Berger and Heisler 1977; Heisler et al. 1983; Heisler and Glass 1985, Ishimatsu et al. 1988). In *Varanus niloticus*, blood O_2 levels are consistently higher in the right aortic arch than in the left. Careful studies of the intracardiac shunts in this species (Ishimatsu et al. 1988), using the radiolabelled microsphere technique combined with blood gas analysis, confirms the presence of a washout shunt (cf. Heisler 1985; Heisler and Glass 1985). At 35 °C and a heart rate of 53 beats/min, separation of oxygenated and hypoxic blood was nearly perfect, with 91% of the right atrial blood directed into the pulmonary artery. Both left-to-right and right-to-left shunts were small (ca. 10%), the right-to-left shunt fraction exclusively entering the left aortic arch (Ishimatsu et al. 1988).

A striking deviation between the varanids and the other noncrocodilian reptiles is the differentiation between intracardiac blood pressures in the pulmonary and systemic circuits (Millard and Jonansen 1974; Burggren and Johansen

1982). A reduction in pulmonary perfusion pressure in the chelonians is due to the presence of an arterial resistance in the outflow tract (see earlier), but in the varanids the pressure generated by the *cavum pulmonale* is lower than in the *cavum venosum/cavum arteriosum*. This is made possible by the well-developed horizontal muscle ridge ("Muskelleiste"), which presses against the opposite wall of the ventricle during systole, effectively sealing off the *cavum pulmonale* to allow pressure gradients in excess of 4 kPa between the two compartments (Burggren and Johansen 1982).

4 Crocodilians

In the crocodilians, the ventricles are completely separated as in birds and mammals, and the free lumen of the ventricles is larger than in the noncrocodilian reptiles (Webb 1979; van Mierop and Kutsche 1985). The main systemic aorta (right aorta, RAo) leaves the left ventricle, and a common pulmonary artery runs from the right ventricle. However, a uniquely crocodilian arrangement is the emergence from the right ventricle of the left aortic arch as a second aorta (left aorta, LAo) (Figs. 4,5). This remarkable anatomical feature immediately suggests the possibility of a pulmonary bypass shunt but, as will be detailed later, such a shunt develops only under special circumstances.

The LAo, RAo and the pulmonary arterial trunk are enclosed in a shared connective tissue sheath within the pericardium, forming a substantial outflow tract. The lumen of the common base of the RAo and carotid and subclavian arteries is quite large, as is the more thin-walled and elastic common pulmonary trunk, while the diameter of the LAo is relatively modest. The two aortae communicate immediately distal to the aortic valves, via a small aperture, the *foramen Panizzae* (Panizza 1833), and again at a second point further distally, where a vascular anastomosis connects the two vessels (Fig. 4). Beyond this point the RAo continues caudally as the dorsal aorta, while the LAo turns off as the coeliac artery and gives off smaller arteries supplying the gut.

4.1 Left Aortic Blood Flow in Crocodilians

The oxygenation of the LAo blood has been shown to be high and almost identical to that in the RAo, suggesting that LAo blood is derived from the RAo via the *foramen Panizzae* (White 1956,1969; Greenfield and Morrow 1961). Pressure recordings made by White (1956, 1968, 1969) demonstrated synchronous pressure events in the LAo and RAo of the alligator, and he concluded that the foramen remains open throughout the cardiac cycle. Much earlier, the same conclusion, based on anatomical observations, was reached by Fritsch (1869).

However, the size of the *foramen* and its location within the pockets formed by the aortic valves (see Webb 1979), led Sabatier (1873) to suggest that the *foramen* is

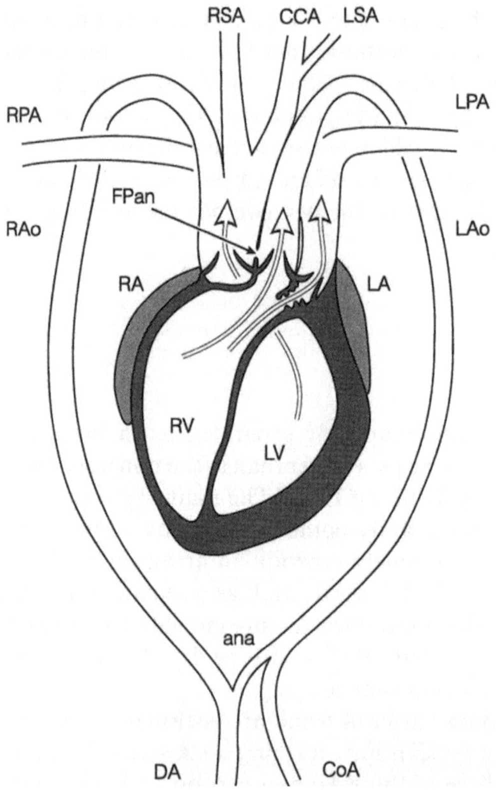

Fig. 4. Schematic diagram of the crocodilian heart, showing the origin of the (common) pulmonary artery (which gives rise to the left, *LPA*, and right, *RPA*, pulmonary arteries), and the left (*LAo*) and right (*RAo*) aorta. The bases of the RAo/carotid/subclavian tract, the Lao and the common pulmonary artery are bound tightly together to form the outflow tract and are enclosed in the pericardium. The two aortae communicate at the *foramen Panizzae* (*FPan*), immediately distal to the aortic valves, and again at the anastomosis (*ana*) further caudally. Distal to the anastomosis, the RAo continues into the tail as the dorsal aorta (*DA*), while the LAo runs as the coeliac artery (*CoA*) to the gut. *Arrows* indicate possible blood pathways, although the LAo valves normally remain closed throughout the cardiac cycle. Note the protrusions at the base of the pulmonary arterial trunk, which fit together like "cogs" (Webb 1979), effectively closing off pulmonary flow during part of the cardiac cycle. *Abbreviations: CCA* common carotid artery; *LA* left atrium; *LSA* left subclavian artery; *LV* left ventricle; *RA* right atrium; *RSA* right subclavian artery; *RV* right ventricle

obstructed by the medial cusp of the aortic values during systole. Indeed, many later studies of anaesthetized as well as unanaesthetized and undisturbed animals showed that the LAo blood pressure trace drops off from that of the RAo during systole (Fig. 6; Greenfield and Morrow 1961; Axelsson et al. 1989; Shelton and Jones 1991). Only during diastole are the pressures identical due to pressure equilibration through the open *foramen Panizzae*. Furthermore, recording of flow patterns in the major vessels of the unanaesthetized caiman, *Caiman crocodylus*

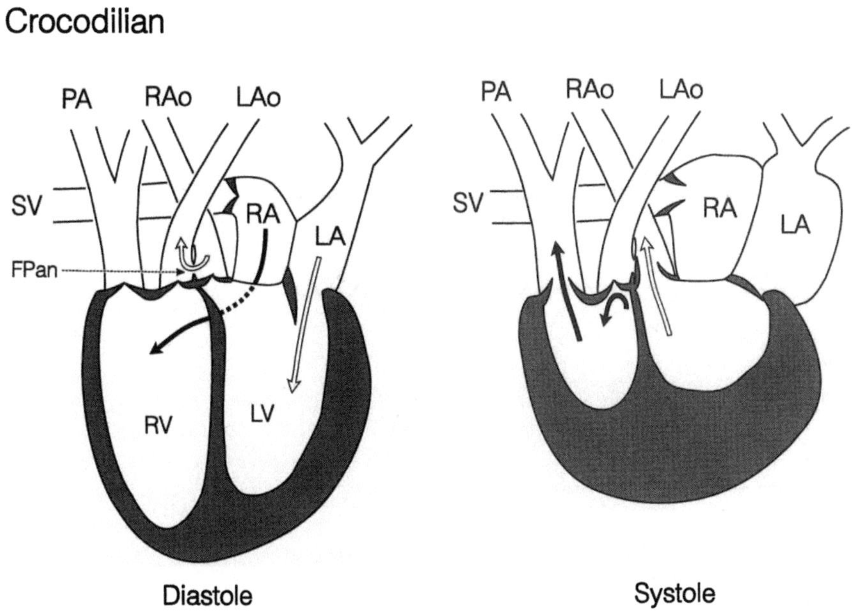

Fig. 5. Diagrammatic representation of the ventricular compartments and major arteries of a crocodilian, showing the flow of hypoxic ("venous"; *filled arrows*) and oxygenated ("arterial"; *open arrows*) blood during diastole (*left*) and systole (*right*). During diastole, the left and right ventricle is filled from its respective atrium. Due to the complete separation of the two ventricles, no mixing of oxygenated blood from the left atrium and hypoxic ("venous") blood from the right atrium occurs. The aortic valves are closed, and blood can therefore flow through the *foramen Panizzae* to equilibrate pressures in the two aortae. During systole, the left ventricle ejects blood into the RAo and during part of the systole the foramen is occluded by the medial cusp of the RAo valve. Blood from the right ventricle enters the pulmonary outflow tract. The LAo blood pressure normally remains higher than right ventricular pressure throughout the cardiac cycle, which prevents opening of the LAo valves (see also Fig. 6). *Abbreviations* as in Fig. 1

(Axelsson et al. 1989), and later in the anaesthetized alligator, *Alligator mississippiensis* (Shelton and Jones 1991), demonstrates a very low net LAo flow that reverses coincidently with asynchrony in the systolic pressure tracings (see also Fig. 6).

An explanation for the discrepancy in the pressure recordings between these later studies and those of White (1956, 1968, 1969) was offered by Grigg and Johansen (1987) in their study of pressure events in unanaesthetized crocodiles (*Crocodylus porosus*). In this study, asynchronous pressure patterns were found to be associated with higher heart rates and higher systemic blood pressures (> 50 mmHg), while a synchronous pattern was recorded during lower blood pressures. To explain the generation of the two patterns, a variable foramen diameter was proposed (Grigg and Johansen 1987). This suggestion wins considerable support from recent observations that the edge of the foramen of *Crocodylus porosus* indeed comprises smooth muscle and is innervated by a rich plexus of nerve terminals containing several neuropeptides as well as markers for adrenergic neurons

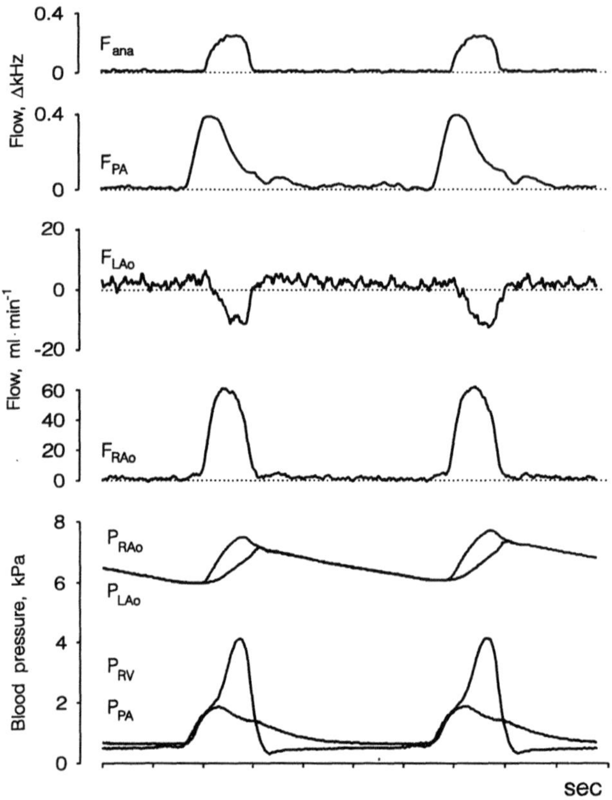

Fig. 6. Recordings of blood flows and pressures in the right ventricle and major arteries in an unanaesthetized 4.7 kg crocodile (*Crocodylus porosus*). Flows in the anastomosis and pulmonary artery were recorded using single crystal Doppler flow probes (Iowa University), and uncalibrated flows are displayed as ΔkHz Doppler shift. Upward pen deflection in the anastomosis flow tracing reflects flow from RAo to LAo. Flows in LAo and RAo were measured using an ultrasonic flow meter (Transonic), and calibrated volume flows are displayed. Blood pressures were recorded via cannulae implanted into the large arteries inside the outflow tract and into the right ventricle. Right ventricular pressure remains below LAo pressure, and the LAo valves remain closed. Note the lack of synchrony in LAo and RAo pressure traces during most of the systole, and the right ventricular pressure peak that exceeds and follows the peak in the pulmonary arterial pressure. F_{ana} Anastomosis flow; F_{PA} left pulmonary arterial flow; F_{LAo} left aortic flow; F_{RAo} right aortic flow; P_{RAo} right aortic pressure; P_{LAo} left aortic pressure. P_{RV} right ventricular pressure; P_{PA} pulmonary arterial pressure. (Courtesy of S. Nilsson, G.C. Grigg, R. Fritsche, C.E. Franklin and M. Axelsson).

(S. Holmgren, P. Karila, I.L. Gibbins unpubl.). Thus, although direct evidence for a variable foramen diameter in vivo is still wanting, the anatomical requisites for a regulated sphincter appear to be present.

The asynchrony in the LAo/RAo pressure patterns terminates abruptly at the end of systole, when the LAo pressure is equilibrated with that in the RAo. This rise, the "foramen spike" (Grigg and Johansen 1987; Grigg 1989), varies in magnitude depending on the pressure difference and the patency of the recording system but, despite claims to the contrary (Shelton and Jones 1991), the "foramen spike" is a normal and very real event in the pressure tracings from the LAo of both caimans and crocodiles (cf. Grigg and Johansen 1987; Axelsson et al. 1989). The size of the foramen spike depends to a large extent on the end-systolic trans-foramen pressure difference, and this in turn depends on how quickly pressure in the LAo builds up due to flow "round the loop" (i.e. from the RAo via the anastomosis and into the LAo). Experimental conditions, such as partial obstruction of the LAo with the recording cannula, may affect the size of the "spike" (Shelton and Jones 1991), and changes in anastomosis diameter will produce the same effects.

4.2 Anastomosis and Coeliac Artery Blood Flow

The vascular anastomosis connecting the two aortae branches off the LAo laterally and runs around to enter the RAo a few centimetres further caudally. In the crocodile, *Crocodylus porosus*, the vessels at this point are embedded in very tough connective tissue, making access to the anastomosis difficult. However, direct blood flow measurements in the anastomosis were made using single crystal Doppler flow probes. The measurements show right-to-left anastomosis flow in the untreated animal, and this contribution from the RAo determines the pressure and flow pattern in the coeliac artery (Fig. 6; Axelsson et al. 1991; S. Nilsson, G.C. Grigg, R. Fritsche, C.E. Franklin and M. Axelsson, unpubl.).

Coeliac artery blood is thus to a large extent derived directly from the RAo, just as in most other vertebrates (Axelsson et al. 1991). There is an additional contribution from the LAo of blood that has entered from the RAo via the *foramen Panizzae*. Feeding appears to lower the vascular resistance of the crocodilian gut (Axelsson et al. 1991). Recent estimates in undisturbed, unanaesthetized *Crocodylus porosus*, in which direct flow measurements in both the coeliac artery and anastomosis were made, suggest that about half of the coeliac artery blood originates from the LAo (S. Nilsson, G.C. Grigg, R. Fritsche, C.E. Franklin and M. Axelsson, unpubl.).

4.3 Pulmonary Blood Flow

The pulmonary outflow tract in crocodilians shows some remarkable anatomical features. The arterial trunk leaves the right ventricle from a funnel-shaped subpulmonary channel, which is lined by fibrous nodules that protrude from the wall and partly occlude the vascular lumen (Fig. 4; Greenfield and Morrow 1961; Webb 1979; van Mierop and Kutsche 1985). These nodules fit together like the "teeth of opposing cogs" (Webb 1979), and may under some circumstances render a substantial resistance to blood flow, producing a persistent pressure gradient between the right atrium and pulmonary artery (Greenfield and Morrow 1961; Shelton and Jones 1991).

We have observed the tightly fitting "cog-teeth" from within the right ventricle of *Crocodylus porosus* using an endoscope, and closure of the subpulmonary arterial channel within the heart is evident during parts of the cardiac cycle. In our experiments with unanaesthetized *Crocodylus porosus*, the recorded pressures in the right ventricle and pulmonary outflow tract show identical development during the early part of ventricular systole. At mid- systole, a rapid rise in right ventricular pressure, coincident with a drop in pulmonary arterial pressure, suggests closure of the subpulmonary arterial channel (Fig. 6; S. Nilsson, G.C. Grigg, R. Fritsche, C.E. Franklin and M. Axelsson, unpubl.).

4.4 Cardiovascular Control in Crocodilians

Despite the fact that application of modern flow and pressure measurement techniques makes the central crocodilian cardiovascular physiology less bewildering, there is little or no information about the in vivo conditions that lead to a right-to-left shunting (opening of the LAo valves), or other major changes in central hemodynamics under natural conditions. As in other diving reptiles, a very substantial bradycardia occurs during submersion (Wilber 1960; Andersen 1961; White 1969; Grigg and Johansen 1987; Axelsson et al. 1991; S. Nilsson, G.C. Grigg, R. Fritsche, C.E. Franklin and M. Axelsson, unpubl.). Reptilian diving/submersion bradycardia depends largely on the cholinergic vagal tonus to the heart (see also Burggren 1975). Our estimates of adrenergic and cholinergic influence on the heart of *Crocodylus porosus* show a doubling of heart rate after atropine injection (from 20 to 40 beats$\cdot$min^{-1}). Heart rate returns almost exactly to the initial value after subsequent injection of the β-adrenoceptor antagonist sotalol (S. Nilsson, G.C. Grigg, R. Fritsche, C.E. Franklin and M. Axelsson, unpubl.).

Completely undisturbed crocodilians on land show a much lower heart rate than those studied in the presence of humans and there appears to be a large psychogenic component in heart rate regulation in the crocodilians (Gaunt and Gans 1969; Huggins et al. 1969; Wienheimer et al. 1982). In the free-ranging *Alligator mississippiensis* equipped for heart rate telemetry, submersion bradycardia does not always occur (Smith et al. 1974).

Early experiments with forced dives in crocodilians produced right-to-left shunts (White 1969), and it was concluded that diving (submersion) was the natural cause of an increase in pulmonary vascular resistance, and therefore a rise in right ventricular pressure large enough to exceed the LAo pressure and thus open the LAo values (Fig 4.) There is good evidence for the development of a right-to-left shunt also during voluntary aerobic dives in relatively undisturbed crocodiles (Grigg and Johansen 1987).

Various compounds such as acetylcholine (Axelsson et al. 1989), substance P (Axelsson et al. 1991) and bombesin (Holmgren et al 1989) produce a marked pulmonary vasoconstriction, and within a few heart beats a right-to-left shunt develops. Evidence for the occurrence of such pulmonary vasoconstriction in vivo does, however, remain equivocal and it is possible that reduced systemic blood pressure is the major cause of a right-to-left shunt (Shelton and Jones 1991). Indeed, observations in the crocodile *Crocodylus porosus* suggest that right-to-left shunting is the cause of a lowered systemic blood pressure in "relaxed" animals, or animals that have been injected with the α-adrenoceptor antagonist phentolamine (which reduces systemic blood pressure) (S. Nilsson, G.C. Grigg, R. Fritsche, C.E. Franklin and M. Axelsson, unpubl.).

These observations, combined with those demonstrating low heart rate in the completely undisturbed animal (see earlier), hint that a pulmonary bypass is a feature of the completely relaxed crocodilian. Clearly, further studies employing techniques that reduce the interference with the animal as far as possible, e.g. telemetry techniques, are necessary to resolve the question of under which circumstances pulmonary bypass occurs in vivo.

5 Concluding Remarks

Although a high degree of separation between oxygenated and hypoxic blood within the heart is evident already in lungfish and amphibians (see Johansen and Burggren 1980), the division of the respiratory and systemic flows have been taken several steps further in the reptiles. The presence of septa and ridges within the ventricle aids distribution of the ventricular ejection into the appropriate outflow vessels. Among the noncrocodilian reptiles, the varanid heart shows the best ventricular separation of left and right atrial blood, thanks to the large valves that obstruct the interventricular canal during diastole, the small *cavum venosum* and the complete systolic separation by the horizontal septum ("Muskelleiste") of the *cavum pulmonale*.

A difference in pulmonary compared to systemic perfusion pressures is evident in many reptiles. This occurs due to high resistance segments in the pulmonary outflow tract and, in varanids and crocodilians, also to the isolation of the *cavum pulmonale* (during systole) or right ventricle, respectively, from the higher pressures generated by the dorsal cava (*cavum venosum/cavum arteriosum*) or left ventricle.

Further properties of the "ideal" heart of an intermittently breathing animal include the retained option for right-to-left (and left-to-right) shunts in the heart of all reptiles, and the very substantial variations in heart rate (and thus cardiac pumping capacity) between periods of apnoea (notably in the diving species) and periods of breathing. Taken together, the properties of the reptilian heart and central vascular arrangement, particularly in varanids and crocodilians, make it very tempting to agree with the view expressed by Grigg (1989) that "the crocodilian heart and cardiovascular system [is] the best 'designed' of all".

Acknowledgements. I wish to thank Drs. Michael Axelsson and Susanne Holmgren for reading the manuscript. Our own research on reptilian circulatory control has been supported by the Swedish Natural Science Research Council, the University of Queensland and the Royal Society for Science and Arts of the University of Göteborg.

References

Andersen HT (1961) Physiological adjustments to prolonged diving in the American alligator *Alligator mississippiensis*. Acta Physiol Scand 53: 23–45

Axelsson M, Holm S, Nilsson S (1989) Flow dynamics of the crocodilian heart. Am J Physiol 256: R875–R879

Axelsson M, Fritsche R, Holmgren S, Grove DJ, Nilsson S (1991) Gut blood flow in the estuarine crocodile, *Crocodylus porosus*. Acta Physiol Scand 142: 509–516

Berger PJ, Heisler N (1977) Estimation of shunting, systemic and pulmonary output of the heart and regional blood flow distribution in unanaesthetized lizards (*Varanus exanthematicus*) by injection of radioactively labelled microspheres. J Exp Biol 71: 111–121

Burggren WW (1975) A quantitative analysis of ventilation tachycardia and its control in two chelonians, *Pseudemys scripta* and *Testudo graeca*. J Exp Biol 63: 367–380

Burggren WW (1977) The pulmonary circulation of the chelonian reptile: morphology, haemodynamics and pharmacology. J Comp Physiol 116: 303–323

Burggren WW (1982) Pulmonary blood plasma filtration in reptiles: a wet vertebrate lung? Science 215: 77–78

Burggren WW (1985) Hemodynamics and regulation of central cardiovascular shunts in reptiles. In: Johansen K, Burggren WW (eds) Cardiovascular shunts: phylogenetic, ontogenetic and clinical aspects. Alfred Benzon Symp 21, Munksgaard, Copenhagen, pp 121–142

Burggren WW (1987) Form and function in reptilian circulations. Am Zool 27: 2–19

Burggren WW, Johansen K (1982) Ventricular hemodynamics in the monitor lizard *Varanus exanthematicus*: pulmonary and systemic pressure separation. J Exp Biol 96: 343–354

Burggren WW, Shelton G (1979) Gas exchange and transport during intermittent breathing in chelonian reptiles. J Exp Biol 82: 75– 92

Campbell G (1971) Autonomic innervation of the pulmonary vascular bed in a toad (*Bufo marinus*). Comp Gen Pharmacol 2: 287–294

Davies DG (1989) Distribution of systemic blood flow during anoxia in the turtle, *Chrysemys scripta*. Respir Physiol 78: 383–390

De Saint-Aubain ML (1982) Vagal control of pulmonary blood flow in *Ambystoma mexicanum*. J Exp Zool 221: 155–158

De Saint-Aubain ML, Wingstrand KH (1979) A sphincter in the pulmonary artery of the frog *Rana temporaria* and its influence on blood flow in skin and lungs. Acta Zool (Stockholm) 60: 163–172

Fritsch G (1869) Zur vergleichenden Anatomie des Amphibienherzens. Arch Anat Physiol 6: 654–758

Gaunt AS, Gans C (1969) Diving bradycardia and withdrawal bradycardia in *Caiman crocodilus*. Nature 223: 207–208

Greenfield LJ, Morrow AG (1961) The cardiovascular hemodynamics of Crocodilia. J Surg Res 1: 97–103

Grigg G (1989) The heart and patterns of cardiac outflow in Crocodilia. Proc Aust Physiol Pharmacol Soc 20: 43–57

Grigg GC, Johansen K (1987) Cardiovascular dynamics in *Crocodylus porosus* breathing air and during voluntary aerobic dives. J Comp Physiol 157: 381–392

Heisler N (1985) Mechanism of intracardiac shunting in reptiles. In: Gilles R (eds) Circulation, respiration and metabolism. Springer, Berlin Heidelberg New York, pp 125–138

Heisler N, Glass ML (1985) Mechanisms and regulation of central vascular shunting in reptiles. In: Johansen K, Burggren WW (eds) Cardiovascular shunts: phylogenetic, ontogenetic and clinical aspects. Alfred Benzon Symp 21, Munksgaard, Copenhagen, pp 334– 353

Heisler N, Neumann P, Maloiy GMO (1983) The mechanism of intracardiac shunting in the lizard *Varanus exanthematicus*. J Exp Biol 105: 15–32

Holmgren S, Axelsson M, Jensen J, Aldman G, Sundell K, Jönsson A-C (1989) Bombesin-like immunoreactivity and the effect of bombesin in the gut, circulatory system and lung of the caiman, *Caiman crocodylus*. Exp Biol 48: 261–271

Huggins SE, Hoff HE, Pena RV (1969) Heart and respiratory rates in crocodilian reptiles under conditions of minimal stimulation. Physiol Zool 42: 320–333

Ishimatsu A, Hicks JW, Heisler N (1988) Analysis of intracardiac shunting in the lizard, *Varanus niloticus*: a new model based on blood oxygen levels and microsphere distribution. Resp Physiol 71: 83–100

Johansen K (1985) A phylogenetic view of cardiovascular shunts. In: Johansen K, Burggren WW (eds) Cardiovascular shunts: phylogenetic, ontogenetic and clinical aspects. Alfred Benzon Symp 21, Munksgaard, Copenhagen, pp 17–37

Johansen K, Burggren WW (1980) Cardiovascular function in the lower vertebrates. In: Bourne GH (eds) Hearts and heart-like organs. Academic Press, London, pp 61–117

Johansen K, Lenfant C, Hanson D (1970) Phylogenetic development of pulmonary circulation. Fed Proc 29: 1135–1140

Luckhardt AB, Carlson AJ (1921) Studies on the visceral sensory nervous system. VIII. On the presence of vasomotor fibers in the vagus nerve to the pulmonary vessels of the amphibian and the reptilian lung. Am J Physiol 36: 72–112

Millard RW, Johansen K (1974) Ventricular outflow dynamics in the lizard, *Varanus niloticus*: responses to hypoxia, hypercarbia and diving. J Exp Biol 60: 871–880

Milsom WK, Langille BL, Jones DR (1977) Vagal control of pulmonary vascular resistance in the turtle, *Chrysemys scripta*. Can J Zool 55: 359–367

Panizza B (1833) Sulla struttura del cuore e sulla circolazione del sangue del *Crocodilus lucius*. Bibl Ital 70: 87–91

Sabatier A (1873) Etudie sur le coeur et la circulation centrale dans la série des vertebrés. Ann Sci Nat Zool Ser V 18: 1–89

Shelton G, Burggren WW (1976) Cardiovascular dynamics of the chelonia during apnoea and lung ventilation. J Exp Biol 64: 323–343

Shelton G, Jones DR (1991) The physiology of the alligator heart: the cardiac cycle. J Exp Biol 158: 539–564

Smith DG, Campbell G (1976) The anatomy of the pulmonary vascular bed in the toad *Bufo marinus*. Cell Tissue Res 165: 199–213

Smith DG, MacIntyre DH (1979) Autonomic innervation of the visceral and vascular smooth muscle of a snake lung (*Ophidia*: Colubridae). Comp Biochem Physiol 62C: 187–191

Smith EN, Allison RD, Crowder WE (1974) Bradycardia in a free ranging american alligator. Copeia 3: 770–772

Van Mierop LH S, Kutsche LM (1985) Some aspects of comparative anatomy of the heart. In: Johansen K, Burggren WW (eds) Cardiovascular shunts: phylogenetic ontogenetic and clinical aspects. Alfred Benzon Symp 21, Munksgaard, Copenhagen, pp 38–56

Webb GJW (1979) Comparative cardiac anatomy of the reptilia. III. The heart of crocodilians and a hypothesis on the completion of the interventricular septum of crocodilians and birds. J Morphol 161: 221–240

Webb GJW, Heatwole H, De Bavay J (1971) Comparative cardiac anatomy of the reptilia. I. The chambers and septa of the varanid ventricle. J Morphol 134:335–350

White FN (1956) Circulation in reptilian heart (*Caiman sclerops*). Ant Rec 125: 417–432

White FN (1959) Circulation in reptilian heart (Squamata). Anat Rec 135: 129–134

White FN (1968) Functional anatomy of the heart of reptiles. Am Zool 8: 211–219

White FN (1969) Redistribution of cardiac output in the diving alligator. Copeia 3: 567–570

White FN (1970) Central vascular shunts and their control in reptiles. Fed Proc 29: 1149–1153

White FN (1976) Circulation. In: Gans C (eds) Biology of the Reptilia. Academic Press, London, pp 275–334

White FN (1985) Role of intracardiac shunts in pulmonary gas exchange in chelonian reptiles. In: Johansen K, Burggren WW (eds) Cardiovascular shunts. Munksgaard, Copenhagen, pp 296–309

White FN, Ross G (1966) Circulatory changes during experimental diving in the turtle. Am J Physiol 211: 15–18

Wienheimer CJ, Pendergast DR, Spotila JR, Wilson DR, Standora EA (1982) Peripheral circulation in *Alligator mississippiensis*: effects of diving, fear, movement, investigator activities, and temperature. J Comp Physiol 148: 57–63

Wilber CG (1960) Cardiac responses of *Alligator mississippiensis* to diving. Comp Biochem Physiol 1: 164–166

Chapter 8

Central Cardiovascular Function in Amphibians: Qualitative Influences of Phylogeny, Ontogeny, and Season

W.W. Burggren

Contents

1 Introduction

The central cardiovascular function of amphibians has been studied in detail for many decades. Some investigators have been motivated primarily by the search for supposedly simple models which may provide a greater understanding of the circulations of higher vertebrates. Others were interested in physiological problems presented by the cardiovascular systems of amphibians, because these vertebrates are of interest in their own right, as a unique, specialized and highly successful group (see Feder 1992). Whatever the rationale for studying the amphibian cardiovascular system, investigators routinely experience its paradoxical nature. In some respects, the circulation of various amphibians represents a relatively simple system for convective transport of blood (e.g. lack of anatomical division of the ventricle). Yet, in other aspects (e.g. the control of cardiac

Department of Biological Sciences, University of Nevada, Las Vegas, NV 89154, USA

Advances in Comparative and Environmental Physiology, Vol. 21

output distribution), the amphibian circulation rivals that of any other vertebrate class in terms of flexibility and adaptability.

Analysis and synthesis of the literature gain considerable additional complexity because of three highly influential factors that have not been consistently controlled in prior investigations—phylogeny, ontogeny, and seasonality. Phylogeny has obvious influences on cardiovascular anatomy and physiology, because there is no more a typical amphibian than there is a typical reptile. Typology, the heavy dependence upon a few model species, has as little use in amphibian cardiovascular studies as it does in any other aspect of animal biology (Gans 1970; Burggren and Bemis 1990). Yet, data from only a few amphibian species mostly anurans, pervade (and bias) the literature. Ontogeny also exerts profound influences on the cardiovascular system of amphibians. Most amphibians undergo a marked metamorphosis that is accompanied by a shift in the balance of water and air breathing. Not surprisingly, the cardiovascular system shows associated changes in both anatomy and physiology. Finally, as a cardiovascular variable, season has largely been either ignored or, if controlled, not fully exploited as an interesting environmental variable. Yet, scattered reports from the literature suggest that such factors can exert a major influence on cardiovascular function.

The object of this chapter is to discuss the influence of phylogeny, ontogeny, and season on the central cardiovascular function of amphibians, and to show how these topics deserve specific attention in future investigations. Some basic understanding of the amphibian circulation is assumed (for background information, the reader is directed to detailed reviews by Shelton 1976, 1985; Johansen and Burggren 1980; Burggren 1988; West and van Vliet 1992).

2 Phylogeny and Central Cardiovascular Function

Of the amphibians, the cardiovascular function of the anurans (frogs, toads) has been studied far more extensively than that of the urodeles (newts and salamanders) or apodans (caecelians).

2.1 The General Anuran Cardiovascular Pattern

All anurans examined to date share the same basic structure of the heart and central arterial circulation. Anatomically distinct left and right atria receiving blood from the lungs and the systemic venous circulation, respectively, pump blood into a single ventricle (Fig. 1). The ventricle, formed from spongy trabeculate myocardium lacking a coronary blood supply, ejects blood into a large conus arterious. This large, elastic artery bifurcates distally into symmetrical and separate systemic, carotid, and pulmocutaneous arterial arches. The pulmocutaneous arch gives rise distally to a large pulmonary artery and a smaller

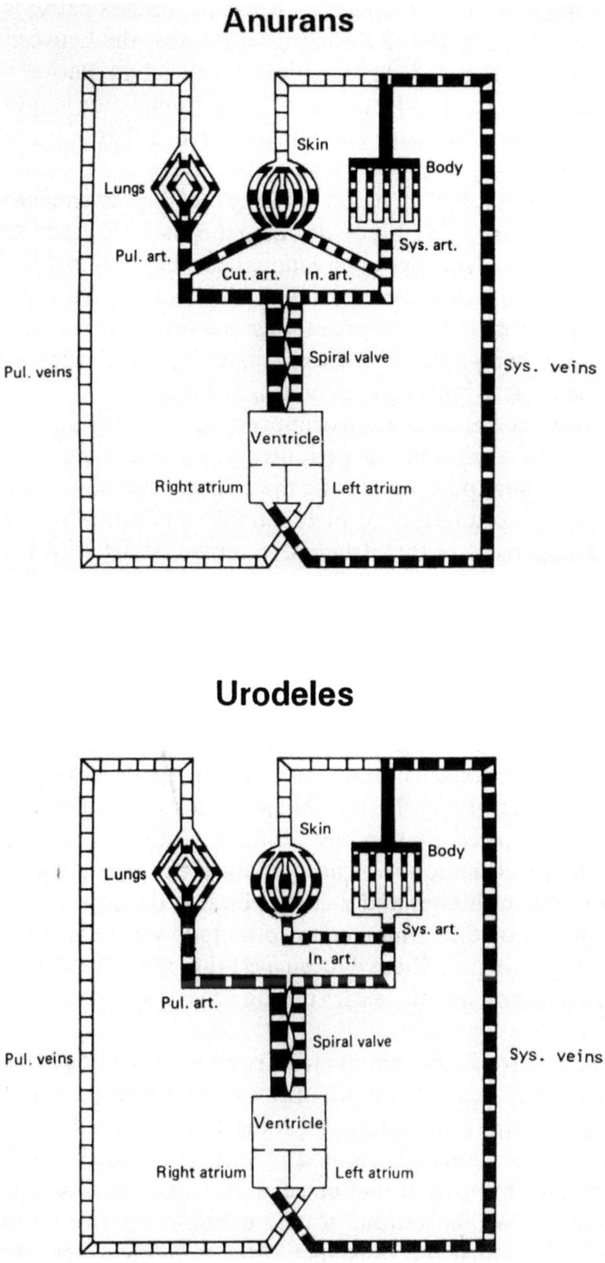

Fig. 1. Diagrammatic representation of the circulation of anurans and urodeles (not lungless forms). Deoxygenated blood is *black*; oxygenated blood is *white*

cutaneous artery, which along with small branches of vertebral arteries perfuses the skin. The orfice between the ventricle and the conus arteriosus, and between the bulbus and the arterial arches, is guarded by semilunar cusp valves. Between these two sets of valves is the single spiral valve, a fold of tissue running the length of the bulbus attached along only one edge (cf. Shelton 1976; Johansen and Burggren 1980).

On the venous side of the circulation, all blood draining the lungs enters into the left atrium via paired pulmonary veins. Blood draining the systemic vascular beds (including the skin) is channeled into systemic venous beds that flow into the right atrium. Thus, even though the skin receives a dual blood supply from the pulmocutaneous and systemic arches, its venous drainage is nonspecialized, and oxygenated blood leaving the skin together with deoxygenated leaving the deeper tissues mixes together in the central veins before entering the heart.

Virtually all studies on the central circulation of anurans agree on several key features. There are no regional variations in pressures generated within the ventricle; it acts as a single pressure pump perfusing the central arterial circulation. Yet, even though there is no anatomical or functional separation of the ventricle, the structural arrangement of the atria and ventricle, along with the highly trabeculate nature of the ventricular wall, can provide a relatively laminar and separate flow of left and right atrial blood through the ventricle and conus arteriosis into the arterial arches. For example, in the anesthetized bullfrog *Rana catesbeiana,* as much 91% of left atrial blood enters the systemic circulation and 84% of right atrial blood enters the pulmocutaneous circulation (Tazawa et al. 1979). Although a highly effective separation of blood flow can be achieved in the anuran heart, the extent of intracardiac shunting varies greatly with pattern of breathing (Shelton 1985; Boutilier 1989). In the classic demonstration of this phenomenon, Shelton (1970) showed that in the South African clawed frog *Xenopus laevis* periods of lung ventilation were accompanied by an increase in blood flow in the pulmocutaneous arches and a corresponding decrease in the systemic arches. During apnea, however, pulmocutaneous blood was reduced to levels below that in the systemic arches. These data suggest that the intracardiac blood shunt increases during apnea, and decreases during actual periods of lung ventilation.

Although the existence of intracardiac shunts within the anuran circulation has been known for more than 20 years, the mechanisms by which these shunts are achieved and controlled is still not completely understood. Since the pulmocutaneous and systemic circulations are essentially located in parallel with respect to the heart, any change in resistance of one circuit relative to the other will cause a redistribution of cardiac output to the circuit with the lower resistance. This mechanism, in which the resistance balance changes between two circuits located in parallel, applies not only between systemic and pulmocutaneous circuits, but also between pulmonary are cutaneous vascular beds within the pulmocutaneous circuit itself. For example, in the marine toad *Bufo marinus,* redistribution of blood flow between pulmonary and cutaneous circuit is achieved by reciprocal changes in the peripheral resistance of these beds. The mechanism triggering these changes appears to involve both O_2- and CO_2-sensitive chemoreceptors in *Bufo marinus* (West and Burggren 1984; see also Burggren 1988; West and Van Vliet 1992). In *Rana catesbeiana*, the relative proportion of blood flow to the skin and lungs can also be adjusted in a reciprocal fashion, in particular to capitalize on a predominance of oxygen in either the air or water (Boutilier et al. 1986).

2.2 Not All Amphibians Are Anurans – The General Urodele Pattern

Frequently, the cardiovascular physiology and anatomy of anurans are assumed representative of amphibians in general. Yet, important distinctions can (and should) be made. For example, urodeles lack a distinct cutaneous artery originating from a pulmocutaneous arch (Fig. 1). The skin of urodeles, therefore, is "just another vascular bed" with respect to its arterial supply, receiving the same blood supply as the other systemic tissues. This difference from the blood supply of the skin of anurans apparently does not prevent the skin of urodeles from serving as an organ of oxygen uptake along with the lungs and gills, as cutaneous O_2 uptake can account for the majority of total O_2 uptake in urodeles (Feder and Burggren 1985).

To what extent the ratio of blood flow between the pulmonary and systemic circuits changes in response to intermittent ventilation has been examined in very few urodeles. In *Amphiuma tridactylum* up to 80% of the blood in the dorsal aorta is derived from the pulmonary veins when blood Po_2's are high during breathing, suggesting a relatively small right-to-left shunt at that time (Toews et al. 1971). Very little left-to-right shunt occurs concurrently. How representative the shunting patterns of *Amphiuma* are of other urodeles is still unclear.

An exceptional modification of the respiratory and cardiovascular systems is found in the urodeles *Chioglossa, Salamandrina, Rhyacotriton,* and the entire family Plethodontidae. As adults, these urodeles completely lack lungs, carrying out all gas exchange using the skin. The pulmonary artery in these salamanders is highly reduced or absent, as is the interatrial septum (see Burggren 1989 for references). This highly specialized mode of life is apparently quite successful, as lungless salamanders are relatively common in both the tropics and temperate climates (Feder 1976).

Another cardiovascular variation that is not typical of anurans is seen in several urodeles. The heart of *Siren intermedia* (Owen 1834; Putnam 1975), *Necturus maculosus* (Putnam and Dunn 1978), and *Cryptobranchus alleganiensis* (Putnam and Parkerson 1985) bears a vertical septum that partially divides the ventricle (Fig. 2). The septum, whose prominence varies with species, superficially resembles that seen in lungfishes, but this structure in urodeles is not homologous with the ventricular septum in the Dipnoi. Whether this unusual cardiac structure alters the performance of the urodele heart, either in pressure generation or blood flow channeling, is unknown. Despite *Siren, Necturus,* and *Cryptobranchus* being large enough for invasive cardiovascular measurements, neither pressure

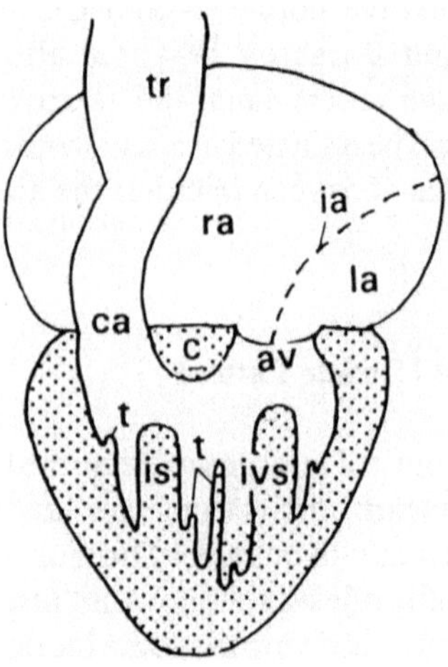

Fig. 2. Ventral view of a frontal section through the ventricle of the urodele *Necturus maculosus.* Note the prominent venticular septum arising from the apex of the ventricle. *av* atrioventricular aperture; *c* anterior ventricular wall; *ca* aperture of the conus arteriosus; *ia* interatrial septum; *is* intraventricular septum; *ivs* interventricular septum; *la* left atrium; *ra* right atrium; *t* muscular trabeculae; *tr* truncus arteriosus (Modified from Putnam 1975)

measurements nor blood gas analyses have been made in the context of assessing the role of the interventricular septum. Incidentally, a number of urodeles have extensive trabecular ridges such as shown for *Necturus* in Fig. 2. Although resembling a true interventricular septum, these structures fail to meet all necessary criteria to be established as such (see Putnam and Dunn 1978, for discussion). Regardless of what these structues actually represent, they may influence blood streaming through the ventricle, and their role in cardiovascular dynamics should also be investivated.

2.3 The Apodan Pattern – Scant Observations

In virtually all aspects of their biology, the apodans (caecilians) are the least well understood of amphibians. Most observations on the cardiovascualr system consist of anatomical observations, with occasional physiological measurements on distantly related species. The central circulation of the semiterrestrial *Siphonops annulatus* from Brazil generally resembles that of urodeles, but this particular species has an incompletely separated and fenestrated interatrial septum (Sawaya 1940; Mendes 1945). The spiral valve is also absent, which together with the abbreviated atrial structure, suggests a circulation with a rather poorly developed ability to separate system and pulmonary venous blood during its passage through the heart and central arterial circulation. However, in *Typhlonectes compressicauda* (also from Brazil), are not only the two atria completely separate, but the leading edge of the spiral valve in the conus arteriosis is fused to the conus wall, forming two distinct channels within the conus (Fig. 3; Toews and MacIntyre, 1978). Arterial blood pressures in these two channels are superimposable, and blood gas analyses indicate that considerable separation is maintained between oxygenated blood being channeled into the systemic arches and deoxygenated blood being channeled into the pulmonary arches.

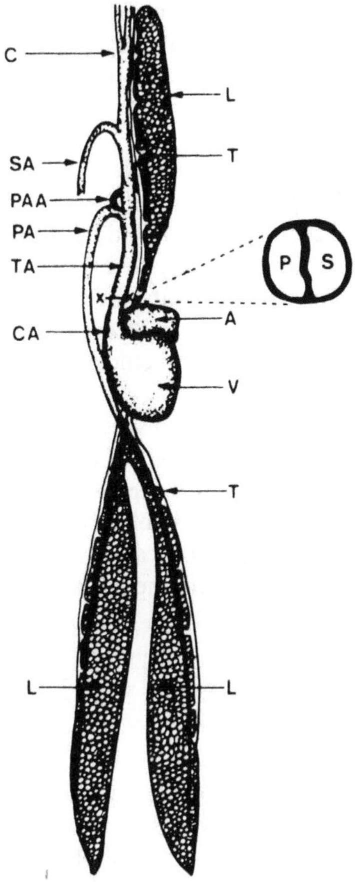

Fig. 3. Central circulation and lungs of the apodan *Typhlonectes compressicauda*. *A* Atrium; *C* carotid artery; *CA* conus arteriosus; *L* lung; *PA* pulmonary artery; *PAA* anterior branch of the pulmonary artery; *SA* systemic arch; *T* trachea; *TA* truncus arteriosus; for expanded view of truncus arteriosus: *P* pulmonary side of truncus; *S* systemic side of truncus. (Toews and McIntyre 1978)

Whether the variation between *Siphonops annulatus* and *Typhlonectes compressicauda* represents two extreme cases of cardiac structure and function in apodans, or whether one or other of the species is representative of most apodans, can only be determined by anatomical and physiological data from a wider range of species.

Interestingly, the ventricle of many apodans has significant ventricular septation produced by muscular trabeculae (see Putnam and Dunn 1978 for references). Although these structures are not homologous with the interventricular septa in some urodeles, they might influence central cardiovascular function and should be investigated in this context.

2.4 Inter- vs Intraorder Variation

The above brief survey of cardiovascular form and function in amphibians reveals considerable variation both within and between orders. There are clearly sufficiently large differences (e.g. the presence or absence of a cutaneous artery) to warrant careful distinction when referring to amphibian circulatory patterns. Anurans, urodeles, and apodans should not be used interchangeably to illustrate "the typical amphibian circulation", since the degree of variability between orders suggests that there is no representative condition.

Although a few species often account for almost all of the published physiological information (e.g. *Rana, Bufo*, and *Xenopus* for anurans; *Ambystoma* for urodeles), sufficient observations exist on other species to indicate that considerable intraorder variation exists in the cardiovascular system. In many instances, the variation has been described only in anatomical traits (the partially divided ventricle of *Siren*, the complete conal separation in *Typhlonectes*), and the physiological implications, if any, of these traits need to be elucidated by experimentation.

Future studies of amphibian cardiovascular systems designed specifically to provide insight into their evolution should be based around a careful choice of species, paying considerable attention to phylogenetic relationships (see Huey 1987; Burggren 1991; Harvey and Pagel 1991; Garland and Adolph 1994 for discussion). Until amphibian biologists understand the patterns of cardiovascular variation due to phylogeny alone, the interpretation of environmental adaptations of the circulation remains problematic.

3 Development of Cardiovascular Function

The ontogeny of most amphibians is characterized by a metamorphic climax that in many species marks a transition from water breathing towards air breathing (see reviews by Burggren and Just 1992; Burggren and Infantino 1994). Associated with the change in respiratory medium is a developmental change in air-breathing organs. While all amphibians breath to some extent with skin as larvae, many also have internal and/or external gills at some point during larval life. Some species develop lungs, which begin to be ventilated well before metamorphosis (see Burggren 1989, 1993). These morphological changes in the respiratory sites involve the addition of new vascular beds and, often, the degeneration of old ones (see Malvin 1985, 1989; Burggren and Just 1992 for reviews). Accordingly, changes in cardiovascular physiology attend these changes in cardiovascular morphology. Progress to date in elucidating these functional changes has occurred in two main areas of interest—regulation of the heart (and especially heart rate) and the ontogeny of central arterial hemodynamics.

3.1 Ontogenetic Patterns of Heart Rate and Its Control

While there have been numerous reports of heart rate in individual stages of amphibian larvae, systematic investigations of how heart rate changes during

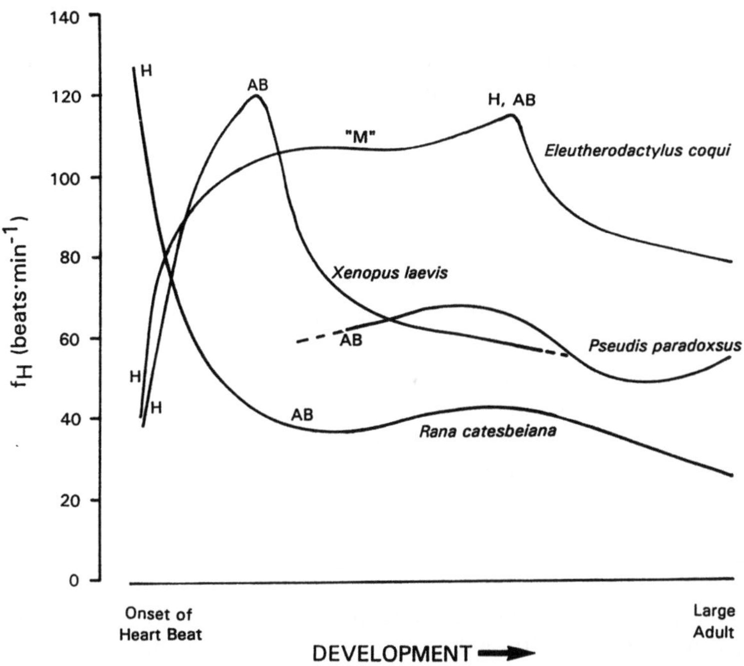

Fig. 4. Changes in resting heart rate (20–25 °C) during development in four species of anuran amphibians. The initial points represent heart rate when the heart first begins to beat, while the terminal points were recorded from the largest adults. Note that this method of plotting "normalizes" the lengths of the total developmental sequence. While this allows plotting of multiple species, it distorts the abscissa such that a vertical line at a given point does not represent the same developmental stage in all species. Because of this, and inherent major differences in the order and timing of developmental events, important developmental landmarks are indicated are indicated by *H*, hatching; *AB*, onset of air breathing; *M*, metamorphosis. Data sources are: *Rana*, Burggren and Doyle (1986a); *Eleutherodactylus* Burggren et al. (1990); *Pseudis*, Burggren et al. (1992); *Xenopus*, Hou (1992) and Bride (1975)

development are few. The existing data are almost exclusively from the following anurans: the bullfrog (*Rana catesbeiana*), the South African clawed frog (*Xenopus laevis*), the coqui (*Eleutherodactylus coqui*), and the paradoxical frog, *Pseudis paradoxus*. Fig. 4 presents data on developmental changes in resting heart rate at 20–25 °C from these anuran species. Clearly, there is no consistent pattern of developmental change when comparing anuran species. Doubtlessly, much variation is attributable to differences in the rate of cardiovascular development and especially in the mode of embryonic/larval development (e.g. direct development into adult vs intermediate larval development). Scaling (allometric) effects due to large changes in body mass may also play a role. Heart rate, like most physiological parameters, has been shown to decrease with increasing body mass, both interspecifically (Stahl 1967) and intraspecifically (see Hou and Burggren 1989 for references). Bullfrogs progress from 40-mg embryos at hatching to 400 + g as adults, representing a 10000-fold increase in body mass. At the same time, resting heart rate at 20 °C falls from nearly 140 between resting heart rate in bullfrog embryos and adults can be predicted almost entirely from allometric equations relating f_H to body mass (W. Burggren, unpubl.). The developmental pattern in the paradoxical frog exhibits an intriguing reversal with respect to body mass changes (extremely large larvae metamorphose into minute frogs). However, the overall change in body mass at metamorphosis results in at most a 30-fold reduction in body mass, which is a relatively small change in terms of producing large heart rate during this developmental transition (Burggren et al. 1992). Similarly, embryonic mass and resting heart rate both rise sharply after the heart first starts beating, which is the opposite of the response expected due purely to allometric influences.

The most complete picture (in a relative sense) of the ontogeny of heart rate control has emerged in *Rana catesbeiana* (Fig. 5), based on experiments with a variety of cholinergic and adrenergic agonists and antagonists as well as normal response to activity and variations in ambient Po_2 (Burggren and Doyle 1986a, b; Kimmel 1990). Description of the individual experiments that generated this broad-ranging scheme for the ontogeny of f_H control in the bullfrog is beyond the scope of this chapter. However, several points are worth emphasizing:

1. In the earliest larval stages following hatching (TK stages I-II), the heart is rather like a metronome. Although small, spontaneous variations in f_H do occur, there is no evidence for reflex adjustment. Heart rate changes occur mainly in response to factors such as temperature change or a degree of hypoxia that is relatively close to the critical Po_2 for oxygen consumption (and thus possibly may represent a direct depressive effect on heart metabolism). Yet, β-adrenergic receptors are present and functional in TK stage I. This suggests that the development of the innervation of the heart, or of endocrine organs releasing catecholamines, may lag behind the development of post-synaptic receptors in cardiac tissue.
2. By TK stage III (the earliest stage examined in this context), both cholinergic inhibitory and adrenergic excitatory nerve fibers exist, but few cardiac reflexes are apparent.
3. By TK stages IV-VII, a noncholinergic, nonadrenergic tachycardia in response to exercise is evident, probably mediated by direct effects (stretching) of increased venous return on the cardiac pacemaker. Even severe hypoxia fails to produce a bradycardia.
4. After the onset of air breathing, resting f_H in larval *R. catesbeiana* reflects a significant vagal inhibition (as revealed by a rise in f_H following administration of atropine) that keeps resting f_H well blow the intrinsic rate. (Interestingly, in *Xenopus laevis* the injection of atropine into the pericardium of NF stage 41–50 larvae causes an increase in f_H from 95 to 170 beats$\cdot$min^{-1}, while the injection of saline causes a corresponding rise of only 25 beats$\cdot$min^{-1} (Hou 1992). These data indicate that a significant vagal tone already occurs in larval *Xenopus* at a developmental stage even earlier than in larval *Rana catesbeiana*.

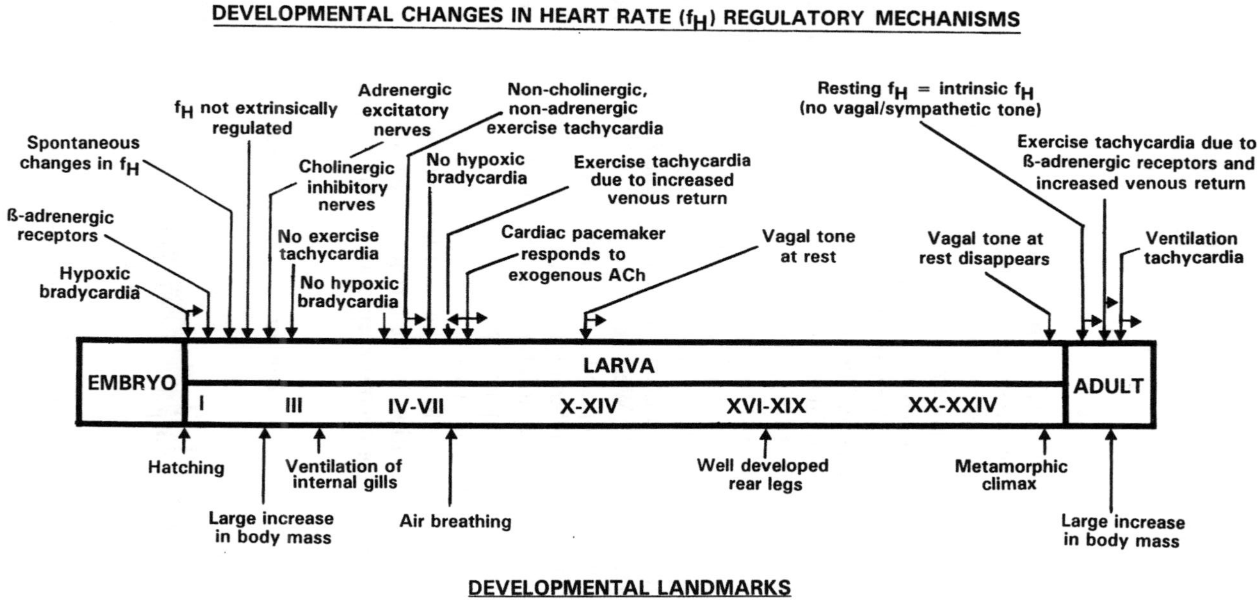

Fig. 5. The ontogeny of cardiac regulation in the bullfrog, *Rana catesbeiana*. The *horizontal bar* represents an individual's life span. Major developmental landmarks as well as the appearence, modification, and/or disappearance of cardiac regulatory mechanisms are shown. *Vertical arrows* represent a single observation of the indicated event. A *vertical arrow combined with a horizontal arrow* indicates the onset of a continuing process

5. After metamorphosis to the adult morph, resting heart rate equals the intrinsic f_H, since neither cholinergic nor β-adrenergic inhibition has any influence on resting f_H. Exercise tachycardia results from a combination of β-adrenergic stimulation and the direct effects (stretching) on the cardiac pacemaker. Ventilation tachycardia (or diving bradycardia) now occurs, largely due to adjustments in vagal tone.

Cholinergic sensitivity of the cardiac pacemaker of *Rana catesbeiana* changes markedly with development (Burggren and Doyle 1986b; Pelster et al. 1993). As larval development progresses towards metamorphic climax, the cholinergic sensitivity of the cardiac pacemaker increases, as determined by the effects of exogenously applied ACh and atropine on resting f_H. After metamorphic climax, however, there is a sharp decrease in cholinergic sensitivity back to the level evident in the earliest larvae. These developmental changes could be caused by changes in the number of ACh receptors in the pacemaker cell membranes and/or changes in the ACh affinity of individual receptors.

Observations of interactions between cardiac and respiratory are few in anuran larvae. Phase coupling of f_H and gill ventilation occurs in some *Xenopus laevis, Pachymedusa dacnicolor*, and *Rana berlandieri* (Wassersug et al. 1981), but its distinct absence has also been reported in *Rana berlandieri* and *Rana catesbeiana* (Burggren et al. 1983; West and Burggren 1983). Most anuran larvae show no response in f_H during episodic lung ventilation, though this is a common cardiac response after metamorphosis (see Burggren and Pinder 1991, for references). Interestingly, ventilation tachycardia does occur in both the giant larvae and the adults of the paradoxical frog, *Pseudis paradoxus* (Burggren et al. 1992). Moreover, larvae of this species occasionally show an increase in heart rate in apparent anticipation of lung ventilation (Fig. 6). This suggests that f_H responses to lung ventilation in the larvae consist of more than either reflexes triggered by mechano- or chemoreceptors only during lung ventilation, or direct

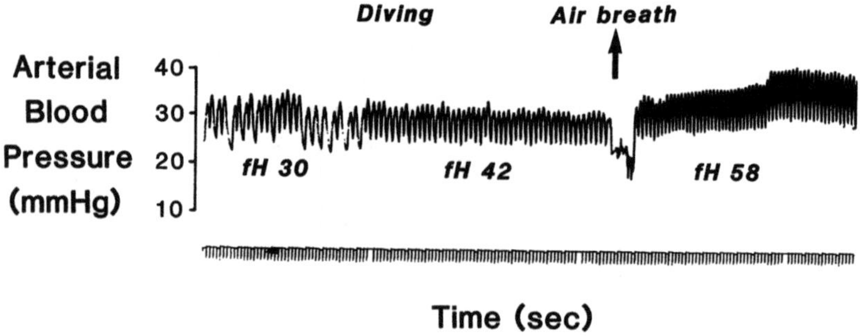

Fig. 6. Arterial blood pressure and heart rate in an unanesthetized 10.3 g, St 45 (Gosner) larva of the paradoxical frog, *Pseudis paradoxsus*. The recording begins towards the end of a voluntary period of apnea. Note the "anticipatory" increase in heart rate about 1 min before the actual termination of apnea by an air breath. (W. Burggren, J. Bicudo, M. Glass, and A. Abe, unpubl. observ.)

neural communication between respiratory and cardiovascular centers in the CNS (central irradiation).

Unfortunately, very little information on developmental changes in the regulation of the heart of urodeles or apodans is available. Important observations have been made, such as the occurrence of ventilation tachycardia in larvae of *Ambystoma tigrinum* (e.g. Health 1980). However, these studies represent only a fragmented view of developmental changes in individual species. There is enormous diversity in life-style, habitat, morphology, and physiology of amphibians (see chapters in Feder and Burggren 1992), and whether development of cardiac regulation is similarly diverse or is more conservative requires more comparative study.

3.2 Development of Central Arterial Hemodynamics

The recent advent of new technology for measuring cardiovascular variables in very small animals (e.g. microelectrodes for pressure measurement, Doppler blood velocity transducers, computer-assisted video analysis of beating hearts) is now allowing investigation of development transformations in cardiovascular function in amphibian larvae.

Our most comprehensive knowledge of the development of central arterial hemodynamics in an amphibian is for *Xenopus*, as a result of recent detailed studies of this species by Hou (1992) and Hou and Burggren (1995a, b). Cardiac stroke volume was determined as the difference between end-systolic and end-diastolic volume of the ventricle, measured by analysis of video images of the heart through the intact body wall of intact embryos and larvae. Blood pressure in the ventricle, conus arteriosus, or truncus arteriosus was determined dynamically with a microelectrode pressure recording system. From stroke volume, f_H and mean arterial pressure, peripheral resistance, cardiac output, and other cardiovascular variables could be calculated. These variables were determined on embryos ranging from as early as NF stage 40, before atrial septation is complete, up to froglets that had completed metamorphosis. Mean arterial pressure increased sharply with development (expressed as body mass). Although f_H showed no significant change with progressive development, mass-specific cardiac stroke volume rose steeply with increasing body mass. As a consequence, so too did mass-specific cardiac output, increasing approximately ten fold from NF stage 40 to NF stage 66 larvae (Fig. 7). A large rise in mass-specific cardiac output in the face of a constant mean arterial pressure indicates a considerable fall in peripheral resistance, and indeed this is the case. Overall, cardiac work increases 10 000-fold between the earliest embryonic and latest larval stages!

Qualitatively, these data on central arterial hemodynamics in *X. laevis* resemble those described for embryos of the domestic chicken, whose circulation is the most commonly studied of all embryonic vertebrates (see Clark 1991). Quantitative differences obsiously exist between the anuran and chick embryo, reflecting differences in metabolic rate and need for nutrient delivery. However,

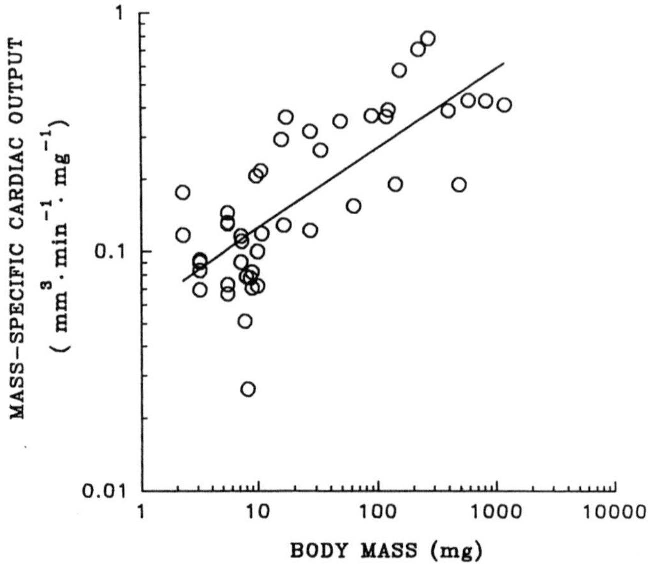

Fig. 7. Mass specific cardiac output as a function of body mass in larval *Xenopus laevis*. The regression line statistics are $r = 0.976$, $P < 0.01$. (Hou 1992)

the broad similarities (especially in early embryonic stages) between two such phyletically distinct vertebrate taxa suggest relatively little intraspecific diversity in the embryonic development of central arterial hemodynamics. If future studies on other amphibians confirm the conservative nature of vertebrate embryonic hemodynamics, then the amphibian embryo may well take up residence beside the chick embryo as an important model in studying hemodynamics.

While *X. laevis* has provided the most comprehensive data on developmental hemodynamics (in terms of pressure/flow/resistance relationships), extensive measurements of central arterial blood pressures have been made in the bullfrog *Rana catesbeiana* (Pelster and Burggren 1991; Pelster et al. 1993). Blood pressure measurements in the ventricle and conus arteriosus of larval bullfrogs completing metamorphosis in a single season indicate that, after the heart completes its S-folding and the formation of pylangial, spiral, and synangial valves (from proximalmost to distalmost), the heart essentially assumes the hemodynamics of the adult bullfrog. This is a highly qualified statement, however, because there is a profound seasonal influence on central arterial hemodynamics. As discussed below, the central hemodynamics of overwintering animals show striking temperature-independent differences.

Central arterial pressures have also been measured in the paradoxical frog, *Pseudis paradoxus*, native to Brazil. The extremely large size of the larvae (up to 150 g at metamorphosis) allows implantation of indwelling catheters to measure arterial blood pressure (Burggren et al. 1992). Mean arterial blood pressure rises from about 5–10 mmHg in stage 37–40 (Gosner scheme) larvae up to about 30

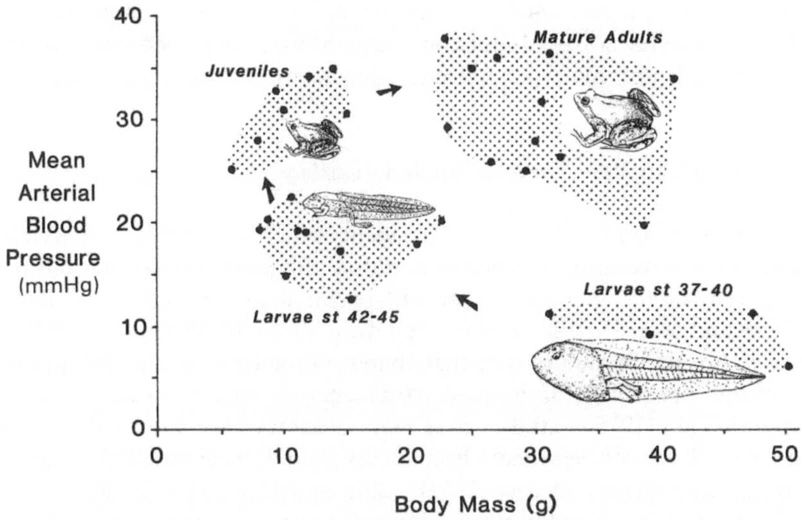

Fig. 8. Mean arterial blood pressure, grouped according to developmental stage, as a function of body mass in unanesthetized larvae, juveniles, and adults in the paradoxical frog, *Pseudis paradoxsus*. (Burggren et al. 1992)

mmHg immediately after metamorphic climax, and stays constant during further growth of the adult (Fig. 8). Definitive statements on how peripheral resistance changes as body mass and developmental stage covary in this species await measurement of other cardiovascular variables (e.g. cardiac output). However, the large size and ready accessibility of the giant larvae of *Pseudis paradoxus* make this species an embodiment of the Krogh principle as applied to studying cardiovascular development in amphibians. Hopefully, measurements of cardiac output in this species can be made in the near future.

4 Seasonal Influences on Cardiovascular Function

As long ago as 1955, Fromm and Johnson noted that "It has been recognized for years that frog tissues show seasonal differences in their physiological behavior, and it is customary to distinguish between summer and winter frogs". Since this time, physiological studies on both anurans and urodeles have repeatedly confirmed this notion (Harlow 1977; Dunlap 1980; Feder 1982; Wolmuth and Crawshaw 1988, among many others). Yet, seasonal effects on amphibian physiology are often ignored in experimental design, and relatively uncommon are studies that actively incorporate seasonality into their hypotheses.

Seasonal changes in an animal's environmental impact upon any physiological process in numerous ways that might be influenced by changes in temperature, nutrition, or photoperiod, no name but a few. It is important to emphasize that seasonal effects are not simply temperature effects. Although temperature is certainly instrumental in their induction, seasonal differences between physiological processes in amphibians can occur independently of body temperature.

4.1 Season and Heart Rate-Temperature Relationships

Heart rate in amphibians is, of course, dependent upon body temperature, as it is in all ectotherms. It has long been appreciated that this fundamental relationship between f_H and body temperature varies with season in anurans (Stier and Bock 1966; Miller and Mizell 1972a, b; Harri and Talo 1975a, b). While some of this difference in f_H may be due to seasonal changes in activity of the autonomic nervous system (specifically of the vagal cardiac nerves; see below), as suggested by Harri and Talo (1975a), studies over many decades have shown that even isolated hearts show this seasonal effect (Barcroft and Izquierdo 1931; Carter 1933; Smith 1951; Harri and Talo 1975b; Chiu and Chu 1989). In *Rana temporaria*, for example, the isolated atria of winter frogs that had been acclimated to either 25 or 5 °C showed a lower spontaneous rate of beating over a range of 5–20 °C than did the isolated atria of summer frogs acclimated to 25 or 5 °C (Fig. 9).

The mechanisms behind these seasonal effects on the f_H-temperature relationship remain unclear. Temperature acclimation affects thyroid hormone

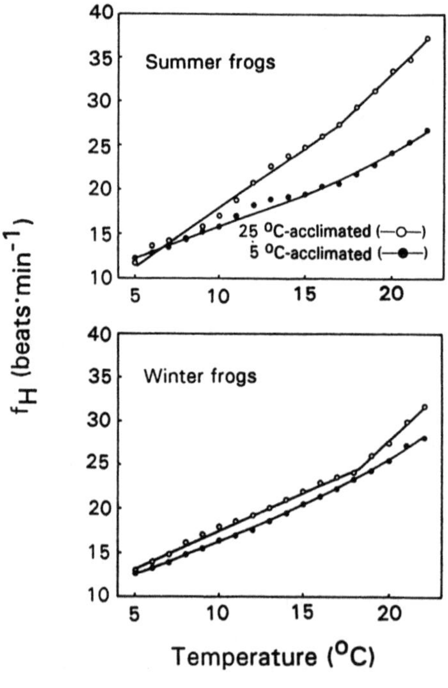

Fig. 9. Seasonal influences on the in vitro heart rate-temperature relationship in the frog *Rana temporaria*. (Harri and Taloa 1975)

output in anurans (Jankowsky 1960; Lagerspetz et al. 1974), and it may be that seasonal variations in thyroid hormone levels condition the heart's pacemaker.

4.2 Season and the Control of Heart Rate in Vivo

Temperature-independent seasonal effects on the control of f_H in amphibians have been well documented. Differences in tonic vagal discharge correlate with season in anurans (Holzapfel 1973; Iriuchijima 1959). Vagal inhibition of the heart of *Rana pipiens* is temperature-dependent, with vagal inhibition more profound at lower temperatures (Lund and Dingle 1968). The temperature at which cardiac inhibition by vagal stimulation will occur depends upon season, with vagal inhibition being reduced at a higher temperature in July than in September. Investigation of f_H changes during diving in *Rana temporaria* and *Bufo bufo* confirms a seasonal component of f_H regulation in vivo (Jones 1968). The rate of adjustment in f_H during diving at 17 °C is slower, and the ultimate f_H reached during diving is higher, in summer *Rana temporaria* and *Bufo bufo* compared with animals acquired during the winter. Seasonal changes in the role of the sympathetic nervous system in controlling f_H do not appear to have been investigated, though related temperature changes have been examined (see below).

The mechanisms behind seasonal changes in control of f_H, like those for the f_H-temperature relationship, remain somewhat enigmatic. More subtle changes than just a change in the relative importance of the cholinergic vs adrenergic regulation of the heart would seem to be involved. For example, a recent study of the adrenergic pharmacology of isolated atria from the tiger frog (*Rana tigrina*) compared the inotropic and chronotropic responses of selective α-and β-agonists and antagonists as a function of temperature and season (Chiu and Chu 1989). The responses were complex, as exemplified by the chronotropic responses of the heart. Phenylephrine produced an increase in the rate of atrial contraction at 24 °C in summer frogs but not in winter frogs. At 6 °C, however, the opposite was true, with phenylephrine increasing the rate in winter frogs but not summer frogs. Chiu and Chu (1989) concluded that in winter frogs at low temperatures the β-adrenergic innervation of the atrium is in quiescence, and that f_H may "by default" become regulated by the α-division of the adrenergic nervous system.

A series of studies by Herman and her colleagues (see Herman 1992 for references) on cardiovascular regulation in *Rana catesbeiana* acclimated to both warm and cold temperatures indicate that temperature alone affects the cardioactive nature of a wide range of substances, ranging from catecholamines to prostaglandins. In cold-acclimated animals, the β-adrenergic receptors in the heart show greatly reduced activity, and prostaglandins likewise have a highly diminished effect on the heart at low temperature. The α-adrenergic component that regulates blood pressure, however, seems undiminished at low temperature. These data, along with those of Chiu and Chu (1989), suggest that the influence of season on regulation of the heart is a complex interaction between seasonal changes in temperature and temperature-independent factors such as hormonally regulated up- or down-regulation of cardiac receptors, for example.

4.3 Season and Hemodynamics

Given the numerous accounts of seasonal effects on f_H, it would seem reasonable to assume that season might also influence hemodynamics (blood pressure, blood flow). Yet, there seem to be very few systematic investigations of seasonal affects on hemodynamics in amphibians. Larval bullfrogs (*Rana catesbeiana*) overwinter up to two times before metamorphosing, providing an ideal opportunity to look at hemodynamics in both winter and summer larvae. Central arterial systolic blood pressure in spring/summer larvae ranges between 10 and 12 mmHg from TK stage II to TK stage XIV (Fig. 10; Pelster and Burggren 1991). In fall/winter larvae, however, the relationship between blood pressure and development, measured at the same temperature as for the fall/winter larvae, changes considerably. Fall/winter larvae at TK stage II have an arterial pressure of only about 2 mmHg, but a six fold increase in pressure up to 12 mmHg occurs between TK stage II and X. Mechanistically, this difference in hemodynamic development with season can be explained by qualitative changes in the cardiac pump. In spring/summer larvae, the central arterial hemodynamics were like that reported for adults (presumably from all seasons), in which the ventricle was the main cardiac pump, and the conus ateriosus additional, lesser impetus to arterial blood after the pylangial valves had closed. In fall/winter larvae, however, ventricular systolic pressure in TK stage II larvae never reached that recorded in the conus or

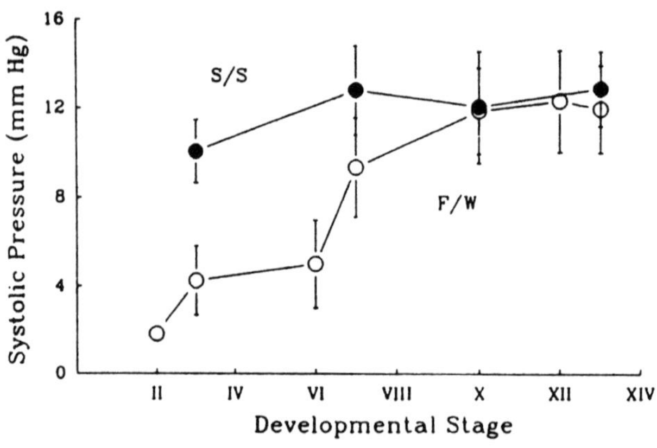

Fig. 10. The seasonal effect on the relationship between systolic arterial blood pressure and developmental stage in larvae of the bullfrog *Rana catesbeiana.* All data were recorded at 20–22 °C. Blood pressure appears constant throughout early larval development when measurements are made in the spring and summer (*S/S*). Blood pressures measured in the fall and winter (*F/W*) indicate that blood pressure rises progressively with development. (Pelster and Burggren 1991)

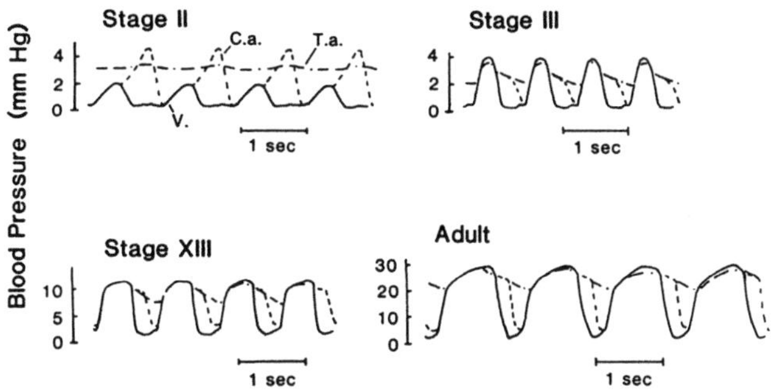

Fig. 11. Blood pressure recorded in the ventricle (*V.*), conus ateriosus (*C.a.*), and truncus ateriosus (*T.a.*) from three larval stages and from the adult of the bullfrog, *Rana catesbeiana*. Measurements were made at 20–22 °C during the winter months. In summer months, the hemodynamics of TK stage II larvae is qualitatively different, closely resembling that measured in older larvae and adults. (Pelster and Burggren 1991)

truncus arteriosus (Fig. 11). In these larvae, the ventricle was acting as an antechamber that primed the conus arteriosus when the ventricle contracted. In this system, the highest pressure in the circulation was measured in the conus, which acted as the main "cardiac" pump in early developmental stages. These qualitative changes in early larval hemodynamics, in which the conus arteriosus assumed the function that would be taken over by the ventricle in later developmental stages, were independent of temperature. How this seasonally related change in central arterial hemodynamics in larval *Rana catesbeiana* is achieved, or even whether it is common to other amphibian species, is unknown at this time.

5 Conclusions and Guiding Principles for Future Studies

Typology, the sometimes irrational dependence upon a particular model animal to illustrate supposedly general principles, is as rampant in comparative cardiovascular physiology as any other discipline. Broadly comparative studies often describe the circulation of the "amphibian", which in actuality is usually a description of the circulation of an adult, ranid frog operating at warm temperatures. Yet, three factors—phylogeny, season, and ontogeny—add considerable complexity to the analysis of cardiovascular function in amphibians, and should not be downplayed.

Phylogeny as a controlled variable has not been a large part of cardiovascular studies in amphibians (or any lower vertebrates) to date. Most often, a species is chosen because considerable background information already exists for it (contributing to a self-fulfilling prophecy?) or because the species is convenient with regards to size or accessibility. Studies are few which have made intraspecific comparisons of two or more species to exploit the power of studies carefully tailored around phylogenetic considerations. Hopefully, in the future, more amphibian cardiovascular studies will compare closely related taxa (not *Rana* vs *Bufo* !) and perhaps even use cladistic analysis or other tools of the evolutionary biologist to understand more clearly cardiovascular adaptations in amphibians.

Season influences processes in amphibians, ranging from thermoregulation (Wolmuth and Crawshaw 1988) to metabolism (Fromm and Johnson 1955) to hemodynamics (Pelster and Burggren 1991). While many investigators carefully control seasonal effects by carrying out their experiments at the same time of year (or working on laboratory-acclimated frogs), others do not take season into account. Doubtlessy, some of the intraspecific variation in measured physiological variables in amphibians that is normally attributed to "biological noise" or simply to "other unknown factors" may well be explained by a failure to control seasonal effects. However, seasonal effects represent more than just an "annoyance" whose effect on experiments is to be controlled. Seasonal effects on physiology represent an important variable (like body mass, size, sex, or species) in amphibians, especially those from temperate climates where seasonal influences may be the most profound. Consequently, investigators should be encouraged to introduce seasonality as a variable into their studies, to better appreciate the flexibility and performance of amphibian physiological systems.

Ontogeny is often ignored in cardiovascular studies (indeed, in most physiological studies). Yet, the adult of a species is simply one of many important developmental stages—studies on adults alone are no more instructive about "how a species works" than are studies on embryos or larvae alone. It is important to recognize that selection pressures operate on all developmental stages of a species. Indeed, the extremely high mortality rate in the immature developmental stages of most lower vertebrates suggests that it is these initial stages, rather than in the terminal adult stage, where natural selection acts most strongly (Burggren 1992). By focusing primarily on adults of a species, we may be missing a great deal of information relating to the evolution of physiological processes. Consequently, more physiological studies should be directed at the investigation of species over their entire developmental continuum, rather than just focusing on a single stage in development.

Finally, it should be emphasized that the influences of phylogeny, ontogeny, and season on central cardiovascular function, while certainly important in amphibians, are not limited to this vertebrate class. Cardiovascular function in fishes and reptiles in all likelihood is similarly influenced by these factors, which should be taken into account in the design of future experiments.

References

Barcroft J, Izquierdo JJ (1931) The relation of temperature to the pulse rate of the frog. J Physiol Lond 71: 145–155

Boutilier RG (1989) Diving physiology: amphibians. In: Wood SC (ed) Comparative pulmonary physiology: current concepts. Dekker, New York, pp 677–695

Boutilier RG, Glass ML, Heisler N (1986) The relative distribution of pulmocutaneous blood flow in *Rana catesbeiana:* effects of pulmonary or cutaneous hypoxia. J Exp Biol 126: 33–39
Bride M (1975) Etablissement de l'innervation dans le coeur du têtard de Xénope et ses répercussions sur le fonctionnement de l'organe. CR Soc Biol 169: 1265–1271
Burggren WW (1988) Role of the central circulation in regulation of cutaneous gas exchange. Am Zool 28: 985–998
Burggren WW (1989) Lung structure and function: amphibians. In: Wood SC (ed) Comparative pulmonary physiology: current concepts. Dekker, New York, pp 153–192
Burggren WW (1991) Does comparative respiratory physiology have a role in evolutionary biology (and vice versa)? In: Woakes AJ, Grieshaber MK, Bridges CR (eds) Physiological strategies for gas exchange and metabolism. Cambridge Univ Press, Cambridge, pp 1–13
Burggren WW (1992) The importance of an ontogenetic perspective in physiological studies: amphibian cardiology as a case study. In: Weber R, Wood SC, Millard R, Hargens A (eds) Strategies of physiological adaptation, respiration, circulation and metabolism. Dekker, New York, pp 235–253
Burggren WW (1993) Respiratory metamorphosis during the water-to-land transitions in developing vertebrates and invertebrates. In: Bicudo E, Glass M, Abé A (eds) The vertebrate gas transport cascade: adaptations to environment and mode of life. CRC Press, Boca Raton, pp 23–34
Burggren WW, Bemis WE (1990) Studying physiological evolution: paradigms and pitfalls. In: Nitecki MH (ed) Evolutionary innovations. Univ Chicago Press, Chicago, pp 191–228
Burggren WW, Doyle ME (1986a) Ontogeny of heart rate regulation in the bullfrog, *Rana catesbeiana.* Am J Physiol 251: R231–239
Burggren WW, Doyle ME (1986b) The action of acetylcholine upon heart rate changes markedly with development in bullfrogs. J Exp Zool 240: 137–140
Burggren WW, Infantino RL Jr. (1994) The respiratory transition from water to air breathing during amphibian metamorphosis. Amer Zool 34: 238–246
Burggren WW, Just JJ (1992) Developmental changes in amphibian physiological systems. In: Feder ME, Burggren WW (eds) Environmental physiology of the amphibians. Univ Chicago Press, Chicago, pp 467–530
Burggren WW, Pinder AW (1991) Ontogeny of cardiovascular and respiratory physiology in lower vertebrates. Annu Rev Physiol 53: 107–135
Burggren WW, Feder ME, Pinder AW (1983) Temperature and the balance between aerial and aquatic respiration in larvae of *Rana catesbeiana.* Physiol Zool 56: 263–273
Burggren WW, Infantino RL Jr, Townsend DS (1990) Developmental changes in cardiac and metabolic physiology of the direct-developing tropical frog *Eleutherodactylus coqui.* J Exp Biol 152: 129–147
Burggren WW, Bicudo JE, Glass ML, Abe AS (1992) Development of blood pressure and cardiac reflexes in the frog *Pseudis paradoxsus.* Am J Physiol 263: R602–R608
Carter G (1933) On the control of the level of activity of the animal body. J Exp Biol 10: 256–273
Chiu KW, Chu JY (1989) Temperature and adrenoceptors in the frog heart. Comp Biochem Physiol 94C; 149–157
Clark EB (1991) Functional characteristics of the embryonic circulation. In: Feinberg RN, Sherer GK, Auerbach R (eds) The develpoment of the vascular system. Karger, Basel, pp 125–135
Dunlap DG (1980) Comparative effects of themal acclimation and season on metabolic compensation to temperature in the hylid frogs, *Pseudacrs triseriata* and *Acris crepitans.* Comp Biochem Physiol 66A: 243–249
Feder ME (1976) Lunglessness, body size and metabolic rate in salamanders. Physiol Zool 49: 398–406
Feder ME (1982) Effects of latitude, season, elevation, and microhabitat on field body temperatures of neotropical and temperate zone salamanders. Ecology 63: 1657–1664
Feder ME (1992) A perspective of environmental physiology of the amphibians. In: Feder ME, Burggren WW (eds) Environmental physiology of the amphibians. Univ Chicago Press, Chicago, pp 1–6

Feder ME, Burggren WW (1985) Cutaneous gas exchange in vertebrates: design, patterns, control and implications. Biol Rev 60: 1–45
Feder ME, Burggren WW (eds) (1992) Environmental physiology of the Amphibia. Univ Chicago Press, Chicago
Fromm PO, Johnson RE (1955) The respiratory metabolism of frogs as related to season. J Cell Comp Physiol 45: 343–359
Gans C (1970) Respiration in early tetrapods—the frog is a red herring. Evolution 24: 740–751
Garland T Jr, Adolph SC (1994) Why not to do two-species comparative studies: limitations on inferring adaptation. Physiol Zool 67: 797–828
Harlow HJ (1977) Seasonal oxygen metabolism and cutaneous osmoregulation in the California newt, *Taricha torosa*. Physiol Zool 50(3): 231–236
Harri MNE, Talo A (1975a) Effect of season and temperature acclimation on the heart rate-temperature relationship in the isolated frog's heart (*Rana temporaria)*. Comp Biochem Physiol 52A: 409–412
Harri MNE, Talo A (1975b) Effect of season and temperature acclimation on the heart rate-temperature relationship in the frog, *Rana temporaria*. Comp Biochem Physiol 52A: 467–472
Harvey PH, Pagel MD (1991) The comparative method in evolutionary biology. Oxford Univ Press, Oxford
Heath AG (1980) Cardiac responses of larval and adult tiger salamanders to submergence and emergence. Comp Biochem Physiol 65A: 439–444
Herman CA (1992) Endocrinology. In: Feder ME, Burggren WW (eds) Environmental physiology of the amphibians. Univ Chicago Press, Chicago pp 40–54
Holzapfel RH (1937) The cyclic character of hibernation in frogs. Q Rev Biol 12: 65–84
Hou P-CL (1992) Devlopment of haemodynamic regulation in larvae of the African clawed toad, *Xenopus laevis*. PhD Diss, Univ Massachusetts, Amherst
Hou P-CL, Burggren WW (1989) Interaction of allometry and development in the mouse *Mus musculus;* heart rate and hematology. Respir Physiol 78: 265–280
Hou P-CL, Burggren WW (1995a) Ventricular and central arterial blood pressures and heart rate during development in early larvae of the anuran amphibian *Xenopus laevis*. Am J Physiol (In Press)
Hou P-CL, Burggren WW (1995b) Cardiac output and peripheral resistance during development in early larvae of the anuran amphibian *Xenopus laevis*. Am J Physiol (In Press)
Huey RB (1987) Phylogeny, history and the comparative method. In: Feder ME, Bennett AF, Burggren WW, Huey RB (eds) New directions in ecological physiology. Cambridge Univ Press, New York, pp 76–98
Iriuchijima J (1959) Sympathetic and vagal discharges in the cardiac nerve of the road. Tohoku J Exp Med 71: 109–119
Jankowsky H-D (1960) Über die hormonale Beeinflussung der Temperaturadaptation beim Grasfrosch (*Rana temporaria*). Z Vgl Physiol 43: 392–410
Johansen K, Burggren WW (1980) Cardiovascular function in the lower vertebrates. In: Bourne GH (ed) Hearts and heart-like organs. Academic Press, New York, pp 61–117
Jones DR (1968) Specific and seasonal variations in development of diving bradycardia in anuran Amphibia. Comp Biochem Physiol 25: 821–834
Kimmel PB (1990) The ontogeny of cardiovascular regulatory mechanisms in the bullfrog (*Rana catesbeiana*). PhD Diss, Univ Massachusetts, Amherst
Lagerspetz KYH, Harri MNE, Okslahti R (1974) The role of the thyroid in the temperature acclimation of the oxidative metabolism in the frog, *Rana temporaria*. Gen Comp Endocrinol 22: 169–176
Lund GF, Dingle H (1968) Seasonal temperature influence on vagal control of diving bradycardia in the frog (*Rana pipiens)*. J Exp Biol 48: 265–277
Malvin GM (1985) Cardiovascular shunting during amphibian metamorphosis. In: Johansen K, Burggren WW (eds) Cardiovascular shunts: phylogenetic, ontogenetic and clinical aspects. Munksgaard, Copenhagen, pp 163–172
Malvin GM (1989) Gill structure and function: amphibian larvae. In: Wood SC (ed) Comparative pulmonary physiology: current concepts. Dekker, New York, pp 148–151
Mendes EG (1945) Contribuicâo para a fisiologia dos sistemas respiratorio e circulatório de

Siphonops annulatus (Amphibia- Gymnophiona). Bol Fac Filos Cienc Let Univ Sa Paulo Ser Zool 5: 283–304

Miller LC, Mizell S (1972a) Seasonal variation in heart rate response to core temperature changes. Comp Biochem Physiol 42A: 733–779

Miller LC, Mizell S (1972b) Seasonal variation in heart-rate-core temperature relationship in *Rana pipiens*: the role of the autonomic nervous system. Comp Gen Pharmacol 3: 434–442

Owen R (1834) On the structure of the heart in the perennibranchiate *Batrachia*. Trans Zool Soc Lond 1: 213–220

Pelster B, Burggren WW (1991) Central arterial haemodynamics in larval bullfrogs (*Rana catesbeiana*): developmental and seasonal influences. Am J Physiol 260: R240–R246

Pelster B, Burggren WW, Petrou S, Wahlqvist I (1993) Developmental changes in the acetylcholine influence on heart muscle of *Rana catesbeiana*: in situ and in vitro effects: J Exp Zool 267: 1–8

Putnam JL (1975) Septation in the ventricle of the heart of *Siren intermedia*. Copeia 1975: 773–774

Putnam JL, Dunn JF (1978) Septation in the ventricle of the heart of *Necturus maculosus*. Herpetologica 34: 292–297

Putnam JL, Parkerson JB (1985) Anatomy of the heart of the Amphibia. II. *Cryptobranchus alleganiensis*. Herpetologica 41: 287–298

Sawaya P (1940) Sôbre o veneo das glandulas cutaneas, a secreçâo e o coraçâo de *Siphonops annulatus*. Bol Fac Filos Cienc Let Univ Sao Paulo Ser Zool 4: 206–270

Shelton G (1970) The effect of lung ventilation on blood flow to the lungs and body of the amphibian, *Xenopus laevis*. Respir Physiol 9: 183–196

Shelton G (1976) Gas exchange, pulmonary blood supply, and the partially divided amphibian heart. In: Spencer Davis P (ed) Perspective in experimental biology. Pergamon Press, Oxford, pp 247–259

Shelton G (1985) Functional and evolution significance of cardiovascular shunts in the Amphibia. In: Johansen K, Burggren WW (eds) Cardiovascular shunts: phylogenetic ontogenetic and clinical aspects. Munksgaard, Copenhagen, pp 100–116

Smith CL (1951) The temperature-pulse rate curve of the isolated frog's heart (*Rana temporaria*). J Exp Biol 28: 141–164

Stahl WR (1967) Scaling of respiratory variable in mammals. J Appl Physiol 22: 453–460

Stier TJB, Bock HC (1966) Seasonal changes of heart rate-temperature relationships in toads. Proc Soc Exp Biol Med 123: 149–151

Tazawa H, Mochizuki M, Piiper J (1979) Respiratory gas transport by the incompletely separated double circulation in the bullfrog, *Rana catesbeiana*. Respir Physiol 36: 77–95

Toews D, MacIntyre DT (1978) Respiration and circulation in an apodan amphibian. Can J Zool 56: 998–1004

Toews D, Shelton G and Randall DJ (1971) Gas tensions in the lungs and major blood vessels of the urodele amphibian, *Amphiuma tridactylum*. J Exp Biol 55: 47–61

Wassersug RJ, Paul RD, Feder ME (1981). Cardio-respiratory synchrony in anuran larvae (*Xenopus laevis, Pachymedusa dacnicolor* and *Rana berlandieri*). Comp Biochem Physiol 70A: 329–334

West NH, Burggren WW (1983) Reflex interactions between aerial and aquatic gas exchange organs in the larval bullfrog. Am J Physiol 244(6): R770–R777

West NH, Burggren WW (1984) Factors influencing pulmonary and cutaneous arterial blood flow in the toad, *Bufo marinus*. Am J Physiol 247: R884–R894

West NH, Van Vliet BN (1992) Sensory mechanisms regulating the respiratory and cardiovascular systems. In: Feder ME, Burggren WW (eds) Environmental physiology of the amphibias: Univ Chicago Press, Chicago, pp 151–182

Wolmuth LP, Crawshaw Ll (1988) The effect of development and season on temperature selection in bullfrog tadpoles. Physiol Zool 61: 461–469

Veno-Lymphatic System

Chapter 9

The Role of the Lymphatic System for Water Balance and Acid-Base Regulation in the Amphibia

D.P. Toews and *L.A. Wentzell*

Contents

1 Historical Perspective

In 1833 the Royal Society of London published an article by Johannes Müller which decribed the existence of "four distinct hearts, having regular pulsations, connected with the lymphatic system, in certain amphibious animals." In this initial publication on the amphibian lymphatic system, Müller described many of the features which characterize this unique network. He found, for example, that frogs have lymph spaces beneath the skin, that there are large accumulations of lymph in these cavities and that the irregular beating of lymph hearts lacks any synchrony with the blood heart. It is perhaps a tribute to the ubiquitous nature and versatility of "the frog" that, before the turn of the 20th century, a wide variety of prominent anatomists, physiologists, embryologists and histologists, such as Panizza (1834), Schiff (1850), Ranvier (1880), von Leydig (1876), Ecker (1889) and Priestley (1878a, b), had published scientific studies elucidating diverse aspects of the amphibian lymphatic system. Of special note is the fact that the first two articles in the first issue of the Journal of Physiology (London) are by John Priestley and detail the anatomy and physiology of amphibian lymph hearts (Priestley 1878a, b).

The initial studies on the amphibian lymphatics were preformed on the European frog, *Rana esculenta,* and as such relate primarily to the physiology and

Biology Department, Acadia University, Wolfville, Nova Scotia B0P 1X0, Canada

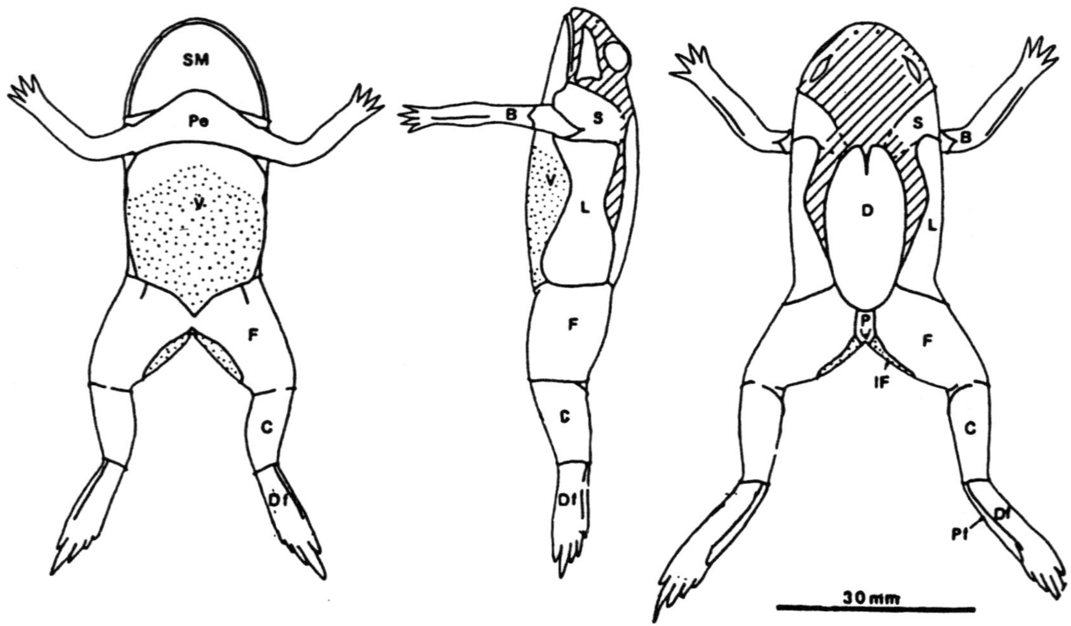

Fig. 1. Ventral, lateral, and dorsal views, respectively, of *Neobatrachus pictus* (representative of terrestrial amphibians including *Bufo marinus*) showing subcutaneous sacs and the septa (*thick outlines*). *B* Brachial sac; *C* crural sac; *D* dorsal sac; *Df* dorsal sac of foot; *F* femoral sac; *If* interfemoral sac; *L* lateral sac; *Pe* pectoral sac; *Pf* plantar sac of foot; *Sm* submandibular sac; *V* ventral sac; *P* pubic sac; *S* scapular sac; *d* dorsal septum; *if* intermediate femoral septum; *v* ventral septum. *Cross-hatching* indicates areas where the skin is closely attached to the muscle, reducing lymph sac area. *Stippling* indicates myointegumental fibres. (Adapted from Carter 1979)

anatomy of the order Anura, the frogs and toads. Anurans characteristically have two to four pairs of lymph hearts located anteriorly in the subscapular region and posteriorly just beneath the skin near the point of attachment of the hindlimb with the body. The salamanders, in the order Urodela, usually have many more pairs of lymph hearts, positioned along the vertebral column. Animals in the order Apoda (some representatives being called caecilians) often resemble large earthworms externally and indeed, like earthworms, show an extensive degree of metamerism of the body musculature. Lymph hearts can be found in each body segment and, as a result, some species have up to 200 pairs of these organs per animal (Kampmeier 1969). Because of their relatively small number of lymph hearts and the abundant availability of experimental animals, species of the order Anura have been the most well studied.

2 Structure of the Lymphatics

Perhaps the earliest detailed description of the subcutaneous anatomy of the lymphatic system is provided by Ecker (1889). He noted that there are many lymph sacs under the skin of *Rana esculenta,* divided by connective tissue septa which allow fluid (lymph) to slowly move from one compartment to the next. It seems reasonable that this compartmentalization alleviates the difficulties which could result posturally if the subcutaneous sacs were a single, large compartment. Carter (1979) clearly described the structure of the lymph sacs in a variety of anurans and suggested that there is a correlation between lymph sac size and habitat, with large lymph sacs being associated with an aquatic life-style and smaller lymph sacs being associated with a more dehydrated environment. Carter also states that these lymph sacs (Fig. 1), hearts and vessels could provide a route for the rapid excretion of water moving into an animal during hydrated conditions. In general terms, it is worth pointing out that until Carter's study, the most prevailing theory for the use by amphibians of subcutaneous lymph sacs was for water storage as a precaution against possible dry conditions (Middler et al. 1968). In effect, these areas could function as subcutaneous bladders. In contrast, most amphibian physiologists today would agree that, because of the high fluid permeability of the amphibian skin, subcutaneous storage of water would most likely result in a greater percentage of water loss during dehydrating conditions, rather than water conservation.

Lymph originates by plasma filtration through a combination of colloid osmotic and blood capillary hydrostatic pressures (Conklin 1930a). Consisting of 97% water, lymph contains concentrations of chloride, sodium, and potassium salts, as well as the blood proteins fibrinogen, albumin and globulins, at levels equal to or slightly below those of the blood (Conklin 1930b; Foglia 1941). The fluid is collected by small lymph capillaries. These vessels are different from those of mammals in that amphibians have relatively short vessels which begin blindly in the muscle tissue, joining together to form large vessels before emptying their

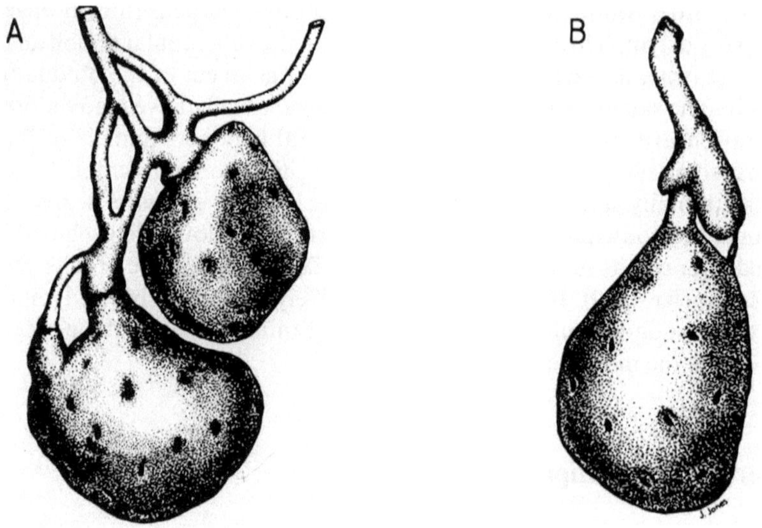

Fig. 2 A,B. Ventromedial view of the left posterior lymph hearts. **A** *Rana pipiens* (after Kampmeier 1969): **B** *Bufo marinus*. The surrounding lymph sinuses are not shown; only the ostia or points of entry in the lymph heart and exit vessels into the venous system are shown

contents into the subcutaneous lymph sacs (Conklin 1930a). Lymph is presumably redistributed within and passed through the lymph sacs by postural movements and by the continuous pumping of (and the probable slight negative pressure created by) the lymph hearts. The posterior lymph hearts collect fluid from the subcutaneous lymph spaces and pump it into the posterior portal vein. The anterior lymph hearts acquire lymph from the viscera and axillary lymph sacs via the suprascapular lymph sinus and drain it into the anterior vertebral vein (Müller 1833; Kampmeier 1969).

The lymph hearts themselves are endothelial-lined chambers composed of skeletal muscle and connective tissue (Fig. 2; Ecker 1889; Deydrup 1964). The surfaces of the lymph hearts usually have numerous openings or ostia for entry of lymph into the hearts. These ostia have one-way valves for fluid entry which close during lymph heart contraction. Additionally, there is usually one efferent valve which remains closed during filling. The lymph hearts are connected to the surrounding body tissues by numerous suspensory ligaments. We believe that contraction of the lymph hearts (systolic phase) provides a degree of tension on the ligaments, and that during relaxation (diastolic phase), a small negative pressure is developed within the heart to facilitate filling.

It appears as though the movement of lymph through the sacs is relatively fast. Carter (1979) injected the dorsal lymph sacs of a variety of frog species with dye and found, for example, in *Litoria caerulea* that the dye quickly moved caudally and appeared in the posterior lymph hearts within 10s of injection. Since there are no discreet routes within the lymph sacs, directional movement must be implemented by the modest aspiration of the lymph hearts. In all anuran amphibians the venous connection will eventually communicate with the renal portal veins and as such will provide a route for lymph flow to the peritubular blood of the kidney.

Baldwin et al. (1990) provide recent evidence which describes the primary routes for lymph return to the circulatory system. Their major concern was the relative importance of the nephrostomes on the kidney which supply direct pores for pritoneal fluid entry to the renal venous circulation. Kampmeirer (1969) viewed the nephrostomes as a "competitor of the lymphatic system," but Baldwin and her coworkers concluded that the nephrostomes are of minor importance in lymph return and that most of the peritoneal fluid is returned to the subvertebral lymph sacs (and anterior lymph hearts) through mesothelial perforations between the two compartments.

Another source of fluid entry into an amphibian, albeit controversial, is the direct influx of ambient water into the lymphatics through the skin. Amphibians do not drink and as a result the uptake of water is across the epidermis when the animals are sitting in water, or for that matter simply in contact with moist soil (Katz 1986). It has always been assumed that this fluid entry is into the capillaries of the skin circulation and hence into the general circulation of the animal. Wentzell et al. (1993) monitored the movement and uptake of tritiated water into the toad *Bufo marinus* in animals which were chronically catheterized in the femoral artery, ureters, bladder, and in the femoral lymph sac. They found that when tritiated water was added to the ambient water, the isotope moved almost simultaneously into the plasma and lymph but within 10 min (and for the following 170 min) levels in the lymph were significantly higher than those in the plasma (Fig. 3). These results suggest a preferential passage of the tritiated water directly into the lymphatic sacs and spaces as opposed to the indirect route into the capillaries and then by filtration into the lymph. Wentzell et al.'s results also show higher levels of tritiated water in the ureteral urine than in the plasma which in itself suggests that an excretory route for water moving across the skin might be from the lymph sacs through the lymph hearts, and then into the renal portal vein which perfuses the peritubular vessels of the kidney. It seems reasonable to suggest that during periods of high water availability this might be an effective route for water elimination. It would be especially advantageous to aquatic species in that rapid fluid entry across the skin could be quickly eliminated by the kidneys without a great deal of hemodilution occurring in the arterial system.

Perhaps the most difficult measure or value to obtain is that of the volume contained within the lymph vessels, interconnecting lymph sacs and peritoneal cavity. The total volume will, of course, be quite variable and will be related to the degree of hydration of the animal. The toad *Bufo marinus* in our laboratory has shown a weight shift of 1.5–2 times when the animal experienced a 24-h bout of dehydration followed by progressive edema which resulted from lymph heart damage or stoppage. It is indeed difficult to establish a "normal" lymph volume

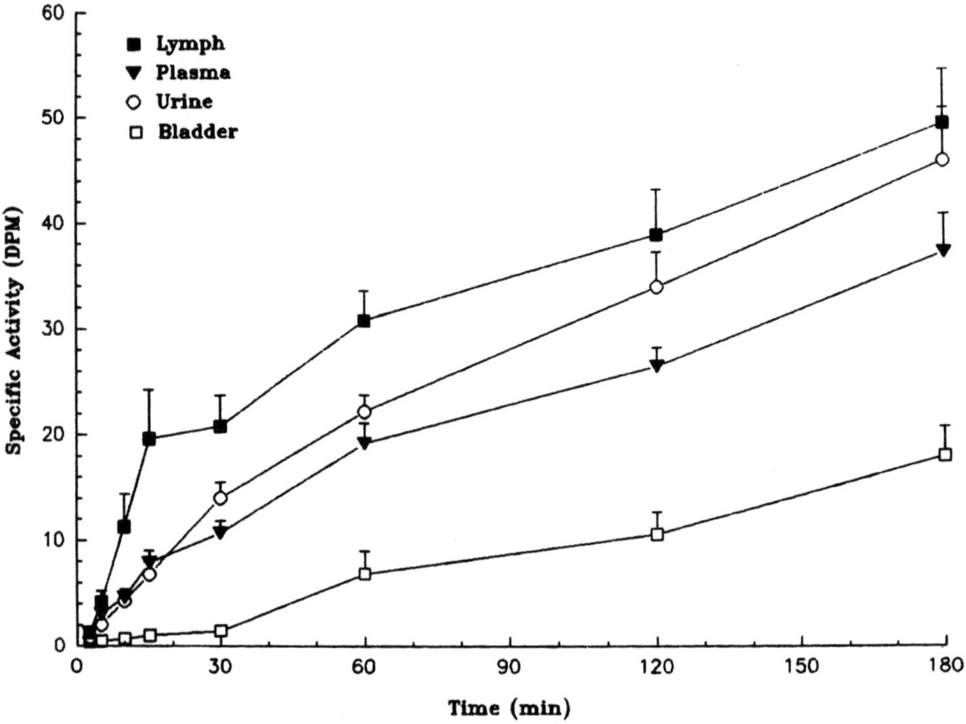

Fig. 3. Uptake of tritiated water from the ambient water into the lymph, plasma, ureteral urine, and bladder urine of 9 *Bufo marinus* over 3 h ($\bar{x} \pm SE$). (Wentzell et al. 1993)

between these two extreme conditions. By modeling fluid movements in all body compartments we have established a "whole animal" interstitial and lymph volume estimate which clearly provides the minimum degree of intercompartmental perturbations when minor lymph pressure and flow variations are introduced into the model system. We estimated this combined lymph and interstitial volume to be about 20% of the body weight of the toads, a value similar to that of Thorson (1964), who calulated the interstitial volume, including lymph, in *Bufo marinus* to be 20%, based on sucrose space measurements. It is simply not possible to limit a definition of lymph, as is done in mammalian physiology, to that fluid contained inside the lymph vessels. As mentioned previously, lymph vessels in amphibians remove fluid from interstitial spaces and usually empty their contents into the lymph sacs and spaces, and, hence, into the lymph hearts. Although at present we do not have a firm estimate for lymph volume, we feel that the fluid contained in the various "lymph compartments" could equal, or possibly exceed, the blood volume in many amphibian species (approximately 7%).

3 Lymph Production and Flow

Müller in 1833 noted "I had long observed that it is very easy to obtain lymph from frogs, because these animals have large lymph-spaces immediately under the skin, and between the muscles. If the skin of the thigh of a frog be cut, the lymph will flow pretty freely, and may be collected in a watchglass; it will continue fluid for ten minutes, and then coagulates; it is at first clear as water". Despite its accessibility and availability, the quantification of this rate of lymph production has nonetheless remained elusive since these early observations. Isayama (1924a, b) provided an indirect estimate of lymph production by destroying lymph hearts in a frog and then estimating the loss of blood being filtered into the lymphatics by noting the increase in erythrocyte counts. His widely quoted estimate and unique unit of measurement was that the volume of lymph collected by the lymphatic system in a frog was equal to about 50 times the plasma volume in a 24-h period. Although we will elaborate upon this in more detail, the more direct methods of lymph flow estimation provided by Jones et al. (1992) suggest that this lymph "turnover" in a 24-h period is closer to 11 times the plasma volume in the toad *Bufo marinus*.

Whatever value is most correct for the individual species, it is nonetheless clear that there is an unusually large volume of lymph which must be continuously reintroduced into the blood circulation. One interesting comparison is the ratio of lymph flow to cardiac output. In a mammal the ratio is about 1/3000 (Eckert et al. 1988) while in the toad we calculated a ratio closer to 1/200 (Jones et al. 1992), indicating that lymph production and circulation are proportionally much greater in amphibians than in mammals.

The importance of the return of the lymph to the bloodstream has been ascertained through experiments involving the destruction of the lymph hearts. Following thermocautery of the anterior and posterior hearts, Foglia (1941) noted a 60% increase in body weight, followed by death within 4 days in *Bufo arenarum* and *Leptodactyllus ocelatus*. The loss of fluid and protein return by the lymph hearts resulted in widespread disruptions throughout the body: loss of blood volume, progressive hemolysis, decrease in mean arterial pressure, reduced urinary output, and decreased glomerular and capillary filtration in the kidney. Baustian (1988) demonstrated that the destruction of the anterior and posterior lymph hearts in *Bufo marinus* resulted in rapid hemoconcentration and death within 1–43 days following hemorrhage, while those animals with intact hearts fully recovered.

The quantification of lymph production has always been attempted through indirect methods, such as Isayama's (1924a, b) erythrocyte count method. Conklin (1930a) found that the immobilization of the lymph hearts in the frog resulted in a 10 to 32% weight gain over 6 h. To provide more accurate values, Jones et al. (1992) developed a surgical technique which not only allowed the direct measurement of lymph production, but also of intralymph heart pressure. An incision is made in the integument of the anaestitized animal directly above the lymph heart. A small cut is then made in the dorsal surface of the heart to allow access to the interior. After spanning the circumference of the portal with a purse-string suture, a length of polyethylene tubing, flared at the insertion end, is positioned within the heart, and suture is tightened and secured. The integument is then closed in such a way that the cannula protrudes vertically between two stitches. After recovery, during which time the cannulae remains occluded to prevent protein depletion, the cannulae may be routed through a fraction collector to determine lymph production, or connected to a transducer to record intralymph heart pressure and heart rate.

Using this procedure, Jones et al. (1992) provided the first direct measurements of lymph production. Interestingly, it was discovered that the flow from the left and right posterior lymph hearts was not always equal. In fact, a "mirror-image" effect was often noted when each was measured separately (Fig. 4). This was not entirely unexpected as Priestley (1878a) noted that the hearts on each side of the body do not beat synchronously with one another, although the anterior and posterior lymph heart on each side do beat together (Pratt and Reid 1939). The asymmetry of flow production between the two sides could account for the great variability in estimates of flow rate prior to this study. The rate of lymph production for both posterior lymph hearts of fully hydrated *Bufo marinus* was determined to be 1.11 ml h^{-1} 100 g^{-1} (Jones et al. 1992). Assuming that blood

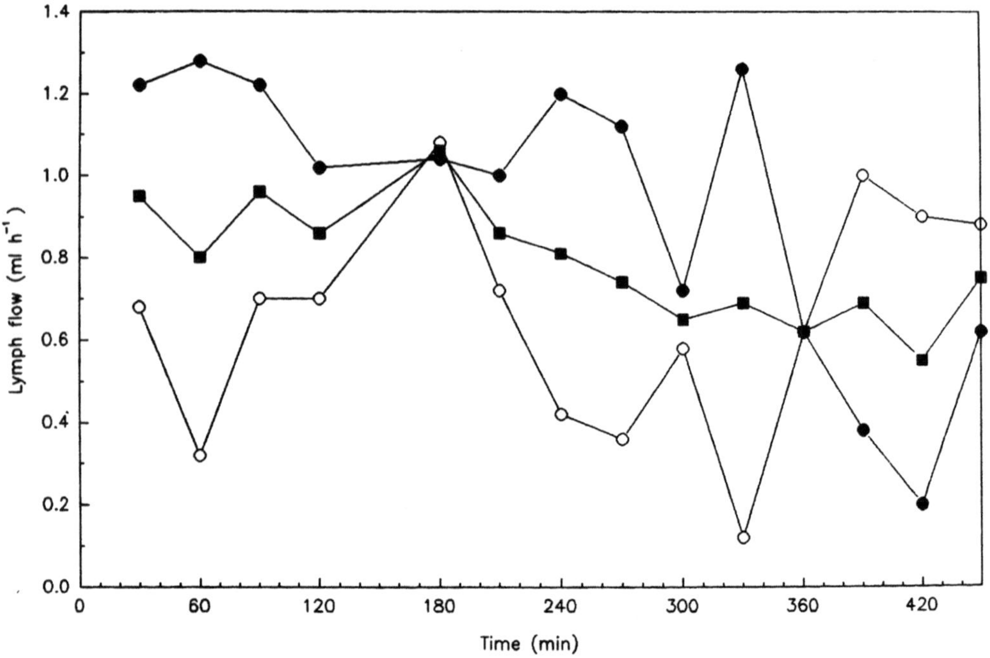

Fig. 4. Lymph flow rates from the posterior lymph hearts of a single *Bufo marinus* under normal hydrated conditions. Flow rates from the left heart (*filled circles*), right heart (*open circles*), and the average of the two hearts (*filled squares*) are shown. (Jones et al. 1992)

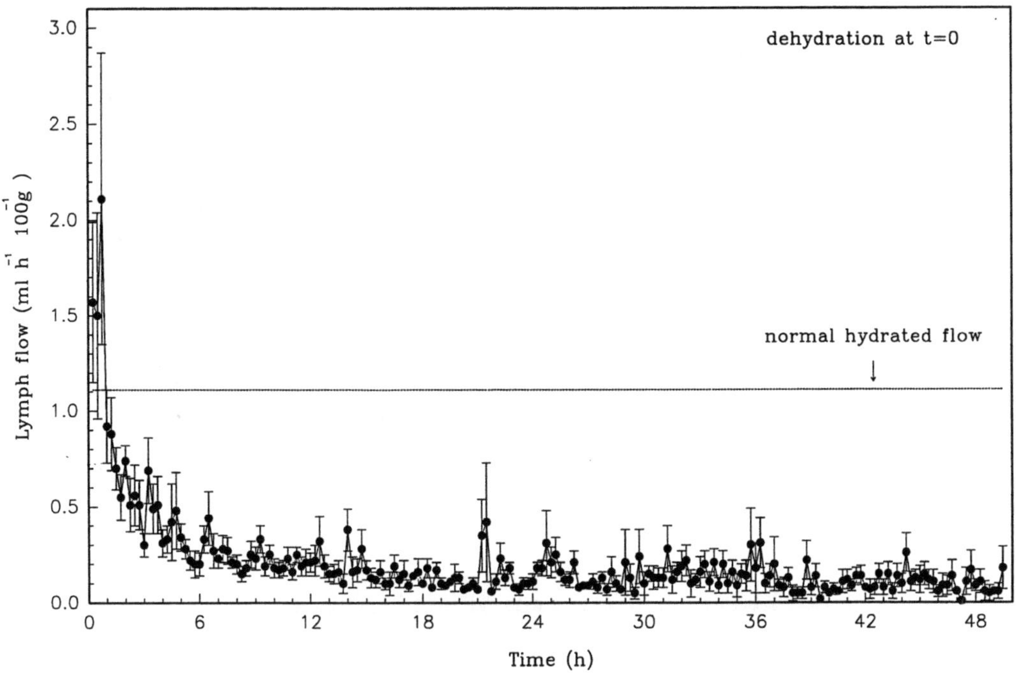

Fig. 5. Lymph flow rates of *Bufo marinus* during dehydration, standardized to the individual mass of the animal (n = 8). Normal hydrated lymph flow is also shown (*dashed line*). (Jones et al. 1992)

volume is equal to 7% of the body mass and hematocrit is 30%, and that the lymph production of the anterior hearts is equal to that of the posterior (Kampmeier 1969), the lymph production over a 24-h period is equal to approximately 11 times the plasma volume, or only about one-fifth the value estimated through indirect methods for the frog (Isayama 1924b).

As would be expected, lymph flow decreases rapidly during dehydration (Fig. 5), although it rarely stops altogether, even during bouts of prolonged water deprivation (Jones et al. 1992). This is presumably due to the continued fluid exchange between the capillaries and the lymph space, which must take place as long as the blood heart continues pumping.

4 Lymph Heart Rate and Pressure

As Müller (1833) noted, the lymph hearts do not exhibit the regular rhythm of the cardiac heart. The rate of lymph heart contraction is, in fact, affected by many outside variables, such as fright, posture, temperature, and activity (Kampmeier 1969). Curare has been shown to slow lymph heart function, while urethane has an acceleratory effect (Conklin 1930a). The lymph hearts will also often cease beating for no apparent reason, only to resume a short time later (Priestley 1878a). Neither do the four hearts beat in unison. It was initially thought that all four hearts were independently innervated (Priestley 1878a; Noble 1932). However, Pratt and Reid (1939) eloquently demonstrated, through smoked drum tracings, that the homolateral anterior and posterior hearts beat in synchrony, and are mutually affected by external stimuli. They therefore concluded that the spinal centres of the anterior and posterior hearts are axially coordinated.

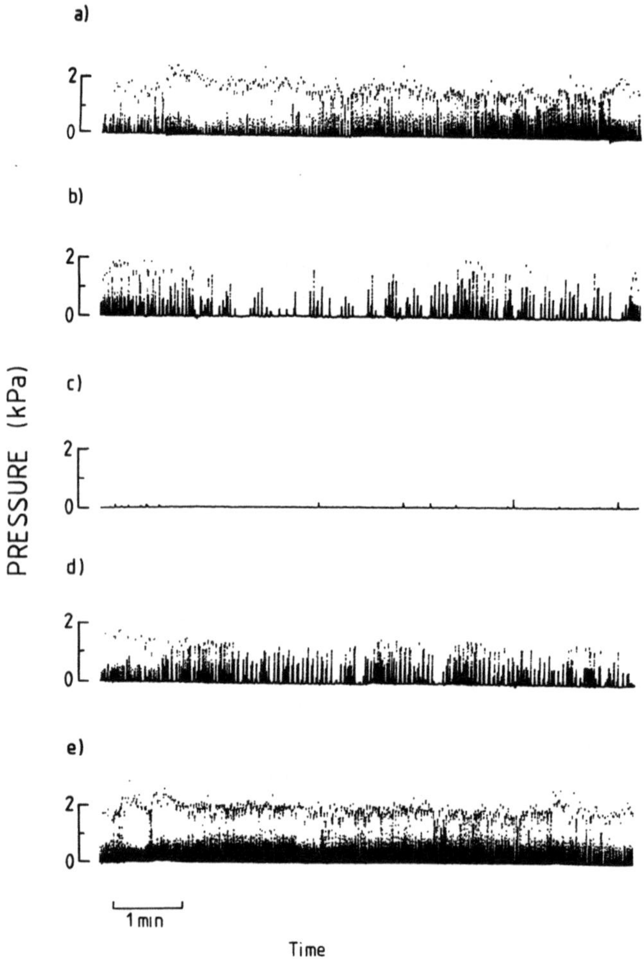

Fig. 6 a–e. Effect of dehydration on lymph heart pressure and rate in a representative *Bufo marinus*. **a** Normal hydrated animal. **b** Detail of a decrease in lymph heart function (3-h dehydration). **c** Detail of almost complete cessation of lymph heart function. This particular stoppage occurred at 7.5-h dehydration and lasted 29.4 min. **d** Detail of lymph heart function at 24-h dehydration with decrease in both pressure and rate when compared to control. **e** Increase in pressure and rate after 30-min rehydration. (Jones et al. 1992)

The resting lymph heart rate is 50–70 beats min^{-1} for the frog (Priestley 1878a), with a stroke volume estimated from heart volume of 0.5–1.0 µl beat^{-1} (Winterstein 1925, in Conklin 1930a; Radwanska 1906, in Kampmeier 1969). However, a lymph heart which receives only a minimal amount of lymph will beat as often as one which is full (Kampmeier 1969), and so lymph volume for each stroke is quite variable. Jones et al. (1992) determined the resting heart rate for fully hydrated *Bufo marinus* to be about 48 beats min^{-1}, decreasing to about 32 beats min^{-1} during prolonged dehydration (t = 22 °C). Reduction or stoppage in lymph heart function, usually for no identifiable reason, was common throughout all the experiments, but particularly numerous during dehydration. In hydrated animals, these pauses typically lasted for up to 30s; during dehydration, the intervals ranged between 3 and 10 min, with occasional stoppages of up to 30 min (Fig. 6b, c, d). Following these periods of near zero lymph production, a temporary increase in both pressure and rate was observed, before the lymph heart production returned to normal.

The first direct values for intralymph heart pressure in *Bufo marinus* were also provided by Jones et al. (1992). In order to open the lymphaticovenous valve in the efferent pore of the lymph heart, the intralymph heart pressure would have to be greater than the pressure in the posterior vertebral vein, which transports the lymph to the renal portal system. This is in fact the case, with the mean pressure in the femoral vein, which is in close proximity to the posterior vertebral vein, measuring 1.4 kPa, and the mean intralymph heart pressure measuring 2.25 kPa in *Bufo marinus* (Fig. 7). The lymph heart pressure, however, is not consistent from one contraction to the next, ranging from almost no pressure to over 4 kPa. As with lymph flow, intralymph heart pressure decreases during dehydration, dropping to about half the hydrated value in *Bufo marinus* (1.01 kPa). Rehydration results in a rapid recovery of lymph heart function (Fig. 6e).

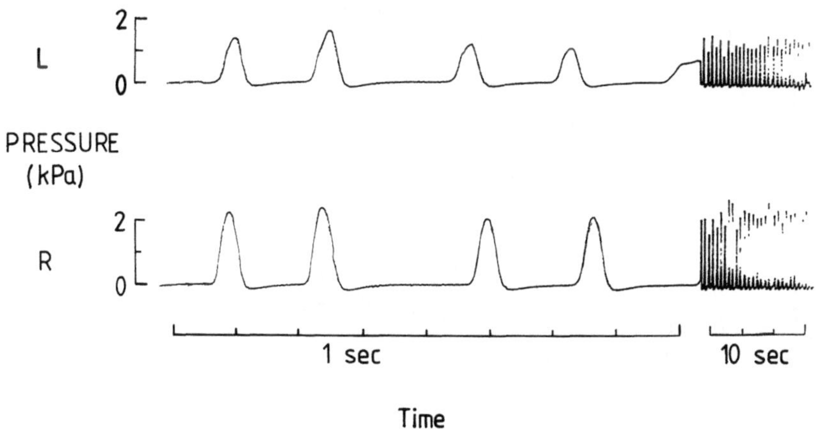

Fig. 7. Intralymph heart pressure of left (*L*) and right (*R*) posterior lymph hearts in a normal hydrated *Bufo marinus*. (Jones et al. 1992)

5 The Lymphatics and Acid-Base Regulation

Perhaps the most significant feature of the amphibian lymphatic system and the potential role this system could play in acid-base regulation relates to the relatively large fluid volume contained in the lymph vessels, sacs and hearts. This "space" could represent a significant expansion of our perceived and previously documented extracellular space estimates for anuran amphibians. Because of the relatively high permeability of amphibian capillaries, the filtered lymph contains all plasma constituents with the exception of red blood cells (Conklin 1930b). The normal protein ratio between lymph and plasma has been reported as 0.5 in *Bufo arenarum* (Zwemer and Foglia 1943) and 0.33 in *Bufo marinus* (Hillman et al. 1987). The high protein content of this large extracellular compartment is significant in that, being anionic in nature, the proteins are buffers to H^+ and other ions, and are therefore important components in the acid-base system.

Garland and Toews (1992), the first to report pH values for the lymph in amphibians, found that lymph collected from the posterior lymph hearts of chronically catheterized *Bufo marinus* was consistently about 0.5 pH units more acidic than the arterial blood (at 20 °C lymph pH was 7.85). Using the connula-

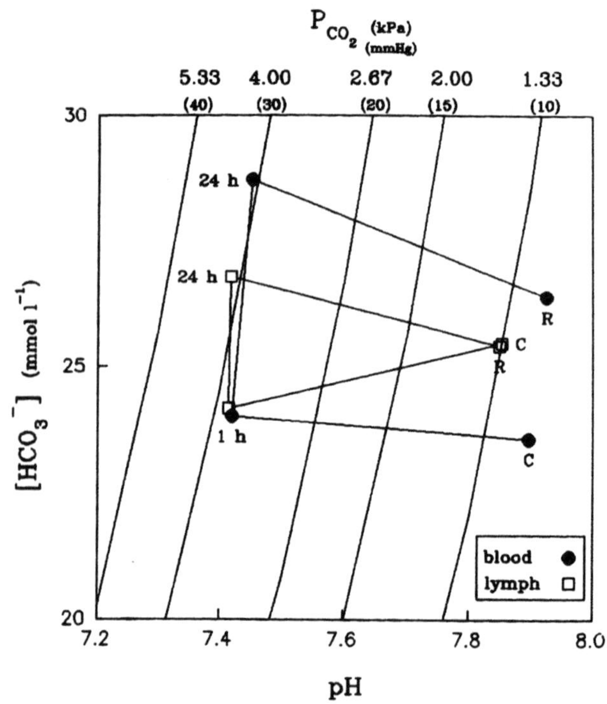

Fig. 8. pH-[HCO_3^-] diagram showing changes in variables in lymph and blood of 11 *Bufo marinus* during control periods (*C*), following 1 and 24 h of exposure to CO_2 and after a recovery period of 24 h from hypercapnia (*R*). (Garland and Toews 1992)

tion technique of Jones et al. (1992) described previously, Garland and Toews (1992) were able to cannulate both the blood and lymph circulatory systems and then expose the large toads to a 24-h bout of 5% CO_2 in air to ascertain the response of each system. They found that, although the lymphatic system had lower protein concentrations and was devoid of red blood cells, the response pattern to an acid-base disturbance was very similar to that of the blood (Fig. 8). Upon exposure to CO_2, the fluids in both compartments showed an initial acidosis (within 1 h) marked by a decrease in pH, followed by a compensatory phase during which $[HCO_3^-]$ increased in the plasma and lymph. Typical of amphibians in general, however, this compensatory phase did not dramatically increase blood or lymph pH values. When the acid-base disturbance was removed, Garland and Toews found that all measured variables in the blood and lymph returned to normal. Their overall conclusion was that the general pattern of changes in acid-base regulation in both compartments was quite similar. They also found that, since the ionic and protein compositions in lymph and plasma were quite similar, the in vitro buffer values for separated lymph and plasma were also not significantly different.

6 General Conclusions

The close relationship between the lymphatic and circulatory systems in amphibians has obviously evolved to ensure survival in a very variable environment. Amphibians are not truly transitional in the evolutionary sense, but most species dwell near the "edges" of the aquatic and terrestrial habitats and, as such, encounter a wide variety of wet and dry conditions. Although it does not appear as though the expansive lymphatics serve a water storage function in most species, the importance of the fluid contained within is evident in its contribution to recovery from such physiological stresses as hemorrhage, dehydration, and hypoxia. The great influx of exogenous water and the continuous filtration of essential blood constituents into the lymphatics necessitate a rapid reintroduction into the excretory and blood circulatory system. It is not surprising, then, that strong pulsatile hearts have evolved in numerous anatomical locations to move large volumes of lymph from one compartment to another. What is most surprising, however, is that even though the potential importance and complex structure of the amphibian lymphatic system were clearly described over 150 years ago in a group of animals that are arguably among the most available for experimental work, many of the physiological secrets of the amphibian lymphatic system still remain intact.

References

Baldwin AL, Rozum JS, Gore RW (1990) Routes of water and small solute transport between blood and lymph in frog. Physiol Zool 63(6): 1141–1156

Baustian M (1988) The contribution of lymphatic pathways during recovery from hemorrhage in the toad *Bufo marinus*. Physiol Zool 61: 555–563

Carter DB (1979) Structure and function of the subcutaneous lymph sacs in the anura (Amphibia). J Herpetoe 13(3): 321–327
Conklin RE (1930a) The formation and circulation of lymph in the frog. I. The rate of lymph production. Am J Physiol 95(1): 79–90
Conklin RE (1930b) The formation and circulation of lymph in the frog. III. The permeability of the capillaries to protein. Am J Physiol 95(1): 98–110
Deydrup IJ (1964) Water balance and kidney. In: Moore JA (ed) Physiology of the Amphibia. Academic Press, New York, pp 251–328
Ecker A (1889) The anatomy of the frog. Clarendon Press, Oxford (transl by Haslem J)
Eckert R, Randall D, Augustine G (1988) Animal physiology. Freeman, New York
Foglia VG (1941) Cause of death of toads after destruction of their lymph hearts. Proc Soc Exp Biol Med 46: 598–601
Garland RJ, Toews DP (1992) Acid-base regulation in response to hypercapnia in the lymphatic and circulatory systems of the toad *Bufo marinus*. Exp Biol 170: 271–276
Hillman S, Zygmunt A, Baustian M (1987) Transcapillary fluid forces during dehydration in two amphibians. Physiol Zool 60(3): 339–345
Isayama S (1924a) Über die Strömung der Lymphe bei den Amphibien. Z Biol 82: 91–100
Isayama S (1924b) Über die Geschwindigkeit des Flüssigkeitsaustausches zwischen Blut und Gewebe. Z Biol 82: 101–106
Jones JM, Wentzell LA, Toews DP (1992) Posterior lymph heart pressure and rate and lymph flow in the toad *Bufo marinus* in response to hydrated and dehydrated conditions. J Exp Biol 169: 207–220
Kampmeier OF (1969) Evolution and comparative morphology of the lymphatic system. Thomas, Springfield
Katz U (1986) The role of amphibian epidermis in osmoregulation and its adaptive response to changing environment. In: Bereiter-Hahn J, Matoltsy AG, Richards KS (eds) Biology of the integument, vol II, Vertebrates. Springer, Berlin Heidelberg New York
Middler SA, Kleeman CR, Edwards E (1968) Lymph mobilization following acute blood loss in the toad *Bufo marinus*. Comp Biochem Physiol 24: 343–353
Müller J (1833) On the existence of four distinct hearts, having regular pulsations, connected with the lymphatic system, in certain amphibious animals. Philos Trans R Soc Lond 1: 89–94
Noble GK (1931) The biology of the Amphibia. McGraw-Hill, New York
Panizza B (1834) Über die Lymphherzen der Amphibien. Arch Anat Physiol: 300
Pratt FA, Reid MA (1939) Synchronism of anuran lymph hearts and the integration of their spinal centers. J Physiol 95: 345–355
Priestley J (1878a) An account of the anatomy and physiology of batrachian lymph-hearts. J Physiol 1: 1–17
Priestley J (1878b) Contributions to the physiology of batrachian lymph-hearts. J Physiol 1: 19–38
Radwanska M (1906) Die vorderen Lymphherzen des Frosches. Bull Int Acad Sci Cracov 213–228 In: Kampmeier OF (1969) Evolution and comparative morphology of the lymphatic system. Thomas, Springfield
Ranvier L (1880) Appareils nerveux terminaux des muscles de lavie organique; coeurs sanguins, coeurs lymphatique; oesophagus; muscles lisses. Lec Anat Gen Faites Coll Fr 7: 350
Schiff M (1850) Vorläufige Bemerkungen über den Einfluß der Nerven auf die Bewegungen der Lymphherzen. Z Rationel Med 9: 259
Thorson TB (1964) The partitioning of body water in amphibia. Physiol Zool 37: 395–399
Von Leydig F (1876) Über die äußeren Bedeckungen der Reptilien und Amphibien. Arch Mikrobiol Anat 12: 119
Wentzell LA, McNeil S A, Toews D P (1993) The role of the lymphatic system in water balance processes in the toad *Bufo marinus*. Physiol Zool 66: 333–347
Winterstein H (ed) (1925) Handbuch der vergleichenden Physiologie I, Jena. In: Conklin RE (ed) The formation and circulation of lymph in the frog. I. The rate of lymph production. Am J Physiol 95: 79–90
Zwemer RL, Foglia VG (1943) Fatal loss of plasma volume after lymph heart destruction in toads. Proc Soc Exp Biol Med 53: 14–17

Chapter 10

Physiological Roles of the Secondary Circulatory System in Fish

A. Ishimatsu[1], *G.K. Iwama*[2] and *N. Heisler*[3]

Contents

1 Introduction

Most studies of the secondary circulation or the "secondary vascular system" (Vogel 1981; Vogel and Claviez 1981) have concentrated on descriptions of morphology, rather than function, among various species (see Steffensen and Lomholt 1992). The scarce available evidence at this time characterizes the secondary vascular system in fishes as being a pool of blood which is low in

[1] Nomo Fisheries Station, Nagasaki University, Nomozaki, Nagasaki, Japan
[2] Department of Animal Science, Faculty of Agricultural Sciences, University of British Columbia, Vancouver, B.C., Canada
[3] Department of Animal Physiology, Humboldt University at Berlin, D-10115 Berlin, and Max-Planck-Institut für experimentelle Medizin, D-37075 Göttingen, Germany

pressure, low in red cell content, absent from the muscle tissues, associated with both the inner and outer surfaces of body including the gills, and of a volume that approaches or exceeds that of the primary circulation. Vogel (1985a) demonstrated the occurrence of an extremely dense capillary network of the secondary system over the scales of trout and tilapia. Satchell (1991) pointed out that similar vascularization on the scales had been observed for both freshwater and seawater teleosts in old anatomical literatures (Jakubowski 1960a,b, 1989; Kaczmarski 1966; Tyszkiewicz 1969), while it is not always clear whether such capillary networks belong to either the primary or secondary vascular system. A similar subepithelial vascular network has also been observed in the Atlantic hagfish *Myxine glutinosa* (Lametschwandtner et al. 1989). Vogel (1985a) also stated that secondary capillary networks are distributed in the mucous membrane of the mouth and the peritoneum, and possibly in the intestinal wall. The central venous sinus (CVS) which is located within the primary lamella of the gill is currently considered to be part of the secondary vascular system (Vogel 1985a).

The secondary vascular system is distinct from the lymphatics of higher vertebrates. It does not have the blind-ended lymphatic vessels, which originate deep in muscle and other tissues and which carry interstitial fluid from those tissues back to the primary venous circulation. Instead, the secondary system in fish is connected to the arterial as well as venous vessels of the primary system, forming a vascular circuit with its capillary network in parallel to that of the primary circuit, thus providing an additional exchange interface. Also, the connection to the primary arterial system ensures higher flow rates and shorter transit time through the secondary circulatory system than would be the case if the periphery terminated in blind ends as in the lymphatic systems. The nature and origin of the fluid carried through the two systems are therefore distinct: lymphatic fluid is composed of almost pure interstitial fluid which has been filtered across the capillary walls, whereas the fluid within the secondary vessels consists, at least in part, of the plasma of the primary circulation.

It is presently unknown whether fish possess a "true" lymphatic system. The secondary vascular system may accordingly perform functions conducted by the lymphatic system in higher vertebrates, such as the regulation of fluid balance, biological defense through immune responses, and transport of proteins, lipids and some enzymes. On the other hand, the peculiar distributional pattern of the secondary circulatory system suggests its unique contribution to transepithelial exchange processes of gases and ions. The main purpose of this chapter is to review the literature pertaining to possible functions of the secondary circulation in fishes, with special emphasis on recent physiological experiments.

2 The Central Venous Sinus (CVS) in the Gill

The CVS of the gill remains something of an enigma to fish physiologists. This fairly ubiquitous structure occupies the central portion of the primary filament in close proximity to where chloride cells are located, spanning nearly the full width of the primary filamental epithelium. It is supplied with blood through specialized anastomoses emanating mainly from the efferent side of the arterioarterial vasculature (Laurent 1984). The effluent blood from the CVS is shunted away from systemic vasculature and returned to the heart via the branchial vein. Thus, this pathway can function as a short circuit with possibly a very rapid recirculation rate. Current hypotheses regarding the possible physiological functions of the CVS have been summarized by Randall (1985). Accordingly, the CVS may act as (1) a plasma reservoir (Girard and Payan 1976; Vogel et al. 1973); (2) a collecting reservoir for interstitial fluid of the gill; (3) a nutritional vasculature for the filamental tissues; (4) a substrate-supplying system for chloride cells; and (5) a degradation system for circulating hormones (Nekvasil and Olson 1985; Olson et al. 1986). Boland and Olson (1979) speculated that the CVS may serve as a reservoir for white cells or as a site of immunological activity.

There is only scarce experimental evidence on the functional role(s) of the CVS, even though a linkage with the function of chloride cell has been repeatedly suggested on the basis of morphological evidence (Laurent 1989). This lack of data is probably related to the paucity of experimental techniques for separate assessments of the physiological roles of the two vascular beds in fish gills, i.e. the arterioarterial circuit associated with the epithelium of the secondary lamellae and the arteriovenous circuit including CVS, underlying the epithelium of the primary filaments. A few attempts in this matter have employed in vivo and in vitro preparations. In vitro preparations such as the isolated head or the isolated gill arch preparations have certain theoretical advantages over in vivo preparations, e.g. the former preparations allow easier manipulation of internal and external environments than in vivo preparations (Perry et al. 1984). However, present techniques for such isolated preparations have not always been successful in mimicking in vivo physiological conditions. Any direct comparison of data obtained in vitro with results from in vivo preparations should therefore be viewed with a certain amount of caution.

2.1 Blood Flow Partitioning in the Gill

In vitro preparations have mainly been used to study blood flow partitioning between primary and secondary systems in the gills. Generally, in vitro estimates of the relative perfusion rate through the CVS are distinctly higher than the few available in vivo data, even with relatively low back-pressures at the efferent arterial outflow often applied in in vitro studies. Recently, published data on the flow partitioning in trout gill using an isolated head preparation (Gardaire et al. 1991) have reported that as much as 34–40% of cardiac output flowed through the CVS, while the dorsal aorta received only 53–62% depending on the adrenaline concentration of the perfusate. About 5% was found to perfuse the 'cephalic tissues'. Isolated gill preparations do not usually allow separation of arterioarterial and arteriovenous outflows. Olson (1984) developed a perfused gill preparation with which one can separately collect arterioarterial and arteriovenous effluents. However, even with this very sophisticated preparation, application of physiological dorsal aortic back-pressure resulted in a very low arterial outflow (72% perfused with Ringer's solution, 55% with blood for rainbow trout) and venous outflow of rather high hematocrit (12%, as compared with 18.4% for arterial outflow, rainbow trout).

In contrast, application of an in vivo microcannulation technique of the branchial vein in the rainbow trout (Ishimatsu et al. 1988) gave substantially lower hematocrit values in branchial vein effluent fluid of only 3.5% with simultaneous systemic values of 23–24%. Relative arteriovenous (CVS) flow, in vivo, was calculated from the hemoglobin concentration ([Hb]) of blood sampled from the dorsal aorta, sinus venosus and branchial vein by the application of the Fick principle to be 7% of the cardiac output. This may represent a slight overestimate, due to the fact that the branchial vein probably returns venous drainage of the nutritional vessels of the gill in addition to the CVS output. Assuming the 'true' CVS [Hb] to be same as the [Hb] of the fluid from the secondary system of the body surface (Iwama et al. 1993; see below) and the [Hb] of the nutritional vasculature to be same as in the dorsal aorta, then the branchial vein blood is estimated to be composed of 86% CVS output and 14% gill nutritional venous return. Taking this into account, the CVS flow would be calculated to be 6% of the total cardiac output, a value little different from the original estimate of 7%.

Based on the above value of blood flow partitioning and blood pressure reported in the literature, it is possible to estimate relative vascular resistance of the lamellar and CVS circuits (R_{LAM}/R_{CVS}) using the formula:

$$R_{LAM}/R_{CVS} = Q_{CVS}/Q_T \cdot (P_{VA} - P_{DA})/(P_{DA} - P_V),$$

where Q_{CVS}/Q_T is the relative blood flow through CVS, P_{VA}, P_{DA}, and P_V represent ventral aortic, dorsal aortic and central venous pressure, respectively. It should be noted that the above equation holds true for fishes with no arteriovenous anastomoses in the afferent side. Afferent connections are known to exist in a few teleost species (Laurent 1984), including rainbow trout. In rainbow trout, however, their physiological importance appears to be quantitatively very minor because of the vanishingly small number of the afferent anastomoses (Vogel 1978). By applying pressure data for rainbow trout reported by Kiceniuk and Jones (1977) (P_{VA}: 53 cmH_2O, P_{DA}: 42 cmH_2O, P_V:2 cmH_2O), R_{LAM}/R_{CVS} is calculated to be 0.02. This estimate of R_{LAM} comprises all vascular resistance between the two pressure-measuring sites in the dorsal and ventral aortae, including afferent filamental vessels, lamellar complex, efferent filamental vessels, and the sphincter resistance at the base of the efferent filamental artery (Bailly and Dunel-Erb 1986). Also, using P_{DA} for the input pressure of the CVS circuit, this calculation assumes an insignificant pressure drop through the efferent filamental sphincter. This appears to be valid, at least for resting conditions, considering a negligible pressure difference between the dorsal aortic pressure and that in the efferent filamental artery obtained by using a micropuncture technique in anesthetized lingcod, *Ophiodon elongatus* (Farrell and Smith 1981). Stagg and Shuttleworth (1984) calculated the vascular resistance of the two branchial pathways, whose ratio (R_{LAM}/R_{CVS}) was equal to 0.18, using data collected in an isolated perfused gill experiment on *Platichthys flesus*. Olson and Meisheri (1989) reported the ratio of R_{LAM} and R_{CVS} of the isolated perfused trout gill to be 0.07 to 0.15. Although these two values are somewhat different from the in vivo estimate given above, all indicate a relatively high vascular resistance of the arteriovenous circuit. A major portion of the resistance possibly resides in the arteriovenous anastomoses leading into the CVS.

According to the parallel arrangement of the systemic vascular bed and CVS downstream from the efferent side of gill circulation, blood flow partitioning between the two circuits is directly determined by their resistance ratio ($Q_{SYS}/Q_{CVS} = R_{CVS}/R_{SYS}$: SYS denotes systemic circuit including the efferent filamental sphincter). Any fluctuations of the systemic vascular resistance downstream of the efferent sphincter must therefore exert a profound effect on gill blood flow partitioning as does the ratio of the vascular resistance of the two circuits within the gills.

Sundin and Nilsson (1992) have recently measured blood flow directly in the inferior jugular vein and ventral aorta in unanesthetized cod *Gadus morhua* using Doppler techniques. Assuming that the vein represents the major drainage pathway of the arteriovenous gill circuit and that the dorsal connection to the anterior cardinal vein is of limited importance in draining the CVS, they evaluated 8% of the cardiac output in resting cod to be CVS flow. This in vivo estimate is in excellent agreement with our in vivo estimate in trout, although the contribution of the systemic venous return from the lower jaw to the inferior jugular vein flow is presently unknown.

2.2 Control of Blood Flow Partitioning

A detailed account of branchial innervation and control of gill blood flow was provided by Nilsson (1984).

2.2.1 Neural Control

Little information is available about the nervous control of gill blood flow partitioning. Relevant studies therefore have utilized perfused in vitro preparations and cautious interpretations of those data are necessary for the reasons discussed above. The perfused head preparation used for a study of sympathetic nervous control of branchial flow partitioning in the cod by Nilsson and Pettersson (1981) suffered from leakage through uncontrolled vessels, which probably led to the highly variable results during nerve stimulation and drug infusion. Overall, the most consistent response was an increase in total branchial resistance with the redistribution of blood in favor of the dorsal aortic flow. This effect was apparently mediated by α-adrenergic receptors, and abolished by the addition of phentolamine to the perfusate.

The recent finding of the serotonergic innervation in the gill may open a new field of research activity, with possible effects of innervation not only on blood flow partitioning but also on ion transfer activities (Bailly et al. 1989; Laurent 1989).

2.2.2 Humoral Control

Adrenaline. In vitro, adrenaline consistently effects a decrease in flow through the CVS, directing a larger proportion of blood flow towards the systemic circuit in both isolated gill preparations (Olson 1984; Stagg and Shuttleworth 1984) and isolated head preparations (Payan and Girard 1977; Claiborne and Evans 1980). This has been interpreted as a result of either dilation of the sphincters at the base of efferent filamental arteries or constriction of A-V anastomoses probably equipped with α-adrenoceptors (Laurent 1989). Sundin and Nilsson (1992) demonstrated almost complete cessation of inferior jugular vein flow in vivo, following a bolus injection of adrenaline. In that preparation, adrenaline must have caused systemic vasoconstriction by a net result of an α-adrenoceptor-mediated mechanism (Nilsson 1983) which would, by itself, increase blood flow through the arteriovenous circuit, in addition to the vascular effects within the gill. Thus, it was postulated that adrenaline has exerted either a relatively larger vasoconstrictory effect on the arteriovenous circuit than that on the systemic circuit, or a considerable dilation of the efferent filamental and/or branchial arteries occurred, resulting in a reduction of overall vascular resistance of the systemic circuit. However, the latter possibility is unlikely, because the blood pressure in the ventral aorta almost doubled whereas cardiac output was elevated only moderately by adrenaline infusion, indicating an overall increase in vascular resistance. Stagg and Shuttleworth (1984) demonstrated an increase in arterioarterial outflow and a nearly equal fall in vascular conductance (= 1/resistance) of the arterioarterial and arteriovenous circuits upon α-adrenoceptor stimulation in isolated flounder gills. Because of the relatively high conductance of the arterioarterial circuit under control conditions (28.5 μl min^{-1} g^{-1} $mmHg^{-1}$) compared to that of arteriovenous circuit (5.2), the observed decline should have redirected flow toward the arterioarterial circuit. Thus, the conductance ratio (G_{LAM}/G_{CVS}) markedly increased from 5.48 to around 45, which is equivalent to a decrease in the resistance ratio (R_{LAM}/R_{CVS}) from 0.18 to 0.02. They also found that a β-adrenoceptor stimulation caused a moderate rise in the conductance of the arterioarterial vasculature without any appreciable effects on flow partitioning. With β-stimulation, the conductance ratio was calculated to have increased only from 5.48 to 6.06.

The extensive use of elevated levels of catecholamines, however, is not generally accepted as physiologically relevant. Laurent (1989) criticized pharmacological, rather than physiological, concentrations often used in in vitro studies and strongly suggested a reassessment of the effects of circulating catecholamines on the partitioning of gill blood flow. Hypercapnia is known to cause a release of catecholamines into the circulatory system (Tang and Boutilier 1988). In our recent study (Iwama et al. 1993), however, no significant change in blood flow partitioning between the systemic and CVS circuits has been observed during hypercapnia. No catecholamine determinations were conducted in those studies of in vivo gill blood flow partitioning.

It is possible that the low relative CVS flow observed in vivo limits further substantial reduction of flow through the CVS circuit. Further investigations are required to elucidate the regulatory roles of the adrenergic system in gill blood flow partitioning in vivo.

Acetylcholine. In contrast to the hemodynamic effects of adrenaline on flow redistribution, acetylcholine is known to divert more blood flow into the arteriovenous circuit. Smith (1977) concluded from an isolated trout gill preparation study that acetylcholine exerted most of its constrictor effect at a site downstream from the secondary lamellae, more specifically downstream from the origin of the vessels to the CVS. Although his preparation suffered from the common limitations of the isolated preparations as described above, his prediction as to the site of acetylcholine action was in good agreement with a more recent anatomical observation by Bailly and Dunel-Erb (1986). They demonstrated that a sphincter at the proximal base of the efferent filamental artery is innervated by the protrematic vagus nerve whose endings are in contact with adrenergic nerve terminals. It was postulated that this pattern of dual innervation represents modulation of the vasoconstrictory effect of the vagus nerve by the sympathetic nerve activity. In addition, the occurrence of a similar innervation pattern has become clarified for the proximal, but not distal, portion of the efferent filamental vasculature, i.e., the efferent filamental arteries and efferent lamellar arterioles (Dunel-Erb et al. 1989).

Farrell and Smith (1981) demonstrated an increase in blood pressure in the efferent filamental artery following acetylcholine administration in their study on a lingcod (*Ophiodon elongatus*) gill preparation by applying a micropuncture technique for the direct pressure recording of those minute vessels. Their results also support the existence of a postlamellar cholinergic constrictor site.

Other Vasoactive Agents. Sundin and Nilsson (1992) also examined the effects of serotonin and two peptides (cholecystokinin-8 and caerulein) on flow partitioning. Serotonin evoked a small increase in cardiac output and a prominent rise in the inferior jugular vein flow in vivo, resulting in approximately a doubling of the partitioning through the vein (equal to CVS flow). The two peptides caused a moderate decrease in both cardiac output and the flow through the jugular vein with no appreciable change in flow partitioning. The effects of atrial natriuretic factor and endothelin on the partitioning of gill blood flow were reported by Olson and Meisheri (1989) and Olson et al. (1991), respectively. The most pronounced effect of atrial natriuretic factor (10^{-8} M) seems to be vasodilation of the arterioarterial circuit that was previously contracted by epinephrine

(10^{-6} M). Endothelin caused an overall increase in vascular resistance, reducing arterioarterial outflow by 65% without affecting arteriovenous flow at 10^{-9} M.

2.3 Ion Transport and the CVS

2.3.1 Localization of the Site of Ion Transport in the Gill

As to the specific site of ion transport in the gill, little is definitely known except for chloride extrusion by chloride cells in saltwater-adapted fish. This finding was made by applying the vibrating probe technique to the opercular epithelia with a high density of chloride cells which are thought to be homologous to chloride cells in the gill. This remains the sole unequivocal demonstration of the ion transport site across fish epithelium with respect to localization (Foskett and Scheffey 1982; Zadunaisky 1984; Foskett and Machen 1985).

We have recently conducted an in vivo experiment to assess the relative roles of the lamellar and CVS circuits in acid-base regulation in rainbow trout (Iwama et al. 1993). Fish were chronically catheterized in the dorsal aorta, sinus venosus and branchial vein and, after recovery, exposed to environmental hypercapnia. During control conditions, the pH value of the effluent fluid from the CVS (samples collected from the branchial vein cannula) was significantly lower than the dorsal aortic pH. The pH difference was attributable to CO_2 addition into the CVS fluid during transit through the CVS (Fig. 1B). In response to the development of a respiratory acidosis, the pH gradient was reversed. This was caused by a significantly higher HCO_3^- concentration in the branchial vein blood compared to that in the dorsal aortic blood (Fig. 1C), which was attributed to transepithelial acid-base relevant ion transport between ambient water and the CVS. The overall rise in [HCO_3^-] was accompanied by a nearly equimolar fall in plasma [Cl^-], while there was no significant difference in Cl^- concentration between sampling sites (Fig. 2). Plasma Na^+ concentration remained nearly constant with no significant change both from control values and between the sites. Calculation indicated that the amount of HCO_3^- gained at the CVS during hypercapnia exceeded the amount accumulated in the extracellular space, suggesting that the filamental epithelium is the main site of acid-base regulation in trout, and the CVS serves as a convective pathway between the ion-transporting site and the general body circulation. The concurrent uniform decline in Cl^-concentration indicates a Cl^--related transfer of HCO_3^- or H^+ in the accumulation of plasma HCO_3^-, but the absence of any [Cl^-] difference between CVS and systemic arterial blood is inconsistent with the idea of a Cl^-/HCO_3^- exchange mechanism in the filamental epithelium. Different locations of the two exchange sites are possible, given that a recently demonstrated electrogenic proton pump (H^+-ATP) that is not directly coupled with transport of a counter ion (Avella and Bornancin 1989; Lin and Randall 1991, 1993) would be the main mechanism for acid-base regulation. Thus, our results can be explained by an electrogenic proton pump mechanism residing in the filamental epithelium, generating the

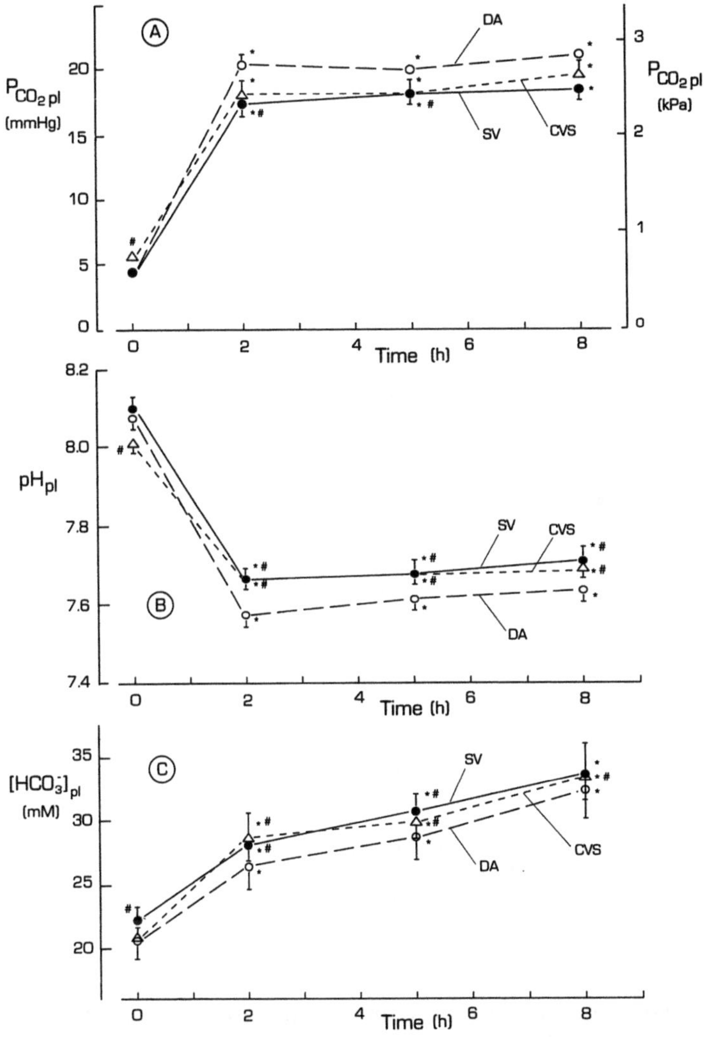

Fig. 1 A–C. Plasma P_{CO_2}, pH and $[HCO_3^-]$ in dorsal aorta (*DA*), sinus venosus (*SV*) and branchial vein (*CVS*) blood during normocapnia and hypercapnia (2% environmental CO_2) of freshwater rainbow trout. *indicates significant difference to the respective average control value; #indicates significant difference between DA and the indicated site at the specific time; CVS and SV are significantly different as well at times 0 and 5 h (mean ± SE, n = 8 for controls, 2 and 5 h; n = 6 for 8-h samples; $P < 0.05$). Data from Iwama et al. (1993)

driving force for passive movement of a negative co-ion (Cl^-), by enhancing efflux and reducing influx, through the location of least resistance. This site will most likely be the respiratory epithelium on the lamellae with its large surface area and short diffusion distances (Heisler 1993). A recent immunohistochemical study has revealed that positive staining for H^+-ATPase was confined to the chloride cells

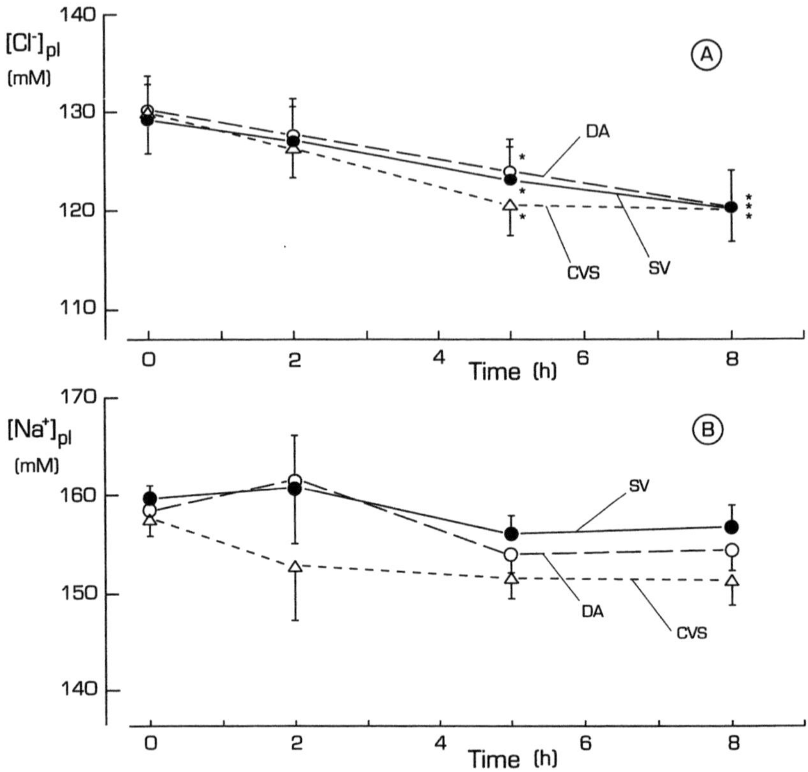

Fig. 2 A–C. Plasma [Cl^-], [Na^+] and osmolarity in dorsal aorta (*DA*), sinus venosus (*SV*) and branchial vein (*CVS*) during normocapnia and hypercapnia (2% environmental CO_2) of rainbow trout. *indicates significant difference to the respective average control value (mean $\pm$SE, n = 8 for controls, 2 and 5 h; n = 6 for 8-h samples; $P < 0.05$). Data from Iwama et al. (1993)

in hypercapnic rainbow trout, while a more uniform distribution was observed for control animals (Lin et al. 1994).

Payan et al. (1984), summarizing their previous studies on ion flux partitioning through the gills of isolated perfused trout head, concluded that Na^+ and Cl^- influxes in freshwater fish could entirely be attributed to lamellar epithelia with no contribution of the filamental surface. For Ca^{2+}, however, they found a large proportion to be transferred across the filamental surface (see also Ishihara and Mugiya 1987).

The cited studies are probably not directly comparable and should be viewed with a certain amount of reservation. Our in vivo preparation has the advantage of allowing direct determination of acid-base parameters of the branchial vein blood, a sampling site representing almost exclusive drainage from the CVS (see above). Differentiation of ion fluxes between the two branchial epithelia by this technique requires application of the Fick principle, including a number of complicating factors such as the relatively long time period of fluid sampling through the very thin branchial cannula (ca. 40 min for flushing and collection of a 0.5-ml sample). This does not allow simultaneous sampling from all cannulation sites which is essential for dynamic isotope influx studies in order to assess unidirectional ion fluxes. In addition, ion transfer as well as fluid flow may not always be distributed homogeneously among all gill arches. Our in vivo cannulation technique can provide information about the acid-base status of fluid in a particular arch, but an assessment of the entire gill vasculature is clearly an extrapolation, which has to be validated by further experimentation. Detailed critiques of isolated perfused head preparations have been published elsewhere (Olson 1984; Perry et al. 1984).

For more detailed information on potential gill ion transport models, the reader is referred to relevant reviews (see e.g. Marshall 1985; McDonald and Prior 1988; Avella and Bornancin 1989; Shuttleworth 1989; Lin and Randall 1991, 1993).

2.3.2 Chloride Cells and Acid-Base Regulation

There is increasing anatomical evidence suggesting a functional role of chloride cells in acid-base regulation. Generally, proliferation and/or enlargement of apical surface area of chloride cells are found as a compensatory response to acid-base disturbance (hypercapnia: Cameron and Iwama 1987; hyperoxia: Laurent and Perry 1991; water acidification: Laurent and Perry 1991; Leino and McCormick 1984). Recently, however, Goss et al. (1992a) reported a sharp reduction in chloride cell fractional surface area in channel catfish (*Ictalurus nebulosus*) exposed to hypercapnia (2% CO_2), which coincided with a rapid drop in Cl^- influx. These authors demonstrated that adjacent epithelial cells extended over the apical openings of the chloride cells and physically restricted their association with the ambient water. Based on these data, they proposed that Cl^-/HCO_3^- exchange mechanism is associated with chloride cells and that morphological modulation of the transport mechanism plays a crucial role in acid-base regulation of the species. Goss et al. (1992 b) reported a similar but less pronounced change in chloride cell morphology in hypercapnic rainbow trout. Paradoxically, Cameron and Iwama (1987) found an opposite change in chloride cell morphology (a rise in chloride cell fractional surface area) in another species of Ictaluridae, *Ictalurus punctatus*, in response to hypercapnia (a stepwise rise up to 8% CO_2, 96 h). The difference in hypercapnia regime and/or water ionic composition ($[Na^+]$ and $[Cl^-]$ were seven to eight times higher in Cameron and Iwama 1987) might have caused the apparently opposite morphological change. Perry and Goss (1994) recently reported an increase in chloride cell fractional area in response to $NaHCO_3$ infusion in rainbow trout.

Water acidification provoked chloride cell proliferation in fathead minnow *Pimephales promelas* (Leino and McCormick 1984), while Laurent and Perry (1991), based on the data on the effects of water acidification on gill morphology of rainbow trout, concluded that acidification alone only stimulates the preexisting chloride cells but is not a sufficient stimulus to provoke proliferation of the cells.

3 Secondary Circulatory System of the Body

According to Vogel (1985a), distribution of the capillary network of the secondary system is restricted to the skin including the fins, the mucus membrane of the mouth, the peritoneum, and possibly the intestinal wall. Kampmeier (1969) presented a scheme of complex arrangements of the systemic secondary veins (or "lymphatics" of his days) in teleosts. The deep and superficial lymphatics were shown to be connected by several large-bore vessels at the level of the shoulder girdle and the posterior end of the abdominal cavity, the latter called "circumanal lymph plexus". There are several major trunks close to the kidney whose posterior ends fuse with the plexus. We are not aware of any anatomical study on the distribution of the secondary vessels within the kidney. If a direct connection were in fact to exist, a functional linkage between the kidneys and the secondary circulatory system may present an entirely new aspect of the functional role of the secondary system.

Secondary vasculature has recently been found in the heat exchangers in skipjack tuna, *Katsuwonus pelamis* (Dewar et al. 1994).

3.1 Hemodynamics of the Secondary Circulatory System of the Body

3.1.1 Volume, Flow and Pressure of the Secondary System

Very few data are available regarding volume, flow rate, flow direction, and pressure of the secondary circulatory system. The recent review by Steffensen and Lomholt (1992) summarizes current knowledge concerning these aspects. The volume of the secondary circulatory system is about 4.9% of body weight in comparison to 3.4% for the primary system in rainbow trout. Fluid flow into the secondary system was found to be very low, being only 0.3% of the total cardiac output. Steffensen and Lomholt (1992) applied a two-compartment model, representing the primary and secondary systems, to analyze the disappearance of labeled plasma protein from the primary circulation. This obviously reflects the overall exchange rate between the two systems. Taking into account the relatively high flow rate found for the CVS in the two in vivo studies (6-8% of the cardiac output), a relatively large fluid pool in the systemic part of the secondary system had to be expected to exchange fluid with the primary circulation at extremely low exchange rates. This suggests that the secondary circulatory system is evidently composed of various subcompartments with largely different or even variable perfusion rates (as well as other hemodynamic characteristics), a diversity similar to that known to exist in the primary system.

Blood pressure in the secondary circulation is largely unknown. The only available data that have been obtained from the lateral subcutaneous vein of rainbow trout are in the range of -1.3 to 3.8 cm H_2O at a dorsal aortic pressure of 41 cmH_2O (Ishimatsu et al. 1992). Nothing is known about blood pressure in the secondary arteries.

3.1.2 Caudal heart

The caudal heart, or more traditionally the "lymph heart (propulsor)", has been the subject of scientific investigation for centuries, but our knowledge has been restricted only to purely morphological aspects while physiological aspects and its control mechanisms have hardly been studied. This organ is prevalent among fishes but it is said to be characteristically lacking in the Acanthopterygii (see Kampmeier 1969 for review of anatomical literature). In the course of a reappraisal of the fish vascular system, Vogel (1985b) demonstrated that this peculiar organ is a component of the secondary circulatory system pumping the drainage of the secondary venous fluid of the caudal region into the caudal vein of the primary system. The beating rate of the caudal heart is surprisingly high. Chan (1971) reported 165–230 beats min^{-1} at 18 °C in *Anguilla japonica.* Davie (1981) found a relationship between beating rate and temperature for *Anguilla australis schmidtii.* He found a lower rate (90 beats min^{-1}) than Chan (1971) at comparable temperature, although still much higher than systemic heart rate (40–60 beats min^{-1}, Hipkins 1985).

The vascular anatomy of the tail region of *Anguilla* was studied by Chan (1971) with emphasis on the vascular connection of the caudal heart and the urophysis. Although a complete picture of venous drainage pathways from this neurosecretory organ could not be given, he found that the blood in the urophysial plexus was directed posteriorly, and ran ventrally into the space between the heads of the caudal fin rays. From this description, it seems possible that the caudal secondary system serves in the transport of urophysial hormones. This notion is supported by data indicating that the beating rate of the caudal heart is stimulated in a dose-dependent manner by an urophysial extract (Chan 1971). Davie (1981) demonstrated that the caudal heart beat of the eel is neurogenic and the segment of spinal cord which sends off a nerve at the penultimate vertebra is necessary for the maintenance of heart beat. A urophysial hormone, urotensin II, thought to be released in response to hypovolemia, elicits an increase in caudal heart rate and exerts a hypertensive effect on the caudal vein pressure (Chan 1975). These data may indicate the existence of a regulatory system initiating fluid shifts from the secondary circulatory system to the primary system in response to a fall in blood pressure. The diuretic effect of urotensin II is probably related to the hypertensive effect of the hormone.

3.2 Physiological Functions

3.2.1 Ion Transport Through the Body Surface

Experimental evidence for an involvement of the secondary circulation in acid-base regulation in rainbow trout has been presented by Ishimatsu et al. (1992). During hypercapnia, pH and HCO_3^- concentrations of secondary venous fluid were significantly higher than the respective parameters of primary arterial plasma, whereas no difference could be detected during normocapnia (Fig. 3). Concurrently, the chloride concentration in the secondary plasma became significantly lower than in the primary plasma (Fig. 4). The finding of identical acid-base status and ionic composition under control conditions between the secondary and primary plasma agrees well with the lack of any potential difference across rainbow trout skin (Fromm 1968) and brook trout opercular epithelium (Marshall 1985) in vitro. Because chloride cells are known to occur in the body skin of several fish species (Henrikson and Matoltsy 1968; Nonnotte et al. 1979; Stiffler et al. 1986; Whitear 1986; Whitear and Mittal 1986; Hwang 1989; Pfeiler and Lindley 1989), the possibility of direct ionic exchange between ambient water and the secondary capillary fluid beneath the epithelium deserves scrutiny (see also Satchell 1991). On the other hand, it is possible that the observed changes in ionic composition of secondary fluid during hypercapnia do actually reflect renal ionic regulation, on the basis of fluid conveyed into the secondary system via fairly large pathways between lateral cutaneous vessels and deeper vasculature within the kidney. The lack of any information as to the flow patterns within the secondary vascular system prohibits any further speculation on this matter. The interesting physiological implications of the close interrelationship between secondary vasculature and excretory organ, however, may stimulate detailed and critical investigation.

The general capability of fish skin for ionic transport has been well demonstrated by Marshall et al. (1992) in studies using isolated cleithrum skin of freshwater-acclimated rainbow trout. When mounted in an Ussing-style chamber, isolated skin showed active Ca^{2+} uptake, judged from the Ussing flux ratio equation. They also demonstrated a moderately close ($r = 0.63$) positive relationship between Ca^{2+} influx and the density of the chloride (mitocholdria-rich) cells on the cleithrum skin. In contrast, their preparation did not provide any evidence for active Na^+ and/or Cl^- transport. The movements of Na^+ and Cl^- through the cleithrum skin were unaffected by either adrenergic or cyclic AMP stimulation. Interestingly, the density of chloride cells in the cleithrum skin decreased in response to cortisol treatment which is known to cause proliferation of gill chloride cells (Laurent and Perry 1990). In connection with the occurrence of chloride cells in the cleithrum skin, it may be of some relevance to indicate that a large pectoral sinus, which presumably represents a component of the secondary system, lies inside the skin at the base of the pectoral fins (Kampmeier 1969).

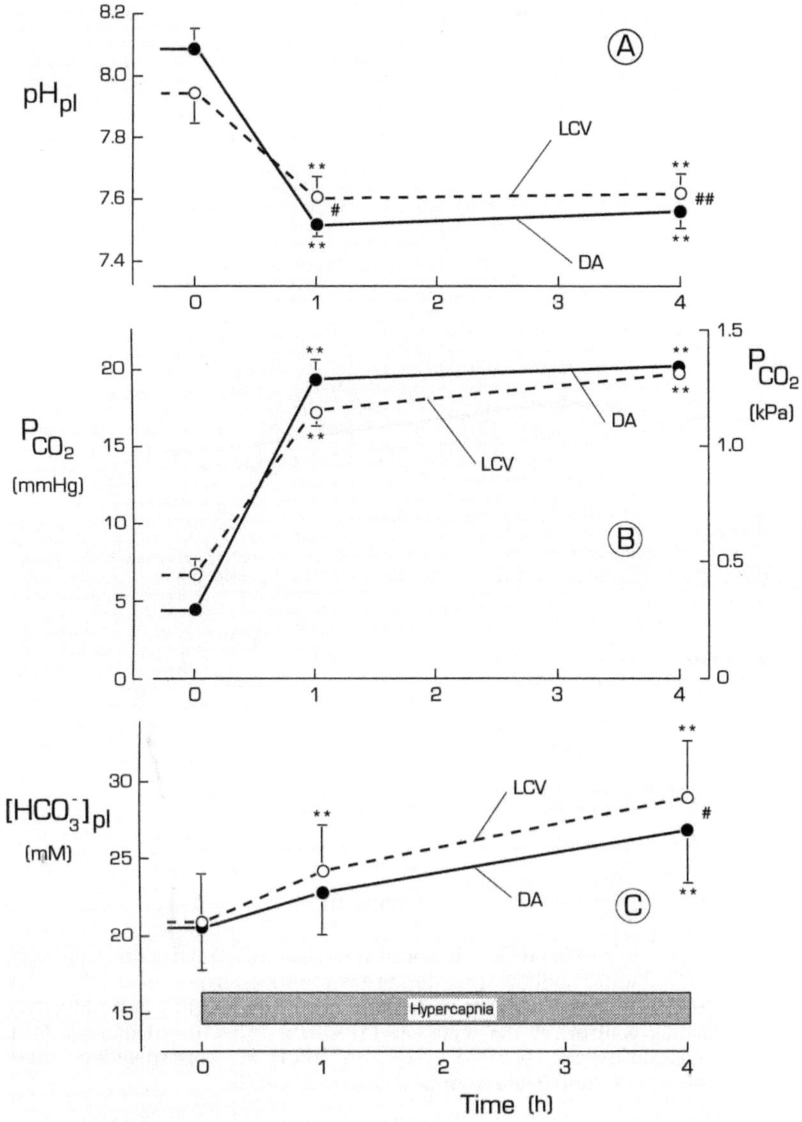

Fig. 3 A–C. Plasma pH, Pco_2 and $[HCO_3^-]$ in dorsal aortic (*DA*) and lateral cutaneous vessel fluid (*LCV*) of freshwater rainbow trout during environmental hypercapnia (2% CO_2). Mean $\pm$SE; n = 6. *Asterisks* denote significantly different values from the corresponding controls at the same site (**$P < 0.01$, Dunnett's test). #denotes a significant difference between the two sites at corresponding times (#$P < 0.05$,##$P < 0.01$, paired t-test). Data from Ishimatsu et al. (1992)

3.2.2 Fluid Balance

The capillary permeability of fishes to proteins is rather high, as demonstrated by Hargens et al. (1974). Their contention was based on the finding of nearly

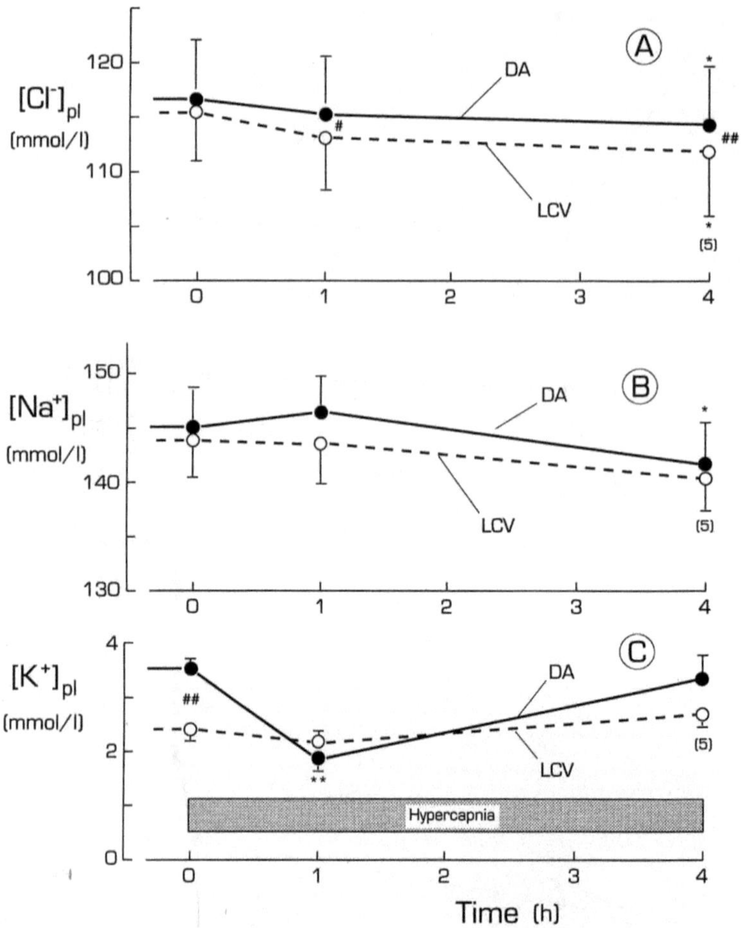

Fig. 4 A–C. Plasma [Cl^-], [Na^+] and [K^+] in dorsal aortic blood (*DA*) and lateral cutaneous vessel fluid (*LCV*) of freshwater rainbow trout during environmental hypercapnia (2% CO_2). Mean ± SE; n = 6 (unless otherwise indicated). *Asterisks* denote significantly different values from the corresponding controls at the same site (*$P < 0.05$, **$P < 0.01$, Dunnett's test). #denotes significant differences between the two sites at corresponding times (#$P < 0.05$, ##$P < 0.01$, paired t-test). Data from Ishimatsu et al. (1992)

identical protein concentrations in extravascular fluid and plasma. This implies that the relative contribution of colloid osmotic forces in fluid balance across fish capillaries is very small. As a consequence, hydrostatic filtration across fish capillary walls is expected to be considerable and fish will definitely "need" a vascular system with a relatively short circulation time to keep fluid balance. However, as pointed out by Satchell (1991), the absence of the secondary system from skeletal muscle seems to make it a poor candidate for the role of returning spilt fluid and proteins back into the primary system. Furthermore, the protrusion of the secondary capillary networks over the scales definitely aggravates the problem by enhancing water exchange with ambient media. Possible interaction between the surface capillary networks of the secondary system and the kidney, as suggested by the anatomical relationship described above, may provide the solution to this problem and deserves further investigation.

3.2.3 Immune Function

There is perhaps a natural tendency to associate immunologic functions and the secondary circulatory system in fishes. There are obvious similarities between the lymphatic system of other vertebrates and the secondary circulatory system. However, little is known about the function of this system in fishes, and definitive statements about the validity of this association cannot be made at this time. There are only a few studies which have addressed this issue. Assuming that the fluid collected from the neural duct of plaice was from the secondary circulation, Ellis and de Sousa (1974) found it to contain all species of white blood cell types, normally found in the primary circulation, only lower in numbers. They reported that there were no erythrocytes in that fluid, and explained the presence of red cells in their samples as being due to contamination from local hemorrhaging. In contrast, D. Alborn and G.K. Iwama (unpubl. observ.) reported leukocyte counts much higher in the secondary circulation of rainbow trout as compared to the primary circulation, when sampled from the lateral vessel according to the cannulation techniques described by Ishimatsu et al. (1992). They also found the lysozyme activity to be higher in the secondary circulation. Because there is a large range in structural characteristics of the secondary circulatory system among fish species, speculation as to its relative importance, compared to the primary circulation and immunologically important tissues, must await further studies.

4 Concluding Remarks

During the preparation of this chapter, we realized that, at present, only very few physiological data are available regarding the functional significance of the secondary circulatory system in fish. Accordingly, a larger portion of this chapter has been highly speculative and we hope that the above discussion of the existing information will stimulate further investigations of the physiology of this interesting vascular system of fish. As for the CVS, the intricate anatomical nature of the arterioarterial and arteriovenous circuits precludes application of many conventional approaches for the separate assessments of exchange processes occurring across the two distinct epithelia of the gill. The application of the recently developed in vivo techniques for the study of the CVS (Ishimatsu et al. 1988; Sundin and Nilsson 1992) will hopefully provide valuable information on this matter. Nevertheless, it is clear that certain studies are only possible with in vitro preparations and further refinements of perfused preparation are very much needed. With some suitable modifications, a perfused single filament preparation

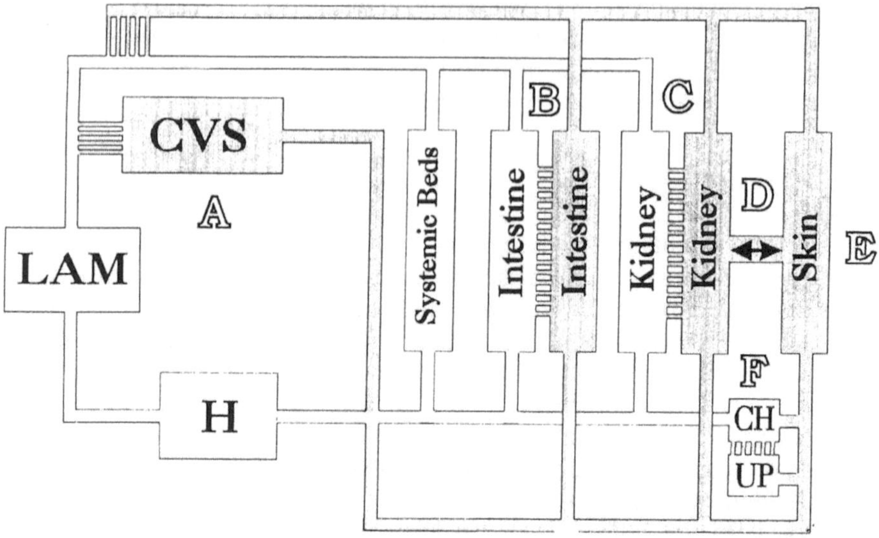

Fig. 5 A–F. Schematic diagram of the secondary circulatory system in fish and open questions regarding the function of the system as discussed in this chapter. **A** Partitioning of ion transport between the lamellar and CVS pathways; **B** distribution and function of the secondary vasculature in the intestinal wall; **C** kidney; **D** renocutaneous functional linkage; **E** importance of the cutaneous secondary system in transepithelial ion and water exchanges; **F** contribution of caudal heart-secondary vasculature complex in the transport of urophyseal hormones

recently reported by Shephard (1992) could prove to be useful in studying CVS functioning. We feel strongly that development of new microsurgical techniques, coupled with microanalytical methods, is necessary to make a breakthrough for the deeper understanding of gill physiology.

Present knowledge about the anatomy of the systemic part of the secondary system is far from being sufficient despite the long history of morphological research into the "lymphatic system" in fishes. This is particularly true for the microcirculation of the system. Scrutiny of both qualitative and quantitative aspects of the anatomy of the secondary circulatory system is overdue, such as electron microscope study, and the application of corrosion casting methods. It goes without saying, hemodynamic aspects of the secondary system, even the most basic ones like blood flow direction, velocity and pressure, have to be carefully investigated before attempts can be made to elucidate the possible functional roles of this vascular system in fishes (Fig. 5).

References

Avella M, Bornancin M (1989) A new analysis of ammonia and sodium transport through the gills of the freshwater rainbow trout (*Salmo gairdneri*). J Exp Biol 142: 155–175

Bailly Y, Dunel-Erb S (1986) The sphincter of the efferent filament artery in teleost gills: I. Structure and parasymphathetic innervation. J Morphol 187: 219–237 aílly Y, Dunel-Erb S, Geffard M, Laurent P (1989) The vascular and epithelial serotonergic innervation of the actinopterygian gill filament with special reference to the trout, *Salmo gairdneri*. Cell Tissue Res 258: 349–363

oland EJ, Olson KR (1979) Vascular organization of the catfish gill filament. Cell Tissue Res 198: 487–500

'ameron JN, Iwama GK (1987) Compensation of progressive hypercapnia in channel catfish and blue crabs. J Exp Biol 133: 183–197

'han DKO (1971) The urophysis and the caudal circulation of teleost fish. Mem Soc Endocrinol 19: 391–412

'han DKO (1975) Cardiovascular and renal effects of urotensin I and II in the eel, *Anguilla rostrata*. Gen Comp Physiol 27: 52–61

'laiborne JB, Evans DH (1980) The isolated, perfused head of the marine teleost fish, *Myoxocephalus octodecimspinosus*: hemodynamic effects of epinephrine. J Comp Physiol 138: 79–85

avie PS (1981) Neuroanatomy and control of the caudal lymphatic heart of the short-finned eel (*Anguilla australis schmidtii*). Can J Zool 59: 1586–1592

ewar H, Brill RW, Olson KR (1994) Secondary circulation of the vascular heat exchangers in skipjack tuna, *Katsuwonus pelamis*. J Exp Zool 269: 566–570

unel-Erb S, Bailly Y (1986) The sphincter of the efferent filament artery in teleost gills: II. Sympathetic innervation. J Morphol 187: 239–246

unel-Erb S, Bailly Y, Laurent P (1989) Neurons controlling the gill vasculature in five species of teleosts. Cell Tissue Res 255: 567–573

llis AE, de Sousa M (1974) Phylogeny of the lymphoid system. I. A study of the fate of circulating lymphocytes in plaice. Eur J Immunol 4: 338–343

arrell AP, Smith DG (1981) Microvascular pressures in gill filaments of lingcod (*Ophiodon elongatus*). J Exp Zool 216: 341–344

oskett JK, Machen TE (1985) Vibrating probe analysis of teleost opercular epithelium: correlation between active transport and leak pathways of individual chloride cells. J Membr Biol 85: 25–35

oskett JK, Scheffey C (1982) The chloride cell: definitive identification as the salt-secretory cell in teleosts. Science 215: 164–166

romm PO (1968) Some quantitative aspects of ion regulation in teleosts. Comp Biochem Physiol 27: 865–869

ardaire E, Avella M, Isaia J, Bornancin M (1991) Measurement of branchial vascular space of trout, *Oncorhynchus mykiss*: effects of adrenaline. J Comp Physiol 161B: 265–269

irard JP, Payan P (1976) Effect of epinephrine on vascular space of gills and head of rainbow trout. Am J Physiol 230: 1555–1560

oss GG, Laurent P, Perry SF (1992a) Evidence for a morphological component in acid-base regulation during environmental hypercapnia in the brown bullhead (*Ictalurus nebulosus*). Cell Tissue Res 268: 539–552

oss GG, Perry SF, Wood CM, Laurent P (1992b) Mechanism of ion and acid-base regulation at the gills of freshwater fish. J Exp Zool 263: 143–159

largens AR, Millard RW, Johansen K (1974) High capillary permeability in fishes. Comp Biochem Physiol 48A: 675–680

leisler N (1993) Acid-base regulation. In: Evans DH (ed) The physiology of fishes. CRC Press, Boca Raton, pp 343–378

lenrikson RC, Matoltsy AG (1968) The fine structure of teleost epidermis. III. Club cells and other cell types. J Ultrastruct Res 21: 222–232

lipkins SF (1985) Adrenergic responses of the cardiovascular system of the eel, *Anguilla australis*, in vivo. J Exp Zool 235: 7–20

lwang PP (1989) Distribution of chloride cells in teleost larvae. J Morphol 200: 1–8

shihara A, Mugiya Y (1987) Ultrastructural evidence of calcium uptake by chloride cells in the gills of goldfish, *Carassius auratus*. J Exp Zool 242: 121–129

shimatsu A, Iwama GK, Heisler N (1988) In vivo analysis of partitioning of cardiac output between systemic and CVS circuits in rainbow trout: a new approach using chronic cannulation of the branchial vein. J Exp Biol 137: 75–88

Ishimatsu A, Iwama GK, Bentley TB, Heisler N (1992) Contribution of the secondary circulatory system to acid-base regulation during hypercapnia in rainbow trout (*Oncorhynchus mykiss*). J Exp Biol 170: 43–56
Iwama GK, Ishimatsu A, Heisler N (1993) Site of acid-base relevant ion transfer in the gills of rainbow trout (*Oncorhynchus mykiss*) exposed to environmental hypercapnia. Fish Physiol Biochem 12: 269–280
Jakubowski M (1960a) The structure and vascularization of the skin of the eel (*Anguilla anguilla* L.) and the viviparous blenny (*Zoarces viviparous* L.). Acta Biol Crac Zool 3: 1–22
Jakubowski M (1960b) The structure and vascularization of the skin of the leathern carp (*Cyprinus carpio* L. var. *nuda*) and flounder (*Pleuronectes flesus luscus* Pall.). Acta Biol Crac Zool 3: 139–162
Jakubowski M (1989) Skin vascularization in fishes compared with that in amphibians. In: Splechtna H, Hilgers H (eds) Trends in vertebrate morphology. Gustav Fischer, Stuttgart, pp 542–545
Kaczmarski F (1966) Structure and vascularization of the skin of the ruff (*Acerina cernua* L.). Acta Biol Crac Zool 9: 165–175
Kampmeier OF (1969) Evolution and comparative morphology of the lymphatic system. Thomas, Springfield
Kiceniuk JW, Jones DR (1977) The oxygen transport system in trout (*Salmo gairdneri*) during sustained exercise. J Exp Biol 69: 247–260
Lametschwandtner A, Weiger T, Lametschwandtner U, Georgieva-Hanson V, Patzner RA, Adam H (1989) The vascularization of the skin of the Atlantic hagfish, *Myxine glutinosa* L. as revealed by scanning electron microscopy of vascular corrosion casts. Scanning Microsc 3: 305–314
Laurent P (1984) Gill internal morphology. In: Hoar WS, Randall DJ (eds) Fish physiology, vol XA. Academic Press, New York, pp 73–183
Laurent P (1989) Gill structure and function: fish. In: Wood SC (ed) Comparative pulmonary physiology: current concepts. Lung biology in health and disease, vol 39. Dekker, New York, pp 69–120
Laurent P, Perry SF (1990) Effects of cortisol on gill chloride cell morphology and ionic uptake in the freshwater trout, *Salmo gairdneri*. Cell Tissue Res 259: 429–442
Laurent P, Perry SF (1991) Environmental effects on fish gill morphology. Physiol Zool 64: 4–25
Leino RL, McCormick JH (1984) Morphological and morphometrical changes in chloride cells of the gills of *Pimephales promelas* after chronic exposure to acid water. Cell Tissue Res 236: 121–128
Lin H, Randall DJ (1991) Evidence for the presence of an electrogenic proton pump on the trout gill epithelium. J Exp Biol 161: 119–134
Lin H, Randall DJ (1993) H^+-ATPase activity in crude homogenates of fish gill tissue: inhibitor sensitivity and environmental and hormonal regulation. J Exp Biol 180: 163–174
Lin H, Pfeiffer DC, Vogl AW, Pan J, Randall DJ (1994) Immunolocalization of H^+-ATPase in the gill epithelia of rainbow trout. J Exp Biol 195: 169–183
Marshall WS (1985) Paracellular ion transport in trout opercular epithelium models osmoregulatory effects of acid precipitation. Can J Zool 63: 1816–1822
Marshall WS, Bryson SE, Wood CM (1992) Calcium transport by isolated skin of rainbow trout. J Exp Biol 166: 297–316
McDonald DG, Prior ET (1988) Branchial mechanisms of ion and acid-base regulation in the freshwater rainbow trout, *Salmo gairdneri*. Can J Zool 66: 2699–2708
Nekvasil NP, Olson KR (1985) Localization of ^3H-norepinephrine binding sites in the trout gill. J Exp Zool 235: 309–313
Nilsson S (1983) Autonomic nerve function in the vertebrates. Springer, Berlin Heidelberg New York
Nilsson S (1984) Innervation and pharmacology of the gills. In: Hoar WS, Randall DJ (eds) Fish physiology, vol XA. Academic Press, New York, pp 185–227
Nilsson S, Pettersson K (1981) Sympathetic nervous control of blood flow in the gills of the Atlantic cod, *Gadus morhua*. J Comp Physiol 144B: 157–163

Nonnotte G, Nonnotte L, Kirsch R (1979) Chloride cells and chloride exchange in the skin of a sea-water teleost, the shanny (*Blennius pholis* L.). Cell Tissue Res 199: 387–396
Olson KR (1984) Distribution of flow and plasma skimming in isolated perfused gills of three teleosts. J Exp Biol 109: 97–108
Olson KR, Meisheri KD (1989) Effects of atrial natriuretic factor on isolated arteries and perfused organs of trout. Am J Physiol 256: R10–R18
Olson KR, Kullman D, Narkates AJ, Oparil S (1986) Angiotensin extraction by trout tissues in vivo and metabolism by the perfused gill. Am J Physiol 250: R532–R538
Olson KR, Duff DW, Farrell AP, Keen J, Kellogg MD, Kullman D, Villa J (1991) Cardiovascular effects of endothelin in trout. Am J Physiol 260: H1214–H1223
Payan P, Girard JP (1977) Adrenergic receptors regulating patterns of blood flow through the gills of trout. Am J Physiol 232: H18–H23
Payan P, Girard JP, Mayer-Gostan N (1984) Branchial ion movements in teleosts: the roles of respiratory and chloride cells. In: Hoar WS, Randall DJ (eds) Fish physiology, vol XB. Academic Press, New York, pp 39–63
Perry SF, Goss GG (1994) The effects of experimentally altered gill chloride cell surface area on acid-base regulation in rainbow trout during metabolic alkalosis. J Comp Physiol 164B: 327–336
Perry SF, Davie PS, Daxboeck C, Ellis AG, Smith DG (1984) Perfusion methods for the study of gill physiology. In: Hoar WS, Randall DJ (eds) Fish physiology, vol XB. Academic Press, New York, pp 325–388
Pfeiler E, Lindley V (1989) Chloride-type cells in the skin of the metamorphosing bonefish (*Albula* sp.) leptocephalus. J Exp Zool 250: 11–16
Randall DJ (1985) Shunts in fish gills. In: Johansen K, Burggren WW (eds) Cardiovascular shunts: phylogenetic, ontogenetic and clinical aspects. Munksgaard, Copenhagen, pp 71–87
Satchell GH (1991) Physiology and form of fish circulation. Cambridge Univ Press, Cambridge
Shephard KL (1992) Studies on the fish gill microclimate. J Comp Physiol 162B: 231–240
Shuttleworth TJ (1989) Overview of epithelial ion-transport mechanisms. Can J Zool 67: 3032–3038
Smith DG (1977) Sites of cholinergic vasoconstriction in trout gills. Am J Physiol 233 R222–R229
Stagg RM, Shuttleworth TJ (1984) Hemodynamics and potentials in isolated flounder gills: effects of catecholamines. Am J Physiol 246: R211–R220
Steffensen JF, Lomholt JP (1992) The secondary vascular system. In: Hoar WS, Randall DJ, Farrell AP (eds) Fish physiology, vol XII. Academic Press, London, pp 185–217
Steffensen JF, Lomholt JP, Vogel WOP (1986) In vivo observations on a specialized microvasculature, the primary and secondary vessels in fishes. Acta Zool (Stockh) 67: 193–200
Stiffler DF, Graham JB, Dickson KA, Stockmann W (1986) Cutaneous ion transport in the freshwater teleost *Synbranchus marmoratus*. Physiol Zool 59: 406–418
Sundin L, Nilsson S, (1992) Arterio-venous branchial blood flow in the Atlantic cod *Gadus morhua*. J Exp Biol 165: 73–84
Tang Y, Boutilier RG (1988) Correlation between catecholamine release and degree of acidotic stress in trout. Am J Physiol 255: R395–R399
Tyszkiewicz K (1969) Structure and vascularization of the skin of the pike (*Esox lucius* L.). Acta Biol Crac Zool 12: 67–79
Vogel WOP (1978) Arteriovenous anastomoses in the afferent region of trout gill filaments (*Salmo gairdneri* Richardson, Teleostei). Zoomorphologie 90: 205–212
Vogel WOP (1981) Struktur und Organisationsprinzip im Gefäßsystem der Knochenfische. Gegenbaurs Morphol Jahrb 127: 772–784
Vogel WOP (1985a) Systemic vascular anastomoses, primary and secondary vessels in fish, and the phylogeny of lymphatics. In: Johansen K, Burggren WW (eds) Cardiovascular shunts: phylogenetic, ontogenetic and clinical aspects. Munksgaard, Copenhagen, pp 143–159
Vogel WOP (1985b) The caudal heart of fish: not a lymph heart. Acta Anat 121: 41–45
Vogel WOP, Claviez M (1981) Vascular specialization in fish, but no evidence for lymphatics. Z Naturforsch 36C: 490–492

Vogel WOP, Vogel V, Kremers H (1973) New aspects of the intrafilamental vascular system in gills of a euryhaline teleost, *Tilapia mossambica*. Z Zellforsch 144: 573–583

Whitear M (1986) The skin of fishes including cyclostomes. In: Bereiter-Hahn J, Matoltsy AG, Richards KS (eds) Biology of the integument 2. Vertebrates. Springer, Berlin Heidelberg New York, pp 8–64

Whitear M, Mittal AK (1986) Structure of the skin of *Agonus cataphractus* (Teleostei). J Zool Lond 210A: 551–574

Zadunaisky JA (1984) The chloride cell: the active transport of chloride and the paracellular pathways. In: Hoar WS, Randall DJ (eds) Fish physiology, vol XB. Academic Press, New York, pp 129–176

Subject Index

Springer-Verlag and the Environment

We at Springer-Verlag firmly believe that an international science publisher has a special obligation to the environment, and our corporate policies consistently reflect this conviction.

We also expect our business partners – paper mills, printers, packaging manufacturers, etc. – to commit themselves to using environmentally friendly materials and production processes.

The paper in this book is made from low- or no-chlorine pulp and is acid free, in conformance with international standards for paper permanency.

MIX
Papier aus verantwortungsvollen Quellen
Paper from responsible sources
FSC® C105338

If you have any concerns about our products,
you can contact us on
ProductSafety@springernature.com

In case Publisher is established outside the EU,
the EU authorized representative is:
Springer Nature Customer Service Center GmbH
Europaplatz 3, 69115 Heidelberg, Germany

Printed by Libri Plureos GmbH
in Hamburg, Germany